581.185
AF744818
Withdrawn from Stock
Lancaster University
Library

Soil Erosion, Conservation, *and* Rehabilitation

BOOKS IN SOILS, PLANTS, AND THE ENVIRONMENT

Soil Biochemistry, Volume 1, edited by A. D. McLaren and G. H. Peterson
Soil Biochemistry, Volume 2, edited by A. D. McLaren and J. Skujiņš
Soil Biochemistry, Volume 3, edited by E. A. Paul and A. D. McLaren
Soil Biochemistry, Volume 4, edited by E. A. Paul and A. D. McLaren
Soil Biochemistry, Volume 5, edited by E. A. Paul and J. N. Ladd
Soil Biochemistry, Volume 6, edited by Jean-Marc Bollag and G. Stotzky
Soil Biochemistry, Volume 7, edited by G. Stotzky and Jean-Marc Bollag
Soil Biochemistry, Volume 8, edited by Jean-Marc Bollag and G. Stotzky

Organic Chemicals in the Soil Environment, Volumes 1 and 2, edited by C. A. I. Goring and J. W. Hamaker
Humic Substances in the Environment, M. Schnitzer and S. U. Khan
Microbial Life in the Soil: An Introduction, T. Hattori
Principles of Soil Chemistry, Kim H. Tan
Soil Analysis: Instrumental Techniques and Related Procedures, edited by Keith A. Smith
Soil Reclamation Processes: Microbiological Analyses and Applications, edited by Robert L. Tate III and Donald A. Klein
Symbiotic Nitrogen Fixation Technology, edited by Gerald H. Elkan
Soil-Water Interactions: Mechanisms and Applications, Shingo Iwata and Toshio Tabuchi with Benno P. Warkentin
Soil Analysis: Modern Instrumental Techniques, Second Edition, edited by Keith A. Smith
Soil Analysis: Physical Methods, edited by Keith A. Smith and Chris E. Mullins
Growth and Mineral Nutrition of Field Crops, N. K. Fageria, V. C. Baligar, and Charles Allan Jones
Semiarid Lands and Deserts: Soil Resource and Reclamation, edited by J. Skujiņš
Plant Roots: The Hidden Half, edited by Yoav Waisel, Amram Eshel, and Uzi Kafkafi
Plant Biochemical Regulators, edited by Harold W. Gausman
Maximizing Crop Yields, N. K. Fageria
Transgenic Plants: Fundamentals and Applications, edited by Andrew Hiatt
Soil Microbial Ecology: Applications in Agricultural and Environmental Management, edited by F. Blaine Metting, Jr.
Principles of Soil Chemistry: Second Edition, Kim H. Tan

Water Flow in Soils, edited by Tsuyoshi Miyazaki
Handbook of Plant and Crop Stress, edited by Mohammad Pessarakli
Genetic Improvement of Field Crops, edited by Gustavo A. Slafer
Agricultural Field Experiments: Design and Analysis, Roger G. Petersen
Environmental Soil Science, Kim H. Tan
Mechanisms of Plant Growth and Improved Productivity: Modern Approaches, edited by Amarjit S. Basra
Selenium in the Environment, edited by W. T. Frankenberger, Jr., and Sally Benson
Plant–Environment Interactions, edited by Robert E. Wilkinson
Handbook of Plant and Crop Physiology, edited by Mohammad Pessarakli
Handbook of Phytoalexin Metabolism and Action, edited by M. Daniel and R. P. Purkayastha
Soil–Water Interactions: Mechanisms and Applications, Second Edition, Revised and Expanded, Shingo Iwata, Toshio Tabuchi, and Benno P. Warkentin
Stored-Grain Ecosystems, edited by Digvir S. Jayas, Noel D. G. White, and William E. Muir
Agrochemicals from Natural Products, edited by C. R. A. Godfrey
Seed Development and Germination, edited by Jaime Kigel and Gad Galili
Nitrogen Fertilization in the Environment, edited by Peter Edward Bacon
Phytohormones in Soils: Microbial Production and Function, W. T. Frankenberger, Jr., and Muhammad Arshad
Handbook of Weed Management Systems, edited by Albert E. Smith
Soil Sampling, Preparation, and Analysis, Kim H. Tan
Soil Erosion, Conservation, and Rehabilitation, edited by Menachem Agassi

Additional Volumes in Preparation

Photoassimilate Distribution of Plants and Crops: Source–Sink Relationships, edited by Eli Zamski and Arthur A. Schaffer

Plant Roots: The Hidden Half, Second Edition, Revised and Expanded, edited by Yoav Waisel, Amram Eshel, and Uzi Kafkafi

Mass Spectrometry of Soils, edited by Thomas W. Boutton and Shin-ichi Yamasaki

Soil Erosion, Conservation, *and* Rehabilitation

edited by

Menachem Agassi
Soil Erosion Research Station
Soil Conservation and Drainage Division
Ministry of Agriculture
Emek-Hefer, Israel

Marcel Dekker, Inc. New York • Basel • Hong Kong

Library of Congress Cataloging-in-Publication Data

Soil erosion, conservation, and rehabilitation / edited by Menachem Agassi.
p. cm. — (Books in soils, plants, and the environment)
ISBN 0-8247-8984-9 (alk. paper)
1. Soil erosion. 2. Rain and rainfall. 3. Runoff. 4. Reclamation of land. 5. Soil conservation. I. Agassi, Menachem. II. Series.
S623.S5735 1995
631.4′5—dc20 95-40716
CIP

The publisher offers discounts on this book when ordered in bulk quantities. For more information, write to Special Sales/Professional Marketing at the address below.

This book is printed on acid-free paper.

Copyright © 1996 by Marcel Dekker, Inc. All Rights Reserved.

Neither this book not any part may be reproduced or transmitted in any form or by any means, electronic or mechanical, including photocopying, microfilming, and recording, or by any information storage and retrieval system, without permission in writing from the publisher.

Marcel Dekker, Inc.
270 Madison Avenue, New York, New York 10016

Current printing (last digit):
10 9 8 7 6 5 4 3 2 1

PRINTED IN THE UNITED STATES OF AMERICA

c.2
95 07129

Preface

Factors affecting processes of runoff and erosion by water in the agricultural field are discussed in the first two parts of this book. Part I emphasizes the importance of rainfall patterns, because estimations of runoff and erosion and the design of flood control, water harvesting, tillage management, and soil conservation plans are all based mainly on precipitation data. Information on rainfall amounts and intensities is available in most parts of the world. These data are probably the only reliable source of information for designing water and soil conservation programs on the basis of probability.

Erosion by water is a secondary process following soil surface sealing by drop impact and runoff initiation. It is established that considerable transportation of detached soil particles outside the field (erosion by water) does not occur without runoff. Collection of erosion data is complicated and expensive and is also considerably affected by the methodology being used. Attempts to correlate these data with rainfall and soil parameters (modeling) have not been very successful. Surface sealing and infiltration processes, on the other hand, are less complicated than erosion processes. They are much easier to measure, and correlating them with rain parameters is much more successful. The effect of soil parameters on surface sealing and infiltration is being studied intensively. However, the quantitative effect of these parameters on sealing and infiltration processes has not yet been established. We can conclude that achieving reliable parameters for successful design of soil conservation programs will be more easily achieved by studying interrelationships between rainfall characteristics, sealing of soil surface, and the ensuing decrease of infiltration rate than by studying and modeling erosion processes, as is currently being done.

In arid and semiarid regions, where water is the main limiting factor for crop production, it is essential to prevent rainwater runoff from arable lands. Therefore, it again will be more meaningful to study the interrelationships among rainfall parameters, soil parameters, and the soil infiltration rate to find better runoff control methods than to study soil erosion processes. We should take it into consideration that by controlling runoff (e.g., by maintaining a relatively high infiltration rate or increasing surface storage), erosion will also be controlled, but the reverse is not necessarily true.

Mulching is a very efficient means to dissipate raindrop impact and to control the ensuing soil surface sealing, runoff, and erosion. Mulching can also reduce evaporation of rainwater and overhead irrigation water. Therefore, mulching can be a vital factor in improving water use efficiency. Three simultaneous processes characterize vast areas of the globe: (1) There is a need to improve water use efficiency as water resources (quantitatively and qualitatively) are being depleted; (2) rapidly increasing amounts of municipal solid wastes are being produced, although awareness of the hazards of these wastes has grown; (3) motivation is increasing to achieve sustainable agriculture, for example, through erosion, runoff, and salinization control. The combined disadvantages of the first two aforementioned processes can be turned to our benefit if we can mulch arable lands and rangelands with composted municipal organic wastes to improve water use efficiency, thus contributing to sustainable agriculture with minimum hazard to the environment and the farming system.

The third part of this book deals with the rehabilitation of soils depleted by human activity. The objective of soil conservation and rehabilitation is to conserve and improve the soil as a healthy habitat for fauna and flora, and to conserve and improve its potential to carry crops.

Menachem Agassi

Contents

Preface *iii*
Contributors *ix*

Part I: Soil Erosion by Water

BASIC FACTORS AND PROCESSES IN SOIL EROSION

1. Infiltration and Seal Formation Processes 1
I. Shainberg and G. J. Levy

2. Rainfall Analysis 23
Joseph Morin

3. Soil Characteristics and Aggregate Stability 41
Yves Le Bissonnais

4. Splash and Detachment by Waterdrops 61
Joe M. Bradford and Chi-hua Huang

5. Slope, Aspect and Surface Storage — 77
Dino Torri

6. The Effect of Surface Cover on Infiltration and Soil Erosion — 107
James E. Box, Jr., and R. Russell Bruce

FORMS OF EROSION

7. Interrill Erosion — 125
Padam Prasad Sharma

8. Rill and Gullies Erosion — 153
Earl H. Grissinger

SOIL LOSS AND RUNOFF ESTIMATION

9. Soil Loss Estimation — 169
Kenneth G. Renard, George R. Foster, Leonard J. Lane, and John M. Laflen

10. Runoff Estimation on Agricultural Fields — 203
Jeffry Stone, Kenneth G. Renard, and Leonard J. Lane

Part II: Soil Conservation

11. Common Soil and Water Conservation Practices — 239
Paul W. Unger

12. Soil Stabilizers — 267
G. J. Levy

Part III: Soil Rehabilitation

13. Reclamation of Gullies and Channel Erosion — 301
Earl H. Grissinger

14. Reclamation of Salt-Affected Soil — 315
J. D. Oster, I. Shainberg, and I. P. Abrol

15. Reclamation of Sodic-Affected Soils 353
Rami Keren

16. Impact of Crop Rotation and Land Management on Soil Erosion and Rehabilitation 375
Jacob Amir

Index *399*

Contributors

I. P. Abrol, Ph.D. Indian Council of Agricultural Research, New Delhi, India

Menachem Agassi, Ph.D. Soil Erosion Research Station, Soil Conservation and Drainage Division, Ministry of Agriculture, Emek-Hefer, Israel

Jacob Amir, Ph.D. Gilat Experiment Station, Agricultural Research Organization, Negev, Israel

James E. Box, Jr., Ph.D. Southern Piedmont Conservation Research Center, Agricultural Research Service, U.S. Department of Agriculture, Watkinsville, Georgia

Joe M. Bradford, M.S., Ph.D. Conservation and Production Systems Research Unit, Subtropical Agricultural Research Laboratory, Agricultural Research Service, U.S. Department of Agriculture, Weslaco, Texas

R. Russell Bruce, M.S., Ph.D. Southern Piedmont Conservation Research Center, Agricultural Research Service, U.S. Department of Agriculture, Watkinsville, Georgia

George R. Foster, Ph.D. National Sedimentation Laboratory, Agricultural Research Service, U.S. Department of Agriculture, Oxford, Mississippi

Earl H. Grissinger, Ph.D. National Sedimentation Laboratory, Agricultural Research Service, U.S. Department of Agriculture, Oxford, Mississippi (Retired)

Chi-hua Huang, Ph.D. National Soil Erosion Research Laboratory, Agricultural Research Service, U.S. Department of Agriculture, and Purdue University, West Lafayette, Indiana

Rami Keren, Ph.D. Department of Soil and Environmental Physical Chemistry, Institute of Soils and Water, The Volcani Center, Agricultural Research Organization, Bet-Dagan, Israel

John M. Laflen, Ph.D. National Soil Erosion Research Laboratory, Agricultural Research Service, U.S. Department of Agriculture, West Lafayette, Indiana

Leonard J. Lane, Ph.D. Southwest Watershed Research Center, Agricultural Research Service, U.S. Department of Agriculture, Tucson, Arizona

Yves Le Bissonnais, Ph.D. Department of Soil Science, Soil Survey Staff of France, National Institute for Agronomic Research, Orleans, France

G. J. Levy, Ph.D. Institute of Soils and Water, The Volcani Center, Agricultural Research Organization, Bet-Dagan, Israel

Joseph Morin, Ph.D. Soil Conservation and Drainage Division, Soil Erosion Research Station, Rupin Institute, Ministry of Agriculture, Emek-Hefer, Israel

J. D. Oster, Ph.D. Department of Soil and Environmental Sciences, University of California, Riverside, California

Kenneth G. Renard, Ph.D. Southwest Watershed Research Center, Agricultural Research Service, U.S. Department of Agriculture, Tucson, Arizona

I. Shainberg, Ph.D. Institute of Soils and Water, The Volcani Center, Agricultural Research Organization, Bet-Dagan, Israel

Padam Prasad Sharma, Ph.D. Land Reclamation Research Center, North Dakota State University, Mandan, North Dakota

Jeffry Stone, Ph.D. Southwest Watershed Research Center, Agricultural Research Service, U.S. Department of Agriculture, Tucson, Arizona

Dino Torri, M.S. Soil Genesis, Classification, and Cartography Research Center, National Research Council, Florence, Italy

Paul W. Unger, Ph.D. Conservation and Production Research Laboratory, Agricultural Research Service, U.S. Department of Agriculture, Bushland, Texas

Soil Erosion, Conservation, *and* Rehabilitation

1

Infiltration and Seal Formation Processes

I. Shainberg and G. J. Levy
Institute of Soils and Water, The Volcani Center, Agricultural Research Organization, Bet-Dagan, Israel

I. INTRODUCTION

A. Infiltration

When water is supplied to the soil surface, whether by precipitation or irrigation, some of the water penetrates the surface and flows into the soil, while some may fail to penetrate and instead accumulate at the surface or flow over it. Infiltration is the term applied to the process of water entry into the soil, usually by downward flow through the soil surface. Hence, "infiltration rate" (IR) is defined as the volume flux of water flowing into the profile per unit of soil surface area under any set of circumstances. Under conditions where water is supplied to the soil without energy, IR and the variation thereof with time are known to depend upon initial water content and suction, texture, structure, and uniformity of the soil profile (Hillel, 1980). Generally, IR is high during the early stages of infiltration, particularly when the soil is initially quite dry, but decreases monotonically to approach a constant rate asymptotically, due to a decrease in the matric suction gradient which occurs as infiltration proceeds.

The decrease in IR from an initially high rate can also result from a gradual deterioration of soil structure, usually caused by the impact energy of waterdrops, which leads to partial sealing of the profile by the formation of a dense surface seal. It is well known that seal formation at the soil surface predominates in the decrease of infiltration during rain (Duley, 1939; Epstein and Grant, 1973;

Morin and Benyamini, 1977) or irrigation with pressurized overhead sprinkler systems (Aarstad and Miller, 1973). The decrease in IR may result in a situation where the application rate of the water (i.e., rain/irrigation intensity) exceeds the IR of the soil. Consequently, water will start running off the field, and soil erosion will commence.

The IR values of sealed soils depend on the hydraulic conductivity (HC) of the seal formed at the soil surface; hence some researchers tried to measure the HC of the seal. McIntyre (1958) found that the HC of the upper and lower layers of the seal of a fine sandy loam soil were 2000 and 200 times lower than the HC values recorded for the undisturbed soils. Bresler and Kemper (1970) measured an HC value of 1.5×10^{-4} mm s^{-1} in the upper 2–3 mm of a sealed clay loam soil. However, since the crust is a very thin layer and it is very difficult to separate it from the rest of the soil profile, it is extremely difficult to determine its HC. Consequently, IR is widely used to characterize water penetration into the soil, particularly when seal formation is involved.

Numerous equations, some entirely empirical and others theoretically based, have been proposed in attempts to express IR as a function of time or total volume of water infiltrated into the soil (i.e., cumulative infiltration). The theoretical equations (Green and Ampt, 1911; Philip, 1957) arise out of mathematical solutions to physically based theories of infiltration. The empirical expressions (Horton, 1940) are not so restrictive as to the mode of water application, since they do not imply surface ponding from time zero on, as do the Green-Ampt and Philip equations (Hillel, 1980). An equation commonly used is a Horton-type equation developed by Morin and Benyamini (1977) which describes the infiltration process as a function of cumulative rain rather than cumulative time:

$$I_t = I_c + (I_i - I_c)e^{-\gamma pt} \tag{1}$$

where

I_t = instantaneous infiltration rate (mm h^{-1})
I_c = asymptotical final infiltration rate (mm h^{-1})
I_i = initial infiltration rate (mm h^{-1})
γ = constant related to stability of soil surface aggregates (mm^{-1})
p = rain intensity (mm h^{-1})
t = time elapsed from beginning of storm (h)

Calculated IR values derived from this equation were in good agreement with values measured with laboratory and field rainfall simulators, such as those developed by Morin et al. (1967) and Miller (1987a), which are a common tool in seal formation studies.

B. Seal Formation

When soil is exposed to the impact of waterdrops either during rainstorms or when overhead sprinkler irrigation systems are used, severe changes in the structure of the soil surface and a reduction in soil permeability is observed (Duley, 1939; Aarstad and Miller, 1973). It has been suggested (Epstein and Grant, 1973) that seal formation due to the beating action of raindrops results from the mechanical breakdown of aggregates, at the soil surface, resulting in the formation of primary particles and microaggregates, which in turn reduce the porosity of the soil surface. Furthermore, raindrop impact causes direct compaction of the soil surface (Epstein and Grant, 1973).

Agassi et al. (1981) suggested that seal formation is due to two complementary mechanisms: (i) physical disintegration of soil aggregates and their compaction caused by the impact of the raindrops and (ii) chemical dispersion and movement of clay particles into a region at 0.1–0.5 mm depth, where they lodge and clog the conducting pores. The first mechanism is the predominant one and is determined primarily by the kinetic energy of the drops. The second mechanism is controlled mainly by the concentration and composition of the cations in the soil and applied water (Agassi et al., 1981). Both mechanisms act simultaneously and the first enhances the latter.

An additional type of seal is that formed by translocation of fine soil particles and their deposition at a certain distance from their original location (Arshad and Mermut, 1988; Chen et al., 1980). When the velocity of the running water exceeds the shear strength of the soil surface, the water erodes the soil and entrains sediments. As water velocity decreases, the sediments in the water deposit and a depositional seal is formed. Formation of a depositional seal is important in surface (furrow and basin) irrigation where there is no drop impact (Kemper et al., 1985).

The objective of this review is to describe the genesis and properties of the seal (only that formed under waterdrop impact) and its effect on IR. This objective is obtained by summarizing studies on the effect of soil and rain properties on seal formation.

II. SOIL PROPERTIES AND SEAL FORMATION

A. Effect of Clay Content and Type on Infiltration Rate

The tendency of soils to form seals depends on the stability of their structure, which tends to increase with increasing clay content. Increasing clay content results in a hyperbolic increase in the wet sieve aggregate stability (Kemper and Koch, 1966). Clay particles act as cementing material binding the particles together in the aggregates (Kemper and Koch, 1966). Thus, the stability of the aggregates against the impact action of the raindrops should also increase with an increase in clay content.

The effect of clay content on the IR of soils was studied by Ben-Hur et al. (1985). Soil samples with clay contents between 3% and 60% were chosen from two different types of soils. The soils were exposed to rain until steady-state infiltration and corresponding seal formation were obtained (Fig. 1). For both types of soils, soils with 10–30% clay were the most susceptible to seal formation and had the lowest infiltration rate (Fig. 1). With increasing clay content, soil structure was more stable and seal formation was diminished. In soils with lower clay contents (<10%), the amount of clay available to disperse and clog the soil pores was limited and, as a result, an undeveloped seal was formed.

Most of the studies on seal formation and runoff have been conducted on soils in which the dominant clay minerals were smectites. These clay minerals are known to be more dispersive than kaolinitic clays (Frenkel et al., 1978; Goldberg and Glaubig, 1987). However, a few kaolinitic and illitic soils from the southeastern United States (Miller, 1987b; Miller and Scifres, 1988) and South Africa (Levy and van der Watt, 1988; Stern et al. 1991) are known to form seals and produce a significant amount of runoff. Stern et al. (1991) studied 19 kaolinitic and illitic soils (alfisols) from South Africa and divided them into stable and unstable (dispersive) soils, based on their susceptibility to seal formation when exposed to simulated rain. The final IR values for the stable soil were >14.5 mm h^{-1}, whereas for the dispersive soils the final IR values were <4.2 mm h^{-1}. Stern, Ben-Hur and Shainberg (1991) observed that the susceptibility of the unstable kaolinitic soils to seal formation was similar to that of smectitic soils (Kazman et al., 1983). The South African stable and unstable soils had similar properties (e.g., texture, dominant clay mineral (kaolinite), organic matter content, pH, etc.). Hence, these properties could not explain the differences in susceptibility to seal formation among the soils studied. However, Stern et al., (1991) noted that the presence of smectite impurities in the soils was correlated with their stability. Kaolinitic soils with small amounts of smectite were dispersive and susceptible to seal formation. Furthermore, the IR of these soils was very sensitive to Phosphosypsum (PG) application. Conversely, kaolinitic soils in which no smectite impurities were detected were less susceptible to sealing. The effect of various other soil properties (e.g., organic matter, sesquioxides and antecedent moisture content, etc.) on soil dispersion and seal formation is discussed in detail in chapters 4 and 5.

B. Effect of Exchangeable Cations on IR

1. *Introduction*

As discussed previously, clay dispersion is one of the two mechanisms responsible for seal formation and reduction in soil IR.

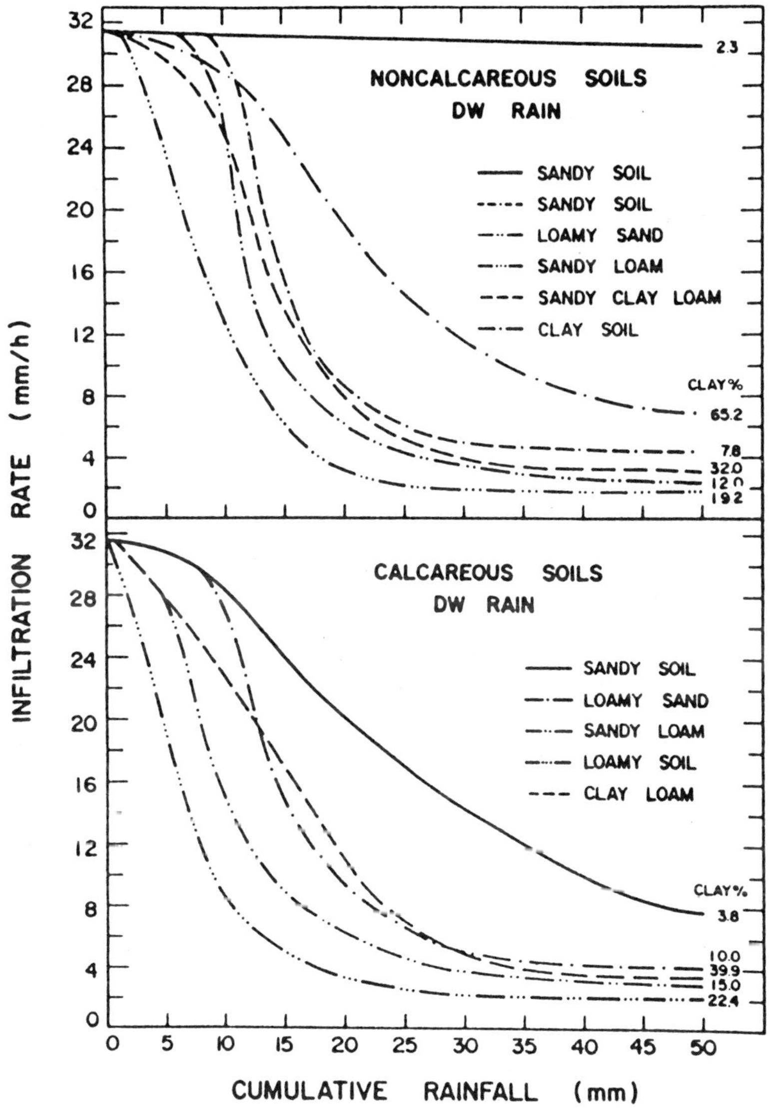

Fig. 1 Infiltration rates of calcareous and noncalcareous soils as a function of clay content and cumulative distilled water (DW) rain. (From Ben-Hur et al., 1985.)

Clay particles are mainly negatively charged because of isomorphic substitutions. Since electroneutrality must be kept, the negative charges are countered by cations present in the soil solution. The behavior of clay particles in aqueous solutions can be explained by the diffuse double layer (DDL) theory, for which a description can be found in, e.g., van Olphen (1977).

Clay dispersion occurs when the charged clay plates, which are moving apart following the penetration of water molecules between adjacent plates, have separated enough so that attractive forces are no longer strong enough to oppose repulsive forces and the plates can move by an external force. According to the DDL theory, the clay dispersion depends greatly on the valency of the cations countering its negative charge (termed exchangeable cations). For example, monovalent cations (Na, K) are considered dispersive, and bivalent cations (Ca, Mg) are considered nondispersive cations.

2. *Effect of Exchangeable Sodium*

The effect of exchangeable sodium percentage (ESP) on IR and seal formation of four smectitic soils varying in texture, clay mineralogy, and $CaCO_3$ content was studied by Kazman, Shainberg and Gal (1983) using a laboratory rainfall simulator and distilled water. In each of the four soils, IR was highly sensitive to low levels of ESP. The results for a sandy loam are presented in Fig. 2. Even at the lowest sodicity (ESP 1.0) a seal was formed and the IR dropped from an initial value of >100 mm h^{-1} to a final IR of 7.0 mm h^{-1}. An ESP value of 2.2 was enough to cause a further drop in the final IR of the sandy loam (final IR of 2.4 mm^{-1}). The amount of rain required to approach the final infiltration rate was also affected by ESP (Fig. 2). As the ESP of the soil increased, the depth of rain required to reach the final IR decreased.

Raindrop impact energy was the same in all the experiments presented in Fig. 2; hence the differences in IR curves for the various soil samples were the result of chemical dispersion of the soil clay caused by sodicity. The high sensitivity of the soil surface to low ESP values is explained by three factors (Oster and Schroer, 1979; Kazman et al., 1983): (i) the mechanical impact of the raindrops, which enhances chemical dispersion; (ii) the absence of surrounding soil matrix (sand particles), which when present slows clay dispersion and movement; and (iii) the almost total absence of electrolytes in the applied distilled water. However, the presence of some minerals ($CaCO_3$ and a few primary minerals) in the soil profile which readily release electrolytes to the percolating solution reduces the dispersion effect of soil sodicity (Shainberg and Letey, 1984).

The mechanical and chemical mechanisms responsible for seal formation are complementary. The mechanical impact of the raindrops has two effects: (i) breakdown of the soil aggregates and (ii) stirring of the soil particles. The latter

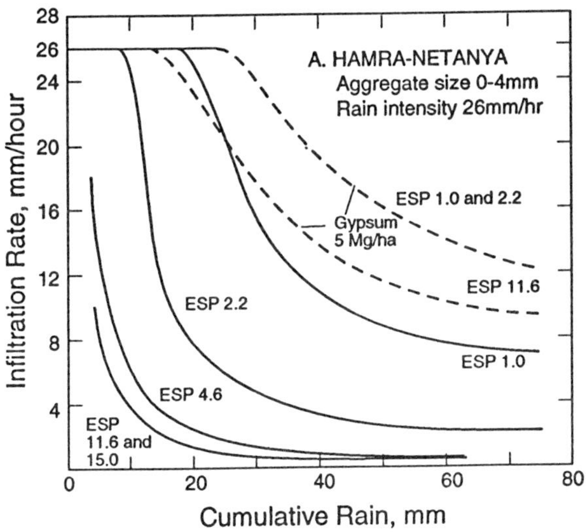

Fig. 2 Effect of soil ESP and phosphogypsum application on the infiltration rate of the Netanya soil as a function of the cumulative rain. (From Kazman et al., 1983.)

enhances the rate of the chemical dispersion. Without the stirring process the rate of chemical dispersion is much slower. In HC measurements, mechanical mixing of the soil is prevented. Thus, the IR is much more sensitive to ESP than is soil HC (Shainberg and Letey, 1984).

Highly weathered soils in the southeastern United States also suffer from dispersion-related degradation of physical properties with additions of NA, either from fertilizers or wastewater sources. The effect of surface application of $NaNO_3$, at fertilizer N rates (0.6 Mg ha^{-1}), on infiltration, runoff, and soil loss of a Greenville soil (clayey, kaolinitic, thermic, Rhodic Paleudult) was studied by Miller and Scifres (1988) and is presented in Fig. 3. Runoff from the untreated soil commenced after 20 mm of rainfall, and the steady-state IR stabilized at approximately 10 mm h^{-1}. Compared to other southeastern U.S. Piedmont soils (Miller, 1987b), the Greenville soil maintained a reasonably high rate of water intake over the rainfall event, indicating a fair resistance to sealing. The $NaNO_3$ treatment (0.6 Mg ha^{-1}) resulted in a nearly immediate surface sealing, as shown by the rapid decline in infiltration reaching a steady-state IR of only 2–3 mm h^{-1} (Fig. 3).

Although the effect of exchangeable Na on the HC of kaolinitic soils is quite small (Frenkel et al., 1978; Chiang et al., 1987), the effects of low ESP values on clay dispersion, sealing, and hydraulic properties of kaolinitic soils are significant. Significant changes in HC will occur only when spontaneous dispersion

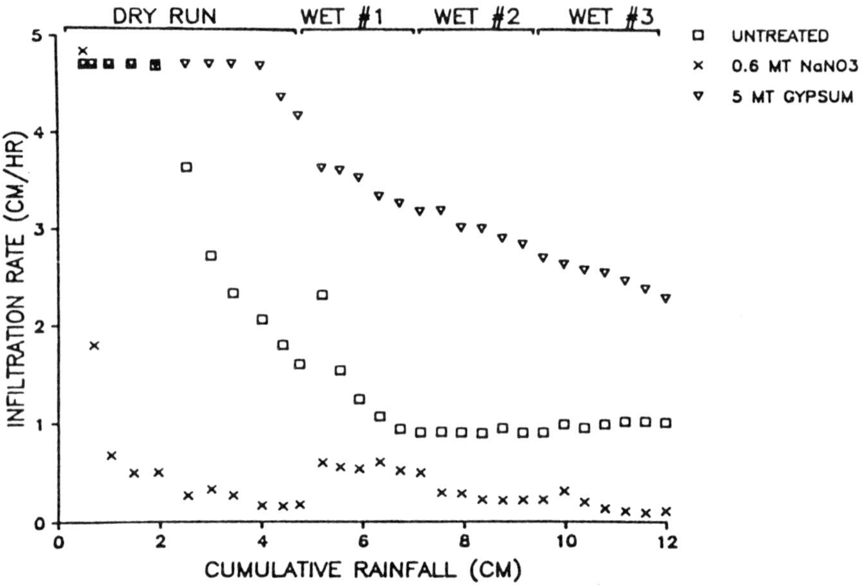

Fig. 3 Infiltration rate of Greenville sandy clay loam soil treated with either 0.6T ha^{-1} $NaNO_3$ or 5 t ha^{-1} $CaSO_4$, compared to untreated control, during 120 mm of simulated rainfall at 45 mm h^{-1}. (From Miller and Scifres, 1988.)

(Emerson, 1967) of aggregates takes place. Kaolinitic aggregates do not spontaneously disperse and thus will not affect soil HC. However, upon mechanical stirring by the beating action of raindrops, aggregates from kaolinitic soils may disperse (in accordance with the level of exchangeable Na) and consequently form a seal with low IR.

3. *Effect of Exchangeable Magnesium*

Exchangeable Mg can cause a deterioration in soil structure and develop a "magnesium solonetz" (Ellis and Caldwell, 1935). In addition, Mg enhances dispersion in montmorillonitic and illitic clays compared to Ca (Bakker and Emerson, 1973). Levy et al., (1988b) compared the effect of Mg to that of Ca as complementary cation to Na on the IR of three South African soils. The results showed that exchangeable Mg had a similar effect to that of Ca on the IR. These researchers concluded that the adverse effect of Mg on clay dispersion was not pronounced in their study because, in IR studies, clay dispersion plays only a secondary role in determining soil sealing and permeability.

Keren (1990) compared the effect of Mg to that of Ca as the complementary cation to Na under a rain kinetic energy range of 3.4–22.9 kJ m^{-3}. At the highest kinetic energy (similar to that used by Levy et al., 1988b), the effect of Mg on the IR was similar to that of Ca. For low to medium kinetic energy (8.0–12.5 kJ m^{-3}) the IR in the Mg-Na treated soil was lower than that in the Ca-Na soil. The adverse effect of Mg on soil permeability could be observed, as previously proposed by Levy et al., (1988b), only under conditions where chemical dispersion is dominant in controlling IR (i.e., raindrops with low to medium kinetic energy).

4. *Effect of Exchangeable Potassium*

The effect of exchangeable K on water transmission properties is controversial (Chen et al., 1983, and references there). Despite that, the effect of K on seal formation and IR has received little attention. Levy and van der Watt (1990) compared the effect of K on the IR of two South African soils to that of Ca and Na. Their results showed that, when K was the complementary cation to Ca, increasing the amount of exchangeable K led to a decrease in the IR. However, when compared to Na, exchangeable K had a more favorable effect on soil IR than Na (Fig. 4). Levy and van der Watt (1990) concluded that K has an intermediate effect, between that of Ca and Na, on IR of soils exposed to rain.

III. RAIN PROPERTIES AND SEAL FORMATION

A. Effect of Rain Properties

1. *Rain Characterization*

Rain is generally characterized by the following parameters: (i) rain intensity, (ii) raindrop median diameter and (iii) final velocity of the median raindrop. The relationships among these parameters were examined (Laws, 1940; Wischmeier and Smith, 1951), and it was found that (i) drops with a large diameter reach a high final velocity, and vice versa, and (ii) the higher the rain intensity the higher the percentage of large drops. The rate of seal formation was found to depend on median drop diameter and rain intensity (Ellison, 1947). On the other hand, Morin and Benyamini (1977) reported that rain intensity had no effect on the final IR of the seals formed.

2. *Effect of Raindrop Kinetic Energy*

Of the various rain properties, raindrop kinetic energy has become in recent years the most common property seal formation is associated with. The effect

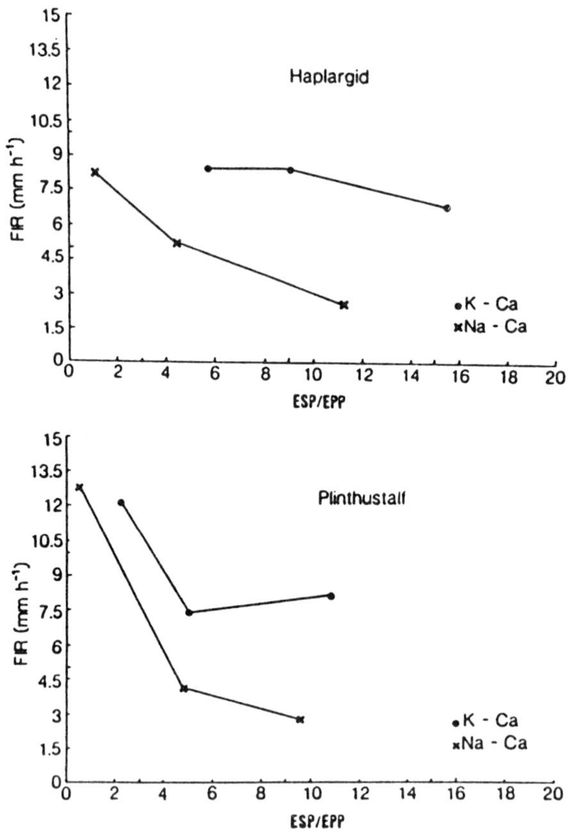

Fig. 4 The effect of exchangeable potassium percentage (EPP) and exchangeable sodium percentage (ESP) on the final IR (FIR) of the (a) Haplargid and the (b) Plinthustalf soils when exposed to distilled water rain. (From Levy and van der Watt, 1990.)

of drop kinetic energy (KE) on seal formation has been studied by many researchers (e.g., Agassi et al., 1985; Thompson and James, 1985; Keren, 1990; Mohammed and Kohl, 1987).

Agassi et al., (1985) showed that when soil was exposed to drops with KE below 0.01 J mm^{-1} m^{-2} (i.e., fog-type rain), no seal was formed. But when the KE was 23.0 J mm^{-1} m^{-2} (a KE typical of high-intensity rainstorms), a seal with a very low permeability was formed. The KE of rain or sprinklers' drops may vary between these two extremes (Hudson, 1971).

A detailed study on the combined effects of raindrop size and impact velocity on IR of soils was carried out by Betzalel et al. (1995). They applied rain of a constant intensity (40 mm h^{-1}) with 2.53- and 3.37-mm-diameter drops falling from 0.4-, 1.0-, 2.0-, 6.0- and 10.0-m heights on two soils and measured the decay in IR. Aggregate disintegration and seal formation increased with an increase in the drops' impact energy (Fig. 5). Differences in the response of the two soils to KE of the drops was limited to height of fall of drops in the range of 0.4–2.0 m; thereafter drop KE resulted in a similar final IR in both soils (Fig. 5). However, the two drop sizes studied had, in most cases, a similar effect on the final IR (Betzalel et al., 1995).

Mohammed and Kohl (1987) plotted the IR for four KE levels against cumulative kinetic energy and observed that the various curves tended to converge during the rapidly decreasing infiltration portion and then diverge as final IR was approached (Fig. 6). Mohammed and Kohl (1987) attributed the divergence in final IR to the differences in drop sizes which resulted from the application intensities not being identical for the different energy levels.

Characterizing rain by KE per unit area and unit time was proposed by Shevbs (1968) as a better term than just using KE, because it incorporates rainfall or sprinkler intensity. This concept was used by Mohammed and Kohl (1987) and Thompson and James (1985), who referred to it as droplet energy

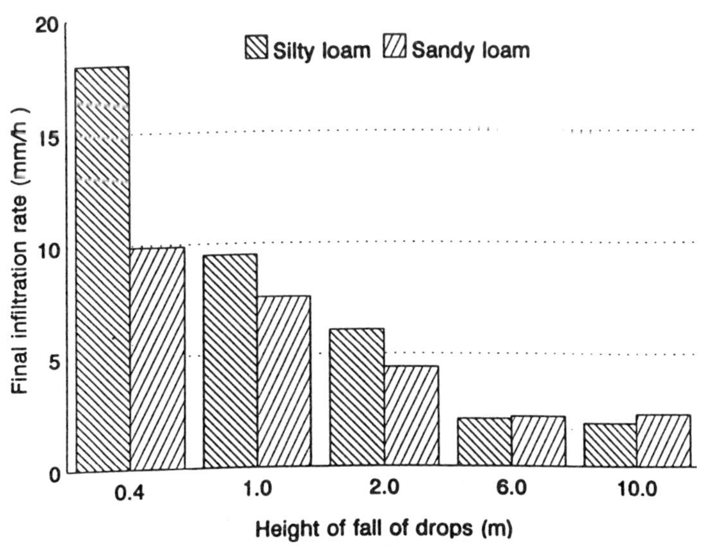

Fig. 5 Final infiltration rate of a sandy loam and a silty loam as a function of drops falling from various heights. (Adapted from Betzalel et al., 1995.)

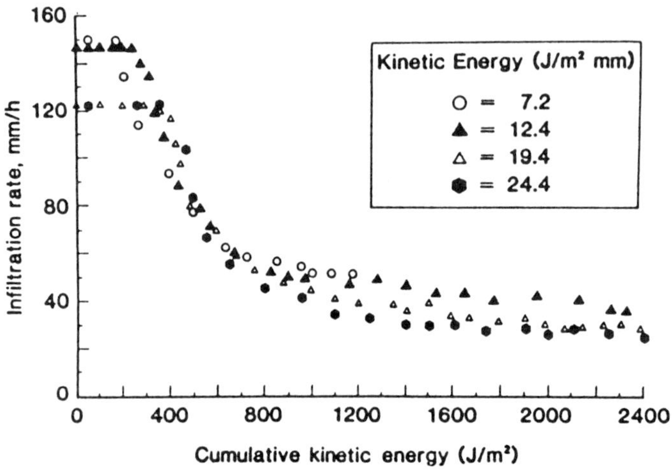

Fig. 6 Infiltration rates as a function of cumulative kinetic energy for four kinetic energy levels on a bare-dry soil surface. (From Mohammed and Kohl, 1987.)

flux (DE_f). These investigators showed that DE_f can be used to determine the amount of water that can be applied to the field without obtaining runoff for different combinations of drop KE and application intensity.

It is expected that the relation between seal formation and drops' KE will depend on the stability of the soil structure. Seal formation in soils with stable aggregate will take place only when the soils are exposed to high-energy rain. Conversely, in soils with unstable structure, the seal may be formed under low-energy rain. This was verified by Shainberg and Singer (1988), who observed that sodic soils were more susceptible to sealing by low impact drops than were Ca soils.

B. Effect of Electrolyte Concentration in the "Rain" on the IR

Clay dispersion is very sensitive to the chemistry (electrolyte concentration and cationic composition) of the applied water (Shainberg and Letey, 1984 and references there). This is particularly true in the case where the soil surface is exposed to the mechanical action of the falling water drops, which enhances clay susceptibility to chemical dispersion. Agassi et al. (1981) studied the effect of electrolyte concentration and soil sodicity on the IR of two loamy soils in Israel, using a laboratory rain simulator. Their results (Fig. 7) revealed that the

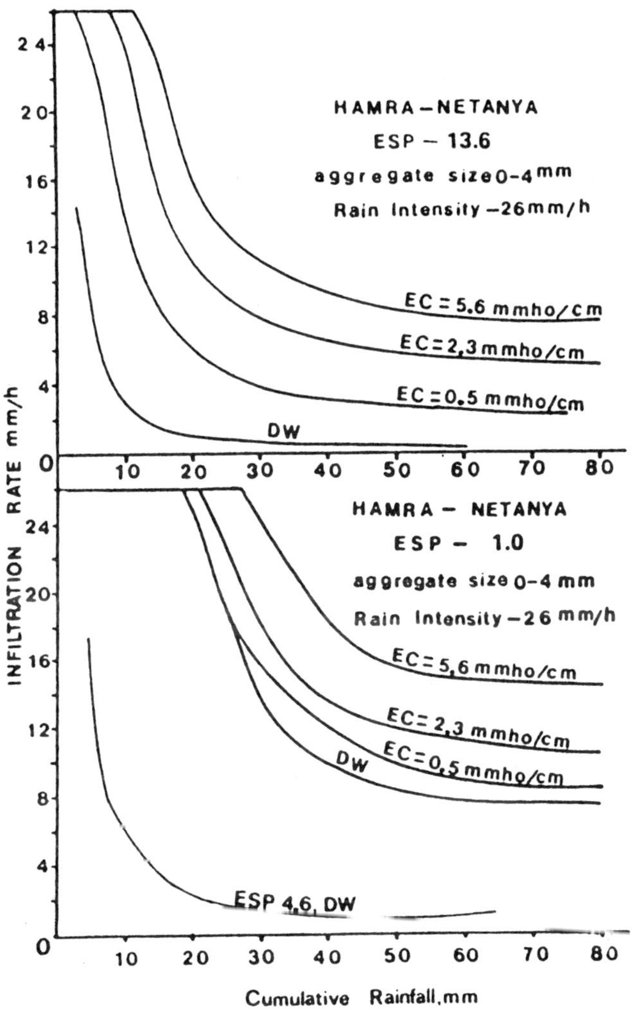

Fig. 7 Effect of electrolyte concentration in rain-simulation experiments on the infiltration rate of Hamra soil. (From Agassi et al., 1981.)

IR was by far more sensitive than the HC to the electrolyte concentration of the applied water (Shainberg and Letey, 1984). Similarly, Oster and Schroer (1979), who studied a Heimdal soil from North Dakota in the United States, found that cationic concentration greatly affected IR even at low sodium adsorption ratios (SAR). They observed an increase in final IR from 2 to 28 mm h^{-1} as cation concentration in the applied water having SAR levels between 2 and 4.6 increased from 5 to 28 meq L^{-1}.

The results in Fig. 7 show that salt concentration had an effect not only on the final IR but also on the rate at which the IR dropped from the initial to its final value. The lower the concentration of the electrolyte solution, the faster the rate at which the IR decreased. Similarly, for the same electrolyte concentration, increasing soil ESP results in a sharper decrease in the IR and a lower final IR.

Seal formation depends on the potential of the soil clays to disperse with increase in soil sodicity and a decrease in the electrolyte concentration. Under conditions favorable for soil clays to disperse, seal formation by raindrops is enhanced.

Chemical dispersion is prevented when electrolytes are present in the soil solution. Hence, under natural rain conditions a readily available electrolyte source should be added to the soil surface to compensate for the lack of electrolytes in the rainwater. It is important that the source of electrolyte added has a solubility that makes a substantial contribution to the ionic strength of the soil solution, yet it must be low enough to allow continued release of salt over a considerable time period. Phosphogypsum was found to be an efficient agent for preventing clay dispersion when added to the surface of smectitic (Kazman et al., 1983) and kaolinitic (Miller, 1987b; Miller and Scifres, 1988) soils. When 5 Mg ha^{-1} PG was spread on the soil surface, the rate of decline in IR and the final IR were both affected (Figs. 2 and 3). Hence, seal formation was slowed down and its permeability was much higher in the PG treatments. PG was effective also in soils of ESP 1.0 (Fig. 2), thus indicating that some chemical dispersion takes place even at very low ESP values, and possibly even in Ca-saturated soils. A detailed review on the effect of soil amendments on IR and soil erosion is given in Chapter 7.

C. Effect of Consecutive Rainstorms

Alternate applications of rain and saline water (by irrigation) predominate in many semiarid regions of the world. The stability and reversibility of the crusts formed were found to depend on the period of drying and water quality (Hardy et al., 1983; Levy et al., 1986). The term *crust* is introduced to distinguish a dried seal from a wet seal. In their experiments, Hardy et al. (1983) studied the effect of subsequent saline water (SW) sprinkling on the stability of crusts formed under distilled water (DW) rain. They found that application of SW in the second storm caused an increase in the final IR of the Netanya soil to 5.6 mm h^{-1} compared to 2 mm h^{-1} in the first storm with DW (Fig. 8). They observed that in this soil the raindrop impact caused a breakdown of the old crust and a formation of a new, more permeable one due to the use of SW. In the loess soil from Nahal-Oz only a slight increase in the final IR was observed in the consecutive SW storm (Fig. 8). Hardy et al. (1983) concluded that in the loess soil the structure of the crust formed in the first storm was stable and did not break during the second storm. In a further study on crust stability, Levy, Shainberg and Morin (1986) observed for the loess soil from Nahal-Oz that, if the crust was sufficiently dry, applying a second storm with SW increased the final IR to 6.4 mm h^{-1} compared to 2.5 mm h^{-1} at the end of the former DW storm. In the case where only partial drying occurred, changing from distilled water in the first storm to saline water in the consecutive storm did not increase the final IR (Levy, Shainberg and Morin, 1986). It was then concluded that sufficient drying of the crust weakens its structure and causes its breakdown by the beating action of the raindrops in the subsequent storm. The quality of the water in that storm will determine the characteristics of the newly formed crust.

The stability and reversibility of seals (i.e., wet crusts) was found to depend on whether the seal was fully or partially developed. Agassi et al. (1988) found that the seal formed by a distilled water rain of high energy and sufficient duration is fully developed. A seal formed by rain with low energy, or by high-energy rain of short duration, or by saline rainwater of high energy and long duration, is a partially developed seal. A fully developed seal is stable and its permeability responds only to simultaneous changes in electrolyte concentration and rain energy. Conversely, a partially developed seal is not stable and its permeability responds even to changes in electrolyte concentration only.

IV. COMBINED EFFECT OF RAIN ENERGY, SOIL SODICITY AND WATER QUALITY ON IR

Studies on the effect of soil sodicity and water quality on seal formation and IR concluded generally that the IR decreases with a decrease in the electrolyte concentration of the applied water and an increase in soil ESP (Shainberg and Letey, 1984 and references there). A quantitative analysis of the relative importance of each of the two factors and possible interaction between them in determining IR has received little attention. Levy et al. (1994) tried to evaluate the relative contribution of soil sodicity and water quality to the final IR of three soil types from Israel, using the stepwise linear regression procedure. Their results showed that both factors had a predominant effect on the final IR, with the electrolyte concentration in the water explaining a greater portion of the variation in the final IR (37.6%) than did soil ESP (30.3%). Conversely, ESP explained a greater portion of the variation in soil loss (14.8%) than did the electrolyte concentration in the applied water (9.7%);

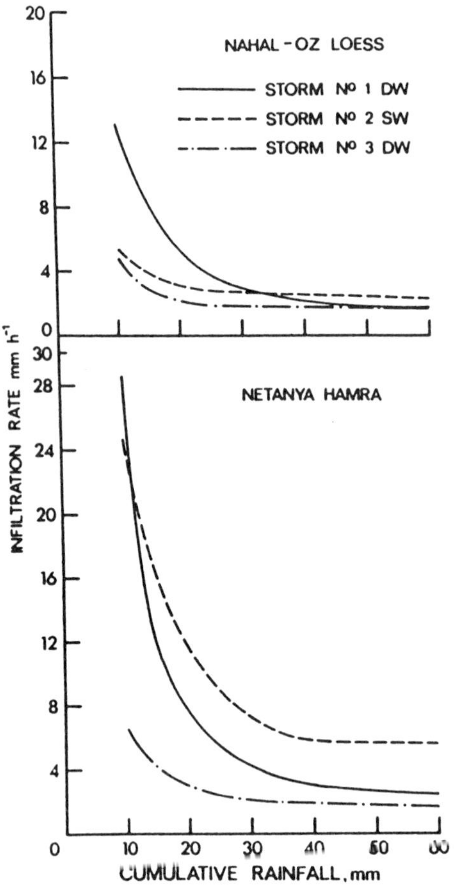

Fig. 8 Infiltration rates of Nahal-Oz and Netanya soils subjected to three consecutive storms of sequence DW-SW-DW. (From Hardy et al., 1983.)

their contribution, however, was secondary to that of clay content (37.5%) (Levy et al., 1994).

Agassi et al. (1994) studied the effect of soil sodicity water quality and drop KE on IR of two kaolinitic soils. They observed that the final IR decreased exponentially as drop KE increased. Agassi et al. (1994) suggested that although all three factors had a significant effect on the IR, drop KE had an overriding effect on the reduction in final IR compared with soil sodicity and water quality.

V. SEAL STRUCTURE: EFFECT OF SOIL SODICITY AND ELECTROLYTE CONCENTRATION

Micromorphological investigations of crust structure (McIntyre, 1958) have shown that it consisted of two distinct parts: (i) an upper skin seal attributable to compaction by raindrop impact; and (ii) a zone of decreased porosity (termed "washed-in zone"), attributed to the accumulation of small particles. The washed-in layer was formed only in soils that were easily dispersed (McIntyre, 1958). Conversely, Chen et al. (1980), studying scanning electron micrographs (SEM) of crust on loessial soil, found only a thin skin seal 0.1 mm thick at the uppermost layer of the soil, with no evidence of accumulation of fine particles in the washed-in zone. In this study, the soil was rained on with tap water whose electrical conductivity was 0.6 dS m^{-1}. This electrolyte concentration generally exceeds the flocculation value of the soil clays and prevents clay dispersion, movement, and the formation of the washed-in layer (Shainberg and Letey, 1984).

Gal et al. (1984) studied the SEM micrograph of a sandy loam with ESP of 1.0 and 11.6, respectively, exposed to distilled water rain. At ESP 1.0, sand grains covered with a skin of clay were observed throughout the upper 2 mm. At this ESP level, both naked sand grains from which the clay skin had been stripped or a layer of accumulated clay was absent. The absence of these phenomena indicates that only a small amount of clay dispersion or clay movement took place. At ESP 1.0, only a thin (~0.1 mm) and compacted layer of broken aggregates sealing the soil surface was observed, resulting in final IR of 7.0 mm h^{-1} (Fig. 2). Conversely, in the SEM micrographs of the sandy loam with ESP 11.6, being exposed to rain, the surface layer (~0.25 mm) consisted of naked sand grains (Gal et al., 1984). An ESP of 11.6 combined with distilled water rain and the mechanical impact of the drops caused clay dispersion and stripping of the clay skins from the sand grains. Following clay dispersion, some downward movement of clay particles occurred, and clay accumulated in the washed-in layer (~0.75 mm thick). Where the washed-in layer was formed the final IR dropped to <2 mm h^{-1} (in spite of the presence of large sand grains at the soil surface).

Onofiok and Singer (1984) performed SEM studies of seals of three different soil types and observed different morphological features in the seals formed. In two of their soils they observed the presence of a the thin compacted surface layer and the washed-in zone. These investigators concluded that differences in the structure of seals are determined by the physical, chemical and mineralogical properties of the soil.

Levy et al. (1988a) observed that the structure of seals is not homogeneous and consists of two distinct microtopographical features. The first are mounds which exhibit high permeability and a structure resembling an unsealed soil. The second are plains with low permeability and a seal that was clearly visible in SEM micrographs. The area occupied by the mounds and their permeability decreased with an increase in soil ESP (Levy et al., 1988a).

The fact that a thin layer, which is readily accessible because it is formed at the soil surface, controls rain infiltration and runoff is very important. Because seal formation is a surface phenomenon it is relatively easy to stabilize the aggregates at the soil surface by application of small amounts of amendments to the surface (organic polymers, gypsum, etc.).

VI. INFILTRATION IN NONARABLE LANDS

Nonarable lands differ from agricultural lands because, among other reasons, they are not subjected to cultivation and mechanical disruption. Hence, unlike cultivated lands, changes in their IR depend on some other mechanisms and not necessarily on seal formation. In sealing soils, IR was reported to either correlate negatively with rainfall intensity (McIntyre, 1958; Farres, 1978) or to be independent of it (Morin and Benyamini, 1977). Conversely, in cases where sealing is not predominant, IR increases with rainfall intensity for two reasons: (1) There is a tendency for higher rainfall intensities to exceed the saturated hydraulic conductivity of larger proportions of the soil surface and thereby to raise the spatially averaged hydraulic conductivity (Hawkins, 1982). (2) Increasing rainfall intensity increases runoff depth. This in turn leads to inundation of a larger fraction of the microtopography and especially the more permeable vegetated mounds, which again increases the spatially averaged hydraulic conductivity (Cook, 1946; Moldenhauer et al., 1960).

It was concluded that the effect of rainfall intensity on IR results from interactions of runoff depth with the hydraulic conductivity of the soil, microtopographic form, vegetation density and hill slope gradient and length (Dunne et al., 1991).

VII. SUMMARY AND CONCLUSIONS

Cultivated soils are structurally unstable and form a seal at the soil surface when exposed to rain or overhead sprinkler irrigation. The formation of a seal determines rain infiltration, runoff, and soil erosion. The rate of seal formation as a function of rain depth is determined by the decrease in IR and may be quantified by an Hortonian type equation which includes three soil parameters: initial IR, final IR, and the rate at which the IR decreases from initial to final value (i.e., rate of seal formation).

The susceptibility of soils to sealing depends on soil properties (clay content and mineralogy and cationic composition of the exchange phase) and rain properties (drop kinetic energy and electrolyte concentration in the applied water). Seal formation depends also on soil management (method of cultivation, type of crops, etc.) and soil slope. Further research on the relation between soil properties and soil sealing is required for the prediction of rain infiltration, successful germination of seeds, etc., from soil properties.

Runoff resulting from seal formation and low IR values may be predicted by combining the rainfall data with the infiltration equation of the soils. Runoff prediction is essential for soil erosion prediction and for a scientific design of water conservation and water harvesting systems.

REFERENCES

Aarstad, J. S., and Miller, D. E. (1973). Soil management practices for reducing runoff under center-pivot sprinkler system. *J. Soil Water Cons. 28*: 171–173.

Agassi, M., Bloem, D., and Ben-Hur, M. (1994). Effect of drop energy and soil and water chemistry on infiltration erosion. *Water Resources Res. 30* (4): 1187–1193.

Agassi, M., Morin, J., and Shainberg, I. (1985). Effect of raindrop impact energy and water salinity on infiltration rates of sodic soils. *Soil Sci. Soc. Am. J. 49*: 186–190.

Agassi, M., Shainberg, I., and Morin, J. (1981). Effect of electrolyte concentration and soil sodicity on infiltration rate and crust formation. *Soil Sci. Soc. Am. J. 45*: 848–851.

Agassi, M., Shainberg, I., and Morin, J. (1988). Effects on seal properties of changes in drops energy and water salinity during a continuous rainstorm. *Aust. J. Soil Res. 26*: 651–659.

Arshad, M. A., and Mermut, A. R. (1988). Micromorphological and physicochemical characteristics of soil crust types in Northwestern Alberta, Canada. *Soil Sci. Soc. Am. J. 52*: 724–729.

Bakker, A. C., and Emerson, W. W. (1973). The comparative effect of exchangeable calcium, magnesium and sodium on some physical properties of red-brown earth subsoils. III: The permeability of Shepperton soil and comparison methods. *Aust. J. Soil Res. 11*: 159–165.

Ben-Hur, M., Shainberg, I., Bakker, D., and Keren, R. (1985). Effect of soil texture and $CaCO_3$ content on water infiltration in crusted soil as related to water salinity. *Irrig. Sci. 6*: 281–294.

Betzalel, I., Morin, J., Benyamini, Y., Agassi, I., and Shainberg, I. (1995). Water drop energy and soil seal properties. *Soil Sci. 159*: 13–22.

Bresler, E., and Kemper, W. D. (1970). Soil water evaporation as affected by wetting methods and crust formation. *Soil Sci. Soc. Am. Proc. 34*: 3–8.

Chen, Y., Banin, A., and Borochovitch, A. (1983). Effect of potassium on soil structure in relation to hydraulic conductivity. *Geoderma 30*: 135–147.

Chen, Y., Tarchitzky, J. T., Brouwer, J., Morin, J., and Banin, A. (1980). Scanning electron microscope observation of soil crusts and their formation. *Soil Sci. 130*: 49–55.

Chiang, S. D., Radcliffe, D. E., Miller, W. P., and Newman, K. D. (1987). Hydraulic conductivity of three southeastern soils as affected by sodium, electrolyte concentration and pH. *Soil Sci. Soc. Am. J. 51*: 1293–1299.

Cook, H. L. (1946). The infiltration approach to the calculation of surface runoff. *Eos. Trans. AGU 27*: 726–747.

Duely, F. L. (1939). Surface factors affecting the rate of intake of water by soils. *Soil Sci. Soc. Am. Proc. 4*: 60–64.

Dunne, T., Zhang, W., and Aubry, B. F. (1991). Effects of rainfall, vegetation and microtopography on infiltration and runoff. *Water Resources Res. 27*: 2271–2285.

Ellis, J. H., and Caldwell, O. G. (1935). Magnesium clay solonetz. *Trans. 3rd Congr. Soil Sci.*, Vol. I, pp. 348–350.

Ellison, W. D. (1947). Soil erosion studies. III: Some effects of erosion on infiltration and surface runoff. *Agric. Eng. 28*: 245–248.

Emerson, W. W. (1967). A classification of soil aggregates based on their coherence in water. *Aust. J. Soil Res. 5*: 47–57.

Epstein, E., and Grant, W. J. (1973). Soil crust formation as affected by raindrop impact. In *Ecological Studies*, 4, *Physical Aspects of Soil Water and Salts in Ecosystems* (A. Hadas, D. Swartzendruber, P. E. Rijtema, M. Fuchs, and B. Yaron, eds.), Springer-Verlag, Berlin-Heidelberg-New York, pp. 195–201.

Farres, P. (1978). The role of time and aggregate size in the crusting process. *Earth Surf. Process. 3*: 243–254.

Frenkel, H., Goertzen, J. O., and Rhoades, J. D. (1978). Effects of clay type and content, exchangeable sodium percentage, and electrolyte concentration on clay dispersion and soil hydraulic conductivity. *Soil Sci. Soc. Am. J. 48*: 32–39.

Gal, M., Arcan, L., Shainberg, I., and Keren, R. (1984). The effect of exchangeable sodium and phosphogypsum on the structure of soil crusts. *Soil Sci. Soc. Am. J. 48*: 872–878.

Goldberg, S., and Glaubig, R. A. (1987). Effect of saturating cation, pH and aluminum and iron oxides on the flocculation of kaolinite and montmorillonite. *Clays Clay Miner. 35*: 220–227.

Green, W. H., and Ampt, G. A. (1911). Studies on soil physics. I: Flow of air and water through soils. *J. Agric. Sci. 4*: 1–24.

Hardy, N., Shainberg, I., Gal, M., and Keren, R. (1983). The effect of water quality and storm sequence upon infiltration rate and crust formation. *J. Soil Sci. 34*: 665–676.

Hawkins, R. H. (1982). Interpretation of source-area variability in rainfall-runoff relationships. In *Rainfall-Runoff Relationships* (V.P. Singh, ed.), Water Resources Publications, Fort Collins, CO, pp. 303–324.

Hudson, N. (1971). *Soil Conservation.* Cornell University Press, Ithaca, NY.

Hillel, D. (1980). *Application of Soil Physics.* Academic Press, New York.

Horton, R. E. (1940). An approach toward a physical interpretation of infiltration-capacity. *Soil Sci. Soc. Am. Proc. 5*: 399–417.

Kazman, Z., Shainberg, I., and Gal, M. (1983). Effect of low levels of exchangeable Na (and phosphogypsum) on the infiltration rate of various soils. *Soil Sci. 135*: 184–192.

Kemper, W. D., and Koch, E. J. (1966). Aggregate stability of soils from western U.S. and Canada. *USDA Tech. Bull.* 1355.

Kemper, W. D., Trout, T. J., Brown, M. J., and Rosenau, R. C. (1985). Furrow erosion and water and soil management. *Trans ASAE. 28*: 1564–1572.

Keren, R. (1990). Water-drop kinetic energy effect on infiltration in sodium-calcium-magnesium soils. *Soil Sci. Soc. Am. J. 54*: 983–987.

Laws, J. O. (1940). Recent studies in raindrops and erosion. *Agric. Engr. 21*: 431–433.

Levy, G. J., Berliner, P. R., du Plessis, H. M., and van der Watt, H. v.H. (1988a). Microtopographical characteristics of artificially formed crusts. *Soil Sci. Soc. Am. J. 52*: 784–791.

Levy, G. J., Levin, J., and Shainberg, I. (1994). Seal formation and interrill soil erosion. *Soil Sci. Soc. Am. J. 58*: 203–209.

Levy, G. J., Shainberg, I., and Morin, J. (1986). Factors affecting the stability of soil crusts in subsequent storms. *Soil Sci. Soc. Am. J. 50*: 196–201.

Levy, G. J., and van der Watt, H. v.H. (1988). Effects of clay mineralogy and soil sodicity on the infiltration rates of soils. *S. Afr. J. Plant Soil 5*: 92–96.

Levy, G. J., and van der Watt, H. v.H. (1990). Effect of exchangeable potassium on the hydraulic conductivity and infiltration rate of some South African soils. *Soil Sci. 149*: 69–77.

Levy, G. J., van der Watt, H. v.H., and du Plessis, H. M. (1988b). Effect of sodium-magnesium and sodium-calcium systems on soil hydraulic conductivity and infiltration. *Soil Sci. 146*: 303–310.

McIntyre, D. S. (1958). Permeability measurements of soil crusts formed by raindrop impact. *Soil Sci. 85*: 185–189.

Miller, W. P. (1987a). A solenoid-operated variable intensity rainfall simulator. *Soil Sci. Soc. Am. J. 51*: 832–834.

Miller, W. P. (1987b). Infiltration and soil loss of three gypsum amended ultisols under simulated rainfall. *Soil Sci. Soc. Am. J. 51*: 1314–1320.

Miller, W. P., and Scifres, J. (1988). Effect of sodium nitrate and gypsum on infiltration and erosion of a highly weathered soil. *Soil Sci. 148*: 304–309.

Mohammed, D., and Kohl, R. A. (1987). Infiltration response to kinetic energy. *Trans. ASAE. 30*: 108–111.

Moldenhauer, W. C., Barrows, W. C., and Swartzendruber, D. (1960). Influence of rainstorm characteristics on infiltration measurements. *Trans. Int. Cong. Soil Sci.*, Vol. 7, pp. 426–432.

Morin, J., and Benyamini, Y. (1977). Rainfall infiltration into bare soils. *Water Resources Res. 13*: 813–817.

Morin, J., Goldberg, S., and Seginer, I. (1967). A rainfall simulator with a rotating disk. *Trans. Am. Soc. Agric. Eng. 10*: 74–79.

Onofiok, O., and Singer, M. J. (1984). Scanning electron microscope studies of soil surface crusts formed by simulated rainfall. *Soil Sci. Soc. Am. J. 48*: 1137–1143.

Oster, J. D., and Schroer, F. W. (1979). Infiltration as influenced by irrigation water quality. *Soil Sci. Soc. Am. J. 43*: 444–447.

Philip, J. R. (1957). The theory of infiltration. 4: Sorptivity and algebraic infiltration equations. *Soil Sci. 84*: 257–264.

Shainberg, I., and Letey, J. (1984). Response of soils to sodic and saline conditions. *Hilgardia 52*: 1–57.

Shainberg, I., and Singer, J. M. (1988). Drop impact energy—soil ESP interactions in seal formation. *Soil Sci. Soc. Am. 52*: 1449–1452.

Shevbs, G. I. (1968). Data on the erosive action of water drops. *Soviet Soil Sci. 2*: 262–269.

Stern, R., Ben-Hur, M., and Shainberg, I. (1991). Clay mineralogy effect on rain infiltration, seal formation and soil losses. *Soil Sci. 152*: 455–462.

Thompson, A. L., and James, L. G. (1985). Water droplet impact and its effect on infiltration. *Trans. ASAE 28*: 1506–1510.

van Olphen, H. (1977). *An Introduction to Clay Colloid Chemistry*, 2nd ed. Wiley, New York.

Wischmeier, W. H., and Smith, D. D. (1951). Rainfall energy and its relation to soil loss. *Trans Am. Geophys. Un. 39*: 285–308.

2

Rainfall Analysis

Joseph Morin
Soil Conservation and Drainage Division, Soil Erosion Research Station, Rupin Institute, Ministry of Agriculture, Emek-Hefer, Israel

I. INTRODUCTION

Estimations of runoff, and erosion, and the design of flood control, water harvesting, tillage management and soil and water conservation plans are all based on precipitation data, properties, soil, and management. Rainfall parameters like annually, monthly and daily amounts, and the intensity of individual rainstorms have to be analyzed for their magnitude and probability of occurrence since engineering and economy decisions are based on them.

Rainfall-runoff relationships for any rainstorm depend on the dynamic interaction between rain intensity, soil infiltration and surface storage. Runoff occurs whenever rain intensity exceeds the infiltration capacity of the soil, providing that there are no physical obstructions to surface flow. Rainfall analysis should consider therefore rain intensities sequence as well as their magnitude.

Information on rainfall amounts and intensities are collected in most parts of the world. Such data is probably the only reliable set of information which allows one to assess water resources planning on a probability basis. Quantitative knowledge of the region rainstorm characteristics is as important as the soil properties information.

II. METHODS AND MODELS FOR RAINSTORM PROBABILITY ANALYSIS

A. Plotting Position Method

The simplest assumption relates the probability (P) directly to the observed frequency. This is known as the California method (Lloyd, 1970) and is given by the general equation

$$P = \frac{m}{n + 1} \times 100 \tag{1}$$

where

n = total number of statistical events
m = rank of events arranged in descending order of magnitude
(m = 1 for largest value, and $m = n$ for smallest value)

The common use of the method is to plot the actual parameter data with their P values on log-probability paper.

A straight line, which has the minimum deviation from plotted points, represents the desired probability relation. Any value of desired probability (X_{p_i}) can be obtained from this line.

B. Lognormal Method

Many phenomena in nature follow the normal distribution. Chow (1954) and others have shown that the logarithms of hydrological and rainfall data follow more closely the normal distributions than the actual data. If y_i is the logarithm (y_i = ln X_i) of a variable X_i normally distributed, then the variable y_i is said to be logarithmic and the set of y_i is normally distributed. Thus,

$$P_{X_i} = \frac{1}{X_i\, \sigma_y \sqrt{2\pi}} \exp - \frac{(y_i - \mu_y)^2}{2\, \sigma^2 y} \tag{2}$$

where μ_y and σ_y are the mean and standard deviation of the natural logarithms of X_i. Assuming that $\mu_y \approx \bar{y}$ and $\sigma_y \approx S_y$ for the rank of the available data, then the size of an event (X_{p_i}) can be estimated for any desired probability p_i by

$$y_{p_i} = Z_{p_i} S_y + \bar{y} \tag{3}$$

where Z_{p_i} is the standard normal deviate for probability P_i and $y_{p_i} = \ln X_{p_i}$. Then

$$X_{p_i} = \exp y_{p_i} \tag{4}$$

and

$$\bar{y} = \frac{\Sigma \ln X_i}{n} \tag{5}$$

and

$$S_y = \sqrt{\frac{\Sigma(\ln X_i)^2 - (\Sigma \ln X_i)^2 / n}{n - 1}} \tag{6}$$

C. Chow Method

The hydrologic or rainfall events can be predicted by the log-probability law proposed by Chow (1954). The parameter value for any desired probability X_{p_i} can be calculated according to the Chow method by

$$X_{p_i} = \overline{X}\,(1 + \mathrm{Cv}\,K) \tag{7}$$

where

$\overline{X}$ = average of recorded data Cv (coefficient of variation) = $\dfrac{S_x}{\overline{X}}$

K = frequency factor

The frequency factor (K) depends on the law of occurrence and the standard deviation of the data. K, can be calculated by the Chow method as:

$$K = \frac{\exp(\sigma_y - Z_{p_i} - \sigma_y^2/2) - 1}{(\exp \sigma_y^2 - 1)^{1/2}} \tag{8}$$

$\sigma_y \approx S_y$ (see Eq. (6))

D. Gumbel Method

For extreme values of rainfall of hydrologic events, such as the annual maximum flood or rainstorm depth, Gumbel (1958) suggested the exponential probability relations. According to his method, the predicted extreme value for any probability can be evaluated by

$$X_{p_i} = \frac{yS_x + \beta}{1.28255} \tag{9}$$

where

$y = -\ln[-\ln(1 - p_i)]$ and
$\beta = \overline{X} - 0.45S_x$.

Detailed information for the different probability analysis and parameters estimation is given by Kinte (1977).

Most of the rainfall data in most parts of the world are limited by the short period of actual measurements. Since all of the prediction methods are based principally on past data, the absolute validity of the predicted values is uncertain.

This is especially true for the most important low-probability events. Since there is no way to be sure which prediction method is superior, our economists and engineers were presented with all of the prediction range as calculated by the different methods.

III. RAIN ANALYSIS

A. Rain Volume

The daily, monthly, and yearly rain volume and rainstorm volume data are obtained from standard daily rain measurements at 8 A.M. In Tables 1 and 2, the yearly and monthly rain volume for one station, as well as the predicted values for some given probabilities, are presented. One of the definitions of a rainstorm is the amount of rainfall in a period of time in which the interval between the rainfall segments does not exceed 24 h. The maximum rainstorm probability for each of the winter months, which is essential for runoff and flood prediction, is presented in Table 3.

B. Rainfall Intensities

A typical recorded rain gauge chart is given in Fig. 1. The graph on the chart is a cumulative rainfall curve, the slope of the graph being proportional to the intensity of the rainfall. Technically, rainstorm analysis starts with recording of each inflection point on the graph by a digitizer.

The analysis of the rainstorm chart of December 19, 1985 from the Ruhama station (Fig. 1), is presented in Fig. 2a–c. The rainstorm was analyzed in segments of rain. Each segment represents a certain period of time with uniform intensity.

In Fig. 2c, the grouping of storm intensities and volume regardless of rank and sequence is presented. For example, segments 19 and 22 have an intensity between 6 and 7 mm h^{-1} with a total volume of $0.63 + 0.69 = 1.32$ mm.

It is obvious that 35.8% of the rainstorm volume is below the intensity of 7 mm h^{-1}. The maximum rain intensity for 30 continuous minutes of rainfall (I_{30}) is 15.65 mm h^{-1}, which falls between segments 19 and 23. The I_{30} value was chosen since it represents the rainstorm erosivity factor R (Wischmeier, 1959).

C. Storm Kinetic Energy and Erosivity

The universal soil loss equation (U.S.L.E.) erosion model (Wischmeier, 1959) is a method for calculating storm kinetic energy E and erosivity R (see Agricultural Handbook 537). The storm kinetic energy can be computed as the sum of each rain intensity group I_i as

$$E = (210 + 89 \log I_i) D_i \text{ (in metric system)} \tag{10}$$

Table 1 Annual Precipitation Data and Predicted Values, Saad Station, Israel

	Annual precipitations		
N	Frequency (%)	Rain (mm)	Year
	(a)		
1	3.4	637.4	1956/57
2	6.9	622.0	1964/65
3	10.3	582.4	1963/64
4	13.8	565.7	1979/80
5	17.2	556.2	1955/56
6	20.7	518.8	1973/74
7	24.1	517.4	1966/67
8	27.6	506.1	1951/52
9	31.0	465.9	1971/72
10	34.5	458.7	1974/75
11	37.9	454.9	1967/68
12	41.4	424.1	1953/54
13	44.8	401.7	1960/61
14	48.3	384.7	1970/71
15	51.7	370.8	1949/50
16	55.2	370.6	1976/77
17	58.6	358.7	1958/59
18	62.1	354.5	1972/73
19	65.5	345.3	1957/58
20	69.0	338.6	1968/69
21	72.4	329.1	1954/55
22	75.9	310.0	1978/79
23	79.3	278.9	1952/53
24	82.8	271.4	1950/51
25	86.2	248.0	1975/76
26	89.7	246.9	1965/66
27	93.1	216.0	1969/70
28	96.6	41.2	1959/60
	(b)		
	5	630.4	
	10	586.3	
	20	526.2	
	30	477.8	
	40	436.4	
	50	377.7	
	60	357.0	
	70	335.7	
	80	277.4	
	90	243.8	
	95	119.7	

Table 1 Continued

CV = 0.343	S = 137.06	AVE = 399.14

(c)

Predicted values of annual rainfall (mm) according to the different methods of rainfall probability analysis.

P_i (%)	Chow	Gumbel	Lognormal
1	820.4	829.1	1220.2
5	653.7	654.9	856.4
10	585.3	577.9	709.4
20	500.0	497.8	564.5
50	377.6	376.6	364.7
80	285.0	286.6	235.6
90	249.5	248.3	187.5
95	218.4	220.2	155.3
99	173.8	174.3	108.1

Table 2 Precipitation for November and Predicted Values, Saad, Israel

(a) Precipitations in November

N	Frequency (%)	Rain (mm)	Year
1	3.3	212.8	1953/54
2	6.7	163.8	1967/68
3	10.0	161.4	1955/56
4	13.3	143.0	1957/58
5	16.7	119.2	1954/55
6	20.0	101.4	1971/72
7	23.3	100.5	1974/75
8	26.7	80.6	1972/73
9	30.0	79.2	1973/74
10	33.3	69.2	1963/64
11	36.7	64.1	1979/80
12	40.0	61.5	1956/57
13	43.3	59.1	1964/65
14	46.7	46.4	1960/61
15	50.0	43.7	1970/71
16	53.3	38.2	1978/79
17	56.7	31.4	1976/77
18	60.0	25.8	1969/70
19	63.3	24.9	1952/53
20	66.7	22.9	1951/52
22	73.3	21.5	1965/66

Table 2 Continued

(a) Precipitations in November

N	Frequency (%)	Rain (mm)	Year
23	76.7	20.8	1968/69
24	80.0	12.8	1950/51
25	83.3	8.0	1959/60
26	86.7	7.4	1949/50
27	90.0	3.5	1958/59
28	93.3	1.5	1977/78
29	96.7	1.1	1966/67

(b)

5	188.2
10	161.4
20	101.4
30	79.2
40	61.5
50	43.7
60	25.8
70	22.9
80	12.8
90	3.5
95	1.3

CV = 0.923 S = 55.66 AVE = 60.30

(c)

Predicted values of monthly rainfall (mm) according to the different methods of rainfall probability analysis

P_i (%)	Chow	Gumbel	Lognormal
1	275.4	234.9	760.4
5	161.3	164.1	304.4
10	120.0	132.9	187.0
20	86.0	100.3	103.6
50	44.2	51.2	33.5
80	22.9	14.6	10.8
90	16.2	0.0	6.0
95	12.2	0.0	3.7
99	7.1	0.0	1.5

Table 3 Maximum Precipitations for a Rainstorm in December and Predicted Values, Saad, Israel

N	Frequency (%)	Rain (mm)	Year	Storm started day	Storm duration days
1	3.2	211.5	1951/572	13	8
2	6.5	169.3	1966/67	17	5
3	9.7	105.7	1963/64	1	3
4	12.9	101.7	1968/69	4	5
5	16.1	98.4	1953/54	14	7
6	19.4	95.7	1977/78	21	3
7	22.6	87.9	1970/71	5	9
8	25.8	86.1	1956/57	5	6
9	29.0	80.4	1980/81	10	8
10	32.3	68.9	1949/50	20	5
11	35.5	64.7	1954/55	5	2
12	38.7	64.7	1971/72	5	4
13	41.9	63.7	1979/80	26	3
14	45.2	61.6	1974/75	3	4
15	48.4	56.9	1955/56	6	2
16	51.6	43.6	1978/79	1	2
17	54.8	40.8	1964/65	12	4
18	58.1	36.6	1972/73	18	4
19	61.3	36.2	1976/77	28	4
20	64.5	33.7	1975/76	8	3
21	67.7	31.7	1957/58	8	2
22	71.0	20.3	1950/51	4	2
23	74.2	19.6	1973/74	16	2
24	77.4	14.0	1952/53	27	2
25	80.6	13.9	1969/70	30	2
26	83.9	11.0	1967/68	29	2
17	87.1	4.8	1958/59	3	1
28	90.3	4.8	1960/61	5	2
29	93.5	3.2	1965/66	19	1
30	96.8	1.5	1959/60	26	1

CV = 0.846 S = 48.84 AVE = 57.76

P_i (%)	Predicted values in mm of rain: Chow	Gumbel	Lognormal
1	243.4	211.0	583.4
5	147.8	148.9	257.0
10	111.5	121.5	166.1
20	82.0	92.9	97.9
50	44.1	49.7	35.6
80	23.8	17.7	12.9
90	17.2	0.0	7.6
95	13.4	0.0	4.9
99	8.0	0.0	2.2

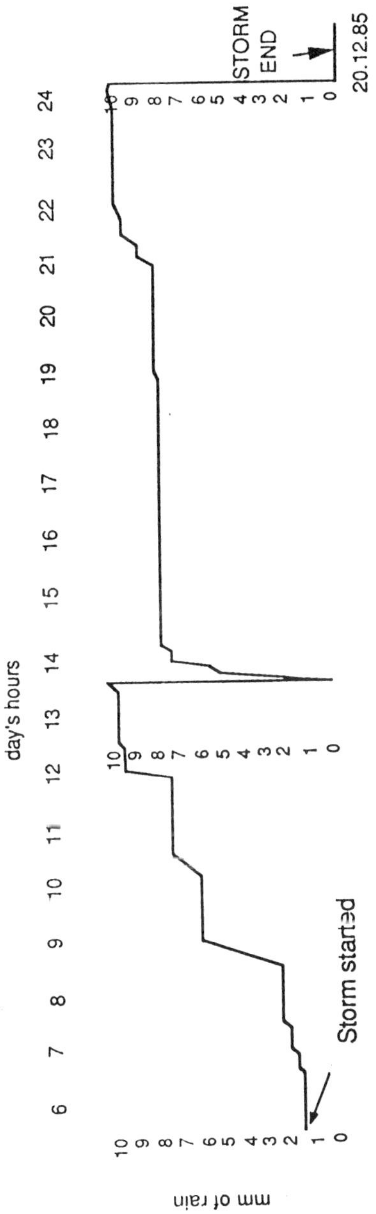

Fig. 1 Rainstorm chart of December 19, 1985, Ruhama, Israel.

(a)

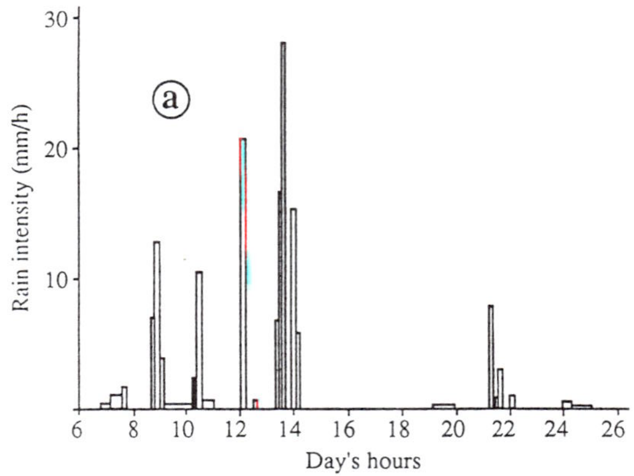

(b)

Segment	From time Hour	to time Hour	Intensity mm/h	Volume mm	Accum.vol. mm	ΔT Hour
1	6.00	6.19	0.07	0.01	0.01	0.19
2	6.19	6.71	0.12	0.7	0.08	0.52
3	6.71	7.11	0.46	0.18	0.26	0.39
4	7.11	7.52	1.04	0.43	0.69	0.41
5	7.52	7.68	1.69	0.26	0.95	0.15
6	7.68	8.68	0.09	0.09	1.05	1.00
7	8.68	8.78	7.06	0.71	1.75	0.10
8	8.78	8.96	12.64	2.27	4.03	0.18
9	8.96	9.08	3.77	0.47	4.50	0.12
10	9.08	10.29	0.27	0.33	4.82	1.21
11	10.29	10.42	2.21	0.21	5.11	0.13
12	10.42	10.51	10.31	0.88	5.99	0.08
13	10.51	11.00	0.59	0.29	6.29	6.28
14	11.00	12.01	0.18	0.18	6.46	1.01
15	12.01	12.11	20.53	2.05	8.51	0.10
16	12.11	12.50	0.10	0.04	8.55	0.38
17	12.50	12.65	0.67	0.10	8.66	0.90
19	13.55	13.64	6.87	0.63	9.37	0.09
20	13.64	13.73	16.30	1.39	10.76	0.08
21	13.73	13.87	27.55	3.86	14.62	0.14
22	13.87	13.97	6.63	0.69	15.31	0.10
23	13.97	14.06	15.23	1.29	16.60	0.08
24	14.06	14.14	5.85	0.50	17.10	0.08
25	14.14	14.28	0.29	0.04	17.14	0.13
26	14.28	17.98	0.11	0.39	17.53	3.70
27	17.98	19.08	0.02	0.03	17.56	1.10
28	19.08	19.94	0.44	0.38	17.94	0.85
29	19.94	21.17	0.05	0.07	18.00	1.23
30	21.17	21.26	7.84	0.67	18.67	0.08
31	21.26	21.42	0.55	0.09	18.76	0.16
32	21.42	21.54	0.91	0.10	18.87	0.11
33	21.54	21.69	3.14	0.47	19.34	0.15
34	21.69	21.95	0.10	0.03	19.36	0.27
35	21.95	22.16	0.96	0.20	19.56	0.20
36	22.16	23.91	0.09	0.16	19.72	1.75
37	23.91	0.30	0.64	0.25	19.96	0.39
38	0.30	1.00	0.34	0.24	20.20	0.70

(c)

Intensity mm/h	Volume mm	Accumulated percentage
-1.00	3.35	16.57%
1.00	0.69	20.00%
2.00	0.29	21.42%
3.00	0.94	26.08%
5.00	0.50	28.54%
6.00	1.32	35.08%
7.00	1.37	41.88%
10.00	0.88	46.21%
12.00	2.27	57.48%
15.00	1.29	63.88%
16.00	1.39	70.74%
20.00	2.05	80.91%
27.00	3.86	100.00%

Storm kinetic energy = 398.31$Jm^{-2}cm^{-1}$ of rainfall
The max. intensity for 30 min was : 15.65 mm/h AT SEGMENT 19-23
Erosivity (R) = 5.23$Jm^{-2}h^{-1}$

Fig. 2 Rainstorm analysis December 19, 1985, Ruhama.

where

E = kinetic energy (J m^{-2} cm^{-1} of rain)
I_i = rain intensity (cm h^{-1})
D_i = rain depth for intensity group I_i of the rainstorm (cm)

The erosivity parameter R according to the U.S.L.E. model presents for a large group of soils the best linear relation between soil erosion and rainstorm erosivity:

$$R = EI_{30} \text{ J m}^{-2} \text{ h}^{-1} \quad \text{(or metric ton per hectare)} \tag{11}$$

where

R = erosivity Jm^{-2} h^{-1}
I_{30} = the maximum rain intensity for a continuous 30 minute period of rainfall cm h^{-1}

1. *Probability Analysis of the Erosivity*

The probability analysis of the sum of the annual rainstorm erosivity for the Yavne station, Israel, is presented in Fig. 3. It was emphasized previously that since the best predicted values are uncertain, it is better to present the range of the values as calculated by the different methods.

Figure 3 presents a sample for this type of probability calculation. In Fig. 3a the calculated annual storm erosivity is presented.

The probability distributions of the annual storm erosivity for the Gumbel, normal, lognormal and Chow methods are presented in Fig. 3b with a rank of given probabilities. The standard deviation of each method represents the differences of the model calculated values from the data when the probabilities of the data were taken as the data frequencies. The asterisk in Fig. 3c represents the actual measured data located in their frequency position according to Eq. (1).

For each model the calculated values are presented for these same frequencies. A chi-square test was conducted between the data and each model value (Fig. 3d). The tests were conducted by comparing the numbers of events in each data group's cells and the predicted values for the same cell's borders. The probability fitness is presented by the cumulative distribution function in the last column of Fig. 3d:

$$p(\chi^2) = \int_0^{\chi^2} f(u)\, du = p_i(X \leq \chi^2) \tag{12}$$

where

(χ^2) = chi-square
$p(\chi^2)$ = probability of chi-square
$Q(\chi^2) = 1 - p(\chi^2) = p_i(X > \chi^2)$

a. Annual Rainstorm Erosivity Data (Jm^2h^{-1})

1958/59	58.2	1959/60	62.3	1960/61	262.7	1961/62	192.5
1962/63	454.6	1963/64	212.7	1964/65	407.8	1965/66	49.4
1966/67	201.6	1967/68	124.3	1968/69	290.0	1969/70	86.5
1970/71	289.4	1971/72	162.7	1972/73	111.7	1973/74	456.0
1974/75	504.8	1975/76	48.7	1976/77	236.7	1977/78	240.8
1978/79	170.3	1979/80	275.3				

$\bar{x} = 22.7$ $\bar{y} = 5.1860$ $S_x = 137.6$ $S_y = 0.7279$ $n = 22$ (S = standard deviation of sample)

b. Probability Distribution (calculated values)

(%)	Gumble	Normal	Log. nor.	Chow
1	654.2	542.7	972.0	744.7
5	479.4	449.0	591.9	476.2
10	402.2	399.0	454.3	379.1
20	321.7	338.5	329.8	291.1
50	200.1	222.7	178.8	184.4
80	109.8	106.9	96.9	126.6
90	71.3	46.4	70.3	107.8
95	43.1	0.0	54.0	96.3
99	0.0	0.0	32.9	81.3
S.DV.	27.19	29.52	55.32	44.85

c. The Annual Rainstorm Erosivity Data. Erosivity ($Jm^{-2}h^{-1}$)

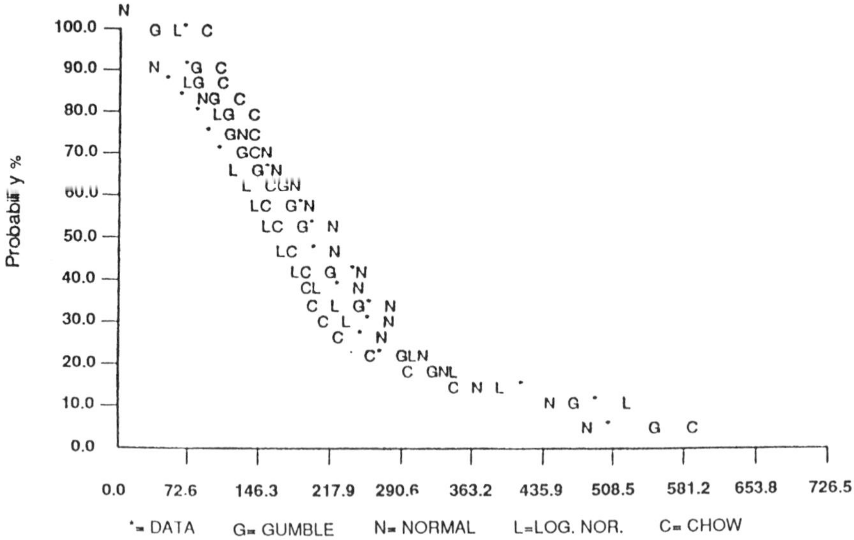

Fig. 3 Probability analysis of the annual rainstorm erosivity.

d. Chi-Square Test (for frequency distribution)

Class Interval	< 86	86–201	201–275	> 275	Probability fitness (%)
DATA	5	6	5	6	
GUMBLE	2.982	8.121	4.498	6.398	57.2
NORMAL	3.545	6.112	4.622	7.721	79.6
Log. NOR.	3.508	8.933	3.478	6.081	51.9
CHOW	0.445	12.084	4.454	5.017	0.0

Fig. 3 Continued

There is no agreement as to whether the chi-square test or the minimum standard deviation is better. However, in general, the range of values might serve as the best guide for economists and engineers. The same system of analysis is applied for other long-term rain and hydrologic data, such as the highest intensity for different time duration, or rainstorm kinetic energy for any desired period.

IV. LONG-TERM EVALUATIONS

Rain record charts are available for most parts of the world. Such data are probably the most reliable long-term bank of information which allows for water resources, flood control and soil conservation projects planning on a probability basis. Some advantages for using this information for long-term evaluations will now be demonstrated.

The highest I_{30} intensity for 32 years at Yavne Station in the southern coastal plain of Israel is presented in Figs. 4 and 5. Figure 4 relates to the first half of the rain season, Fig. 5 to its second half. It is obvious that the intensities for October to January are much higher than those of February to April. Moreover, the higher intensities of the early season occur at a time when the soils are mostly bare. Obviously, runoff and erosion rates are much higher from bare soil than from vegetated soil (Lal, 1974). The high rain intensities during this part of the year should therefore be a key consideration in planning.

A set of the highest daily rain depth, with intensities exceeding 10 mm h^{-1}, is presented in Fig. 6. It demonstrates maximal daily rain depth above 10 mm h^{-1} from the 32-year period. Obviously, any certain rainfall intensity may be selected. The use of this type of analysis is very wide.

Water-harvesting planning for animal and domestic use requires a different approach. In such cases, the main concern is water availability, the duration and sequence of water shortage. The maximum daily rain depths for February to October (the beginning and the end of the dry season in the Mediterranean climate zone) for 32 years is presented in Fig. 7. The analysis shows, for ex-

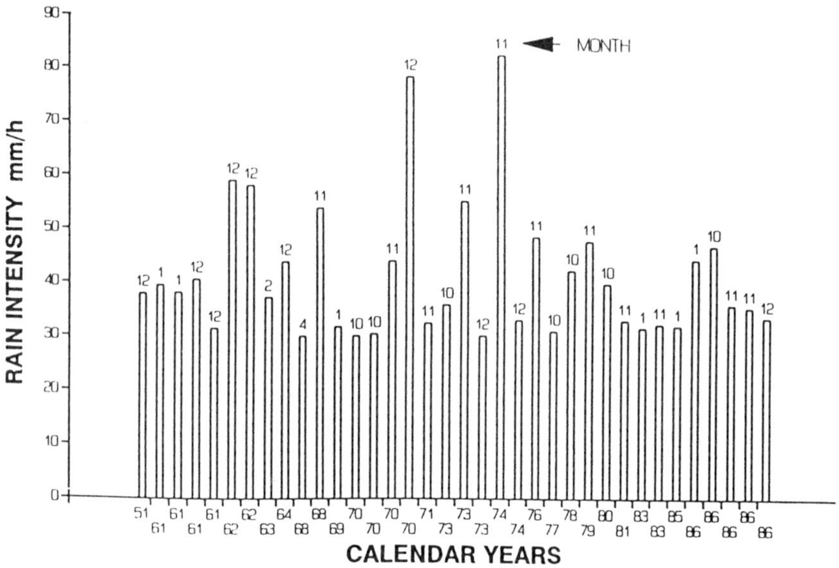

Fig. 4 Highest I_{30} intensities partial series, 1958–1986 for November, December, and January, Yavne, Israel.

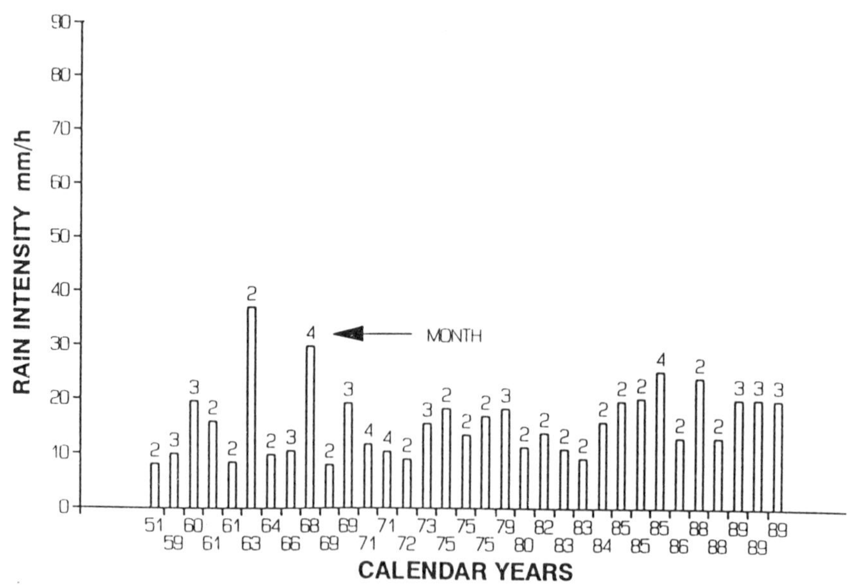

Fig. 5 Highest I_{30} intensities partial series, 1958–1986 for February, March, and April, Yavne, Israel.

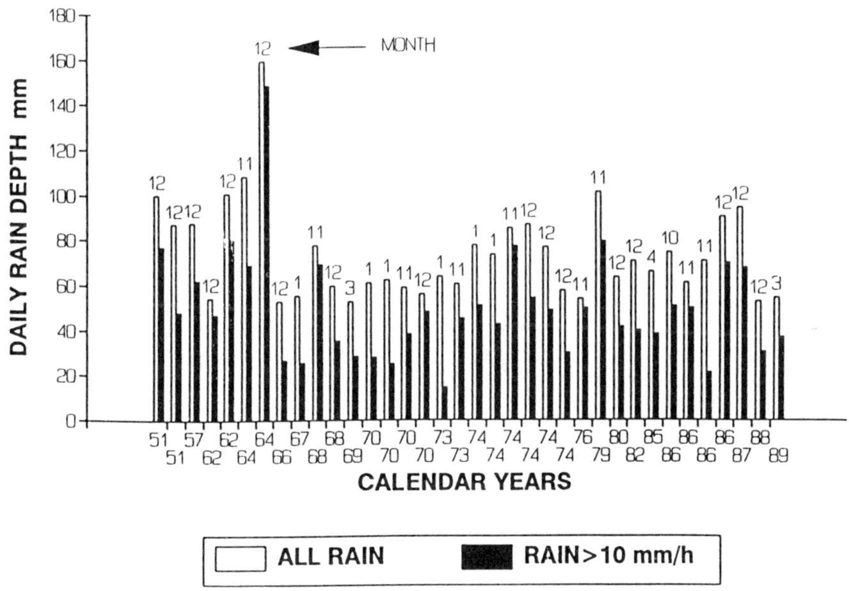

Fig. 6 Maximum daily rain depth for given intensity, partial series 1951–1989, Yavne, Israel.

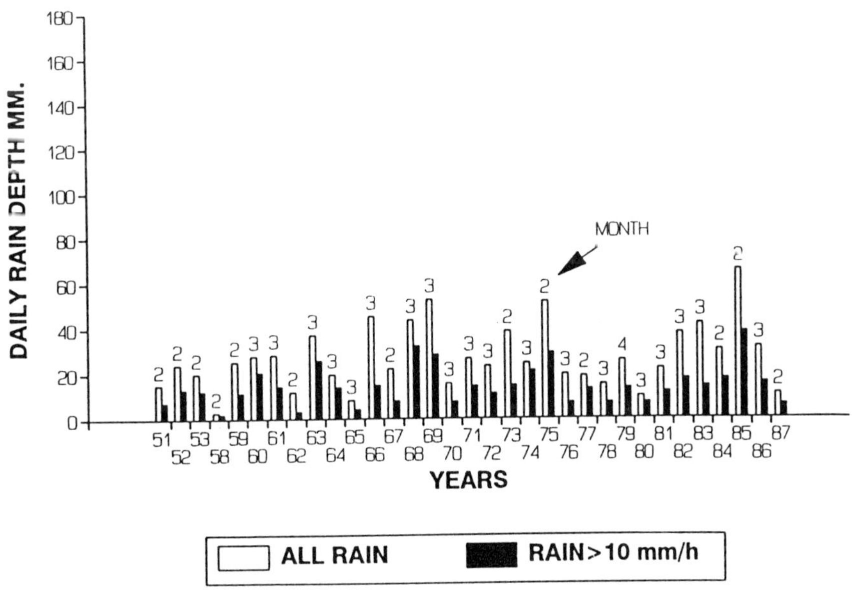

Fig. 7 Maximum daily rain depth for given intensity, annual series 1951–1989, for February, March, and April, Yavne, Israel.

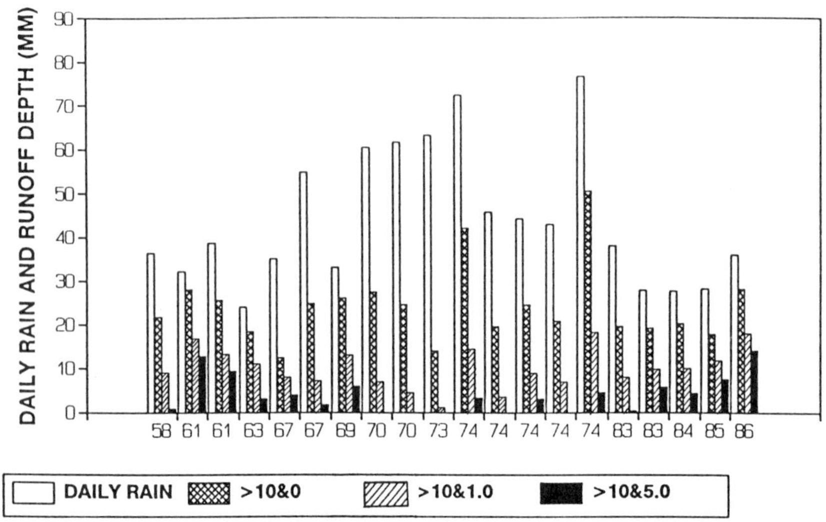

(a)

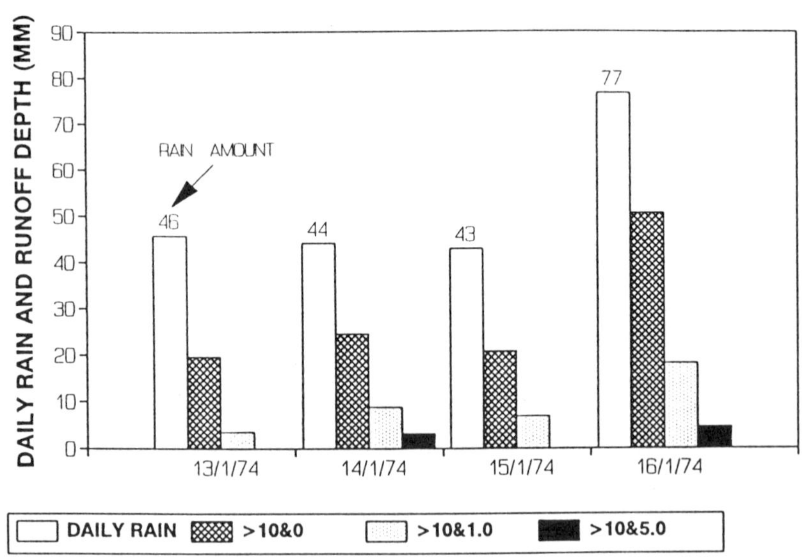

(b)

Fig. 8 Calculated runoff from maximum daily rainfall depth for given infiltration rate and surface storage, partial series, 1958–1986, Yavne. 10 & 0 = 10 mm h^{-1} infiltration rate, and 0 = mm of storage, etc.

ample, that for 1976 to 1981 the maximum daily rain depths exceeding intensities of 10 mm h^{-1} were below 14 mm, while in every second year within this period they were less than 7 mm.

The surface runoff of any storm can be evaluated by combining the appropriate infiltration equation with the actual rain intensity data, taking into consideration not only the excess of the rainfall rate over the infiltration rate but also whether the surface detention was satisfied prior to runoff initiation (Morin and Benyamini 1988). In cases when only the highest runoff events are required, a simpler analysis can be conducted. In this case uniform infiltration rates are assumed for the whole period of the rainstorm. Rainstorm analysis can be performed, assuming a set of relevant infiltration rates and surface storage depths. Figures 8a,b are examples of this type of analysis for Yavne station during January. This month was chosen because in the Mediterranean climate soils of the coastal plain of Israel normally are wet and have a low and stable infiltration rate. In Fig. 8a the highest storms of January for 1958–1986 are presented. It can be seen that surface storage magnitude has a remarkable impact on the production of runoff. At the storm of January 13–16, 1974 (Fig. 8b) the total rain depth was 210 mm. Assuming an infiltration rate of 10 mm h^{-1} and zero storage, the runoff is 109 mm. Adding 1 mm of storage to the infiltration rate of 10 mm h^{-1} gives a runoff of 37 mm. If, however, we have a rougher surface with storage of 5 mm, the runoff will be reduced to only 8 mm. Surface storage is a dynamic parameter which allows maximum infiltration for low-intensity rain segments.

REFERENCES

Chow, V. T. (1954). The log-probability law and its engineering applications. *Am. Soc. Civil Eng.* (separate No. 536), November 1980.

Gumbel, E. J. (1958). *Statistics of Extremes*. Colombia University Press, New York.

Kinte, G. W. (1977). Frequency and risk analysis in hydrology. Water Resources Publ.

Lal, R. (1974). Soil erosion and shifting agriculture. *FAO Soils Bull. 27*: 48–71.

Lloyd, E. H. (1970). Return periods in the presence of persistence. *J. Hydrol. 10*: 91–298.

Morin, J., and Benyamini, Y. (1988). Tillage method selections based on runoff modeling. In *Challenges in Dry Land Agriculture—A Global Prospective*, Proc. Int. Conf. Dry Land Farming, Amarillo TX, 1988, pp. 251–254.

Wischmeier, W. H. (1959). A rainfall erosion index for a universal soil loss equation. *Soil Sci. Soc. Am. Proc. 23*: 246–249.

3

Soil Characteristics and Aggregate Stability

Yves Le Bissonnais
Soil Survey Staff of France, National Institute for Agronomic Research, Orleans, France

I. INTRODUCTION: IMPORTANCE OF SOIL CHARACTERISTICS AND AGGREGATE STABILITY ON SOIL EROSION BY WATER

Soil erosion results from the combined influence of several parameters, the most important of which are soil characteristics as well as topographic, climatic, or land-use parameters. Soil erodibility can be defined as the inherent soil property to react to water action in

1. Reducing infiltration rate and decreasing soil surface roughness due to aggregate breakdown, i.e., increasing the risk of runoff
2. Being detached and transported by the resulting runoff

The *primary physical and chemical soil characteristics* determine soil structure and, therefore, infiltration rate. They also determine the response of the soil surface when it is subjected to rainfall. The *aggregate stability* can be defined as soil structural response to rainfall. It is closely related to primary soil characteristics, but there is no single relationship between aggregate stability and primary soil characteristics. Many attempts were made to find these relationships, but none were found suitable for all soils. In cultivated soils, runoff and erosion often result from surface *sealing and crusting*, which are closely related to aggregate breakdown.

In this chapter, the main soil properties affecting aggregate stability and erosion are reviewed. Then the relationship between aggregate breakdown and erosion is discussed.

II. SOIL PRIMARY CHARACTERISTICS AFFECTING AGGREGATE STABILITY AND ERODIBILITY

The main primary soil properties influencing aggregate stability and erosion have been relatively well known since the early works of Yoder (1936), Hénin (1938) and Ellison (1945). However, numerous studies have reported sometimes contradictory results. We will review and discuss the most recent results concerning the effect of soil texture, clay mineralogy, organic matter, cations, Fe and Al oxides, and $CaCO_3$ on aggregate stability.

A. Soil Texture

It is generally considered that as the silt (0.002–0.05 mm) or silt + very fine sand (0.05–0.10 mm) fraction increases and clay decreases, erodibility increases (Wischmeier and Mannering, 1969). This is because of (i) the aggregation and the bonding effect of clay, (ii) the transportability of fine and nonaggregated particles (i.e. silt), and (iii) the detachability of sand and silt.

In a study on the effect of initial water content, Gollany et al. (1991) found that aggregate stability increases with clay content; and the effect was more pronounced at high water content (Fig. 1). The same result was obtained by Le Bissonnais (1988) with a range of silty soils (Fig. 2). However, with a wider range of soils Le Bissonnais and Singer (1993), as well as Pierson and Mulla (1990), did not find significant correlations between clay content and aggregate stability. Bradford and Huang (1992) obtained a negative correlation between silt and infiltration rate under simulated rainfall. The same result was found by Römkens et al. (1977) and Bradford et al. (1987) with different kinds of soils.

In Africa, working with various sandy soils, Obi et al. (1989) found that runoff and erosion were best predicted by sand percentage with a negative relationship. Trott and Singer (1983) also found a significant negative correlation between coarse sand and erosion rate. Merzouk and Blake (1991) found a sand plus gravel index to be the best erosion indicator in Morocco. These results may be explained by the fact that the clay amount is not sufficient to clog the conducting pores at the soil surface when the fraction of coarser particles increases. Therefore infiltration remains high even if disaggregation occurs. It was shown that surface rock fragments may decrease or increase crusting and erosion, depending on their shape and position (Yair and Lavee, 1976; Poesen, 1986; Van Der Watt and Valentin, 1992).

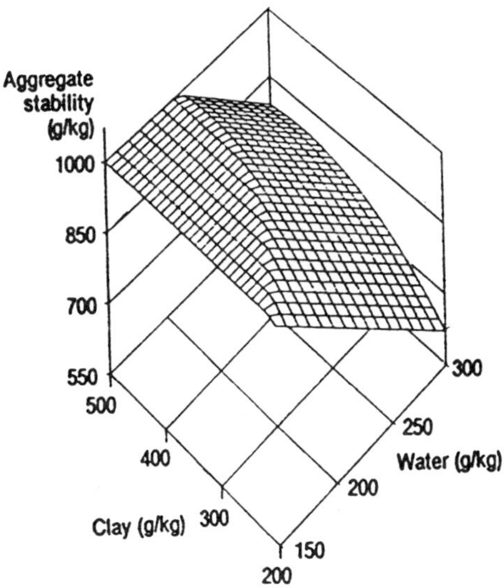

Fig. 1 Effects of clay content and antecedent soil water content on aggregate stability for slightly eroded plots at 260 g kg^{-1} organic C. (From Gollany et al., 1991).

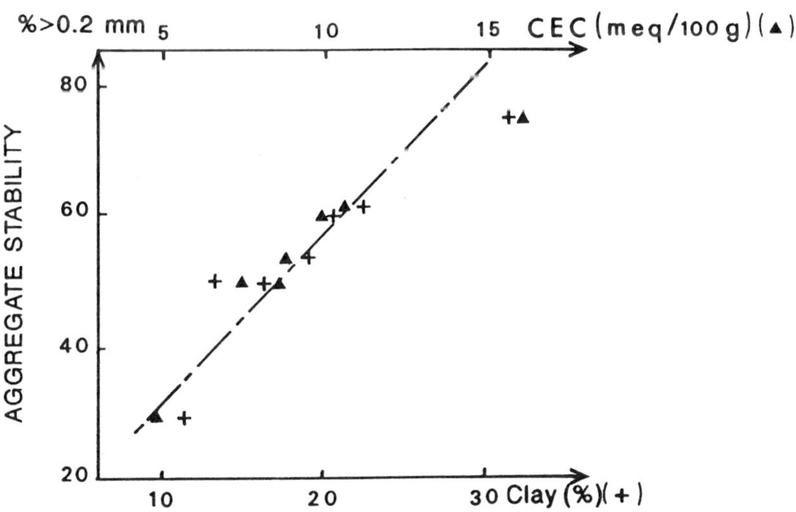

Fig. 2 Effect of clay content and CEC on wet aggregate stability (> 200 μm aggregate ratio) for 7 silty soils. (From Le Bissonnais, 1988).

In conclusion, there is a general agreement that the clay fraction is a positive factor of soil stability because it is a bonding agent of coarser particles. The medium-textured soils (silty and loamy soils) are often the most susceptible to crusting and erosion (Ben-Hur et al., 1985).

However, interactions between texture and other parameters like clay mineralogy and organic matter exist. These interactions influence aggregate breakdown differently, depending on the process. Particularly, clay particles are *aggregating* and *aggregated* particles. Mineralogical and chemical characteristics may modify the effect of clay particles on aggregation and stability. Furthermore, when establishing a correlation between erosion and a single textural class, the other components of texture should also be known. Texture is a complex characteristic in itself, and there is no single parameter giving a clear and complete description of soil texture.

B. Clay Mineralogy

Clay mineralogy influences erosion processes, but the effect is difficult to assess because soils most often contain a mixture of clay minerals. Artificial clay mixtures or aggregates may be good models for studying their effect, but they have not the same behavior as clay in soil. Using pure clay minerals, Emerson (1964) showed that swelling clays like montmorillonite are less subject to slaking than kaolinite or illite because the pressure which is developed by entrapped air is released by swelling; however, fissuring of montmorillonite may occur, due to the combination of stress of entrapped air and swelling of aggregates. As aggregating particles, the smectitic clays should be more efficient than other clays because of their large specific surface and high physicochemical interaction capacity. In accordance with this statement, Young and Mutchler (1977) found that montmorillonite content was highly correlated with aggregate stability and low erodibility. Conversely, El-Swaify and Dangler (1977), studying Hawaiian soils, made a statistical analysis of erodibility and showed that soils with kaolinite were less eroded than soils with montmorillonite. It is probably due to iron-kaolinite interactions producing very strong cohesive forces compared to clay-clay interactions which correspond, for the smectite, to less cohesive forces. Trott and Singer (1983) obtained the same result with California range and forest soils, but again, sesquioxides are controlling the disaggregation and erosion processes.

Another mechanism of aggregate breakdown, chemical dispersion, depends on clay mineralogy together with exchangeable ion composition and electrolyte concentration in the soil solution. Miller and Baharuddin (1986) found a good relationship between soil loss and dispersion on kaolinitic southwest U.S. soils. Comparing soil erosion from various South Africa and Israeli soils, Stern et al. (1991) found that only soils with more than 5% smectite were dispersive and susceptible to seal formation and erosion (Fig. 3). This result is explained by the fact that a small amount of smectite may disperse kaolinite due to prevention of the edge-to-face flocculation of pure kaolinite by the deposition of smectite on the positively charged edge of kaolinite (Frenkel et al., 1978). In this study, the smectitic soils were the most erodible.

The clay mineralogy is related to soil parent material and weathering: Smith (1990) found that stability of kaolinitic soils increases as weathering intensifies. This result can be related to greater amount of kaolinite and sesquioxides. The same effect was observed by Levy and Van Der Watt (1988) and by Collinet (1988) in Africa.

In conclusion, the effect of clay mineralogy on aggregate stability and erosion seems to be ambivalent: clay with low Cation Exchange Capacity-like kaolinite induces less strong aggregates than montmorillonite because of its smaller surface area. However, montmorillonite may disperse more easily than kaolinite. The relationships are complicated because kaolinite is often associated with iron oxides which may induce very strong aggregates. Again, one must not consider one parameter without considering the interactions with the other soil characteristics.

C. Organic Matter

Organic matter is one of the most important and well-known aggregate stabilizing agents in soils. It was considered by Wischmeier and Mannering (1969) as the second most influential property affecting soil erodibility after soil texture. However, this finding is not always consistent in the literature. In addition, there has been considerable experimentation and conjecture on the mechanisms by which soil organic fractions influence aggregate stability and erosion.

Recently Roberson et al. (1991) showed that aggregate stability to slaking increases as the heavy fraction of carbohydrates (polysaccharides) increases, although the total organic content did not change (Fig. 4). On the other hand, in some cases aggregate stability is not affected by organic content variations (Panayiotopoulos and Kostopoulou, 1989). Therefore the total organic matter which is usually measured is probably not the best measurement for predicting aggregate stability or erosion. However, the overall effect of organic matter is positive, and the relationship seems to be logarithmic, although it is often considered as linear within the range of cultivated soils showing moderate organic matter content (Wischmeier et al., 1971). For very high organic matter content, the induced hydrophobic effect may become a problem and reverse the relationship between organic matter content and aggregate stability or erosion.

Ekwue (1990) found a positive relation between organic matter and aggregate stability for soils with grass treatment and a negative relation for those with peat treatment. Splash detachment was reduced for both treatments: grass treatment reduced erosion by increasing aggregate stability, while peat acted as a mulch.

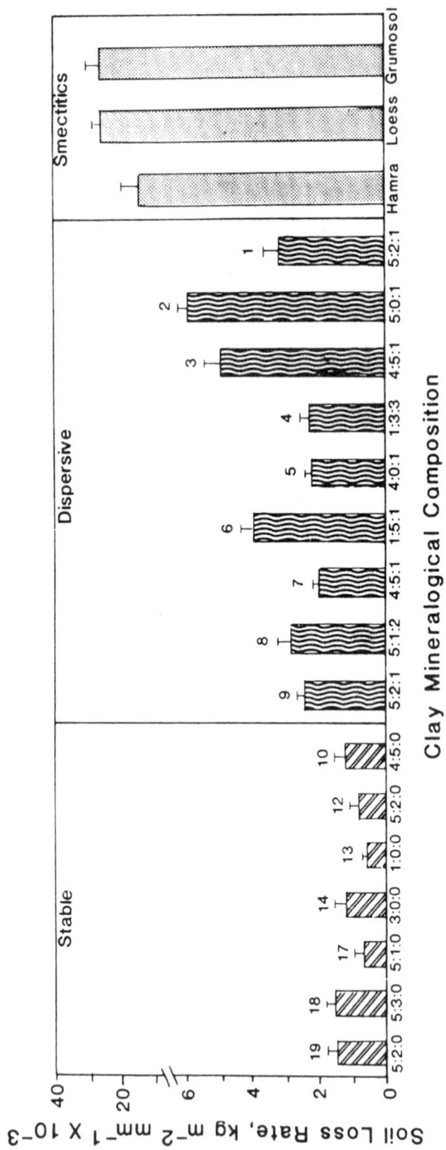

Fig. 3 Soil loss rates of untreated stable, dispersive and smectitic soils. Numbers below the columns represent the mineral ratios of kaolinite, illite, and smectite respectively. (From Stern, Ben-Hur and Shainberg, 1991).

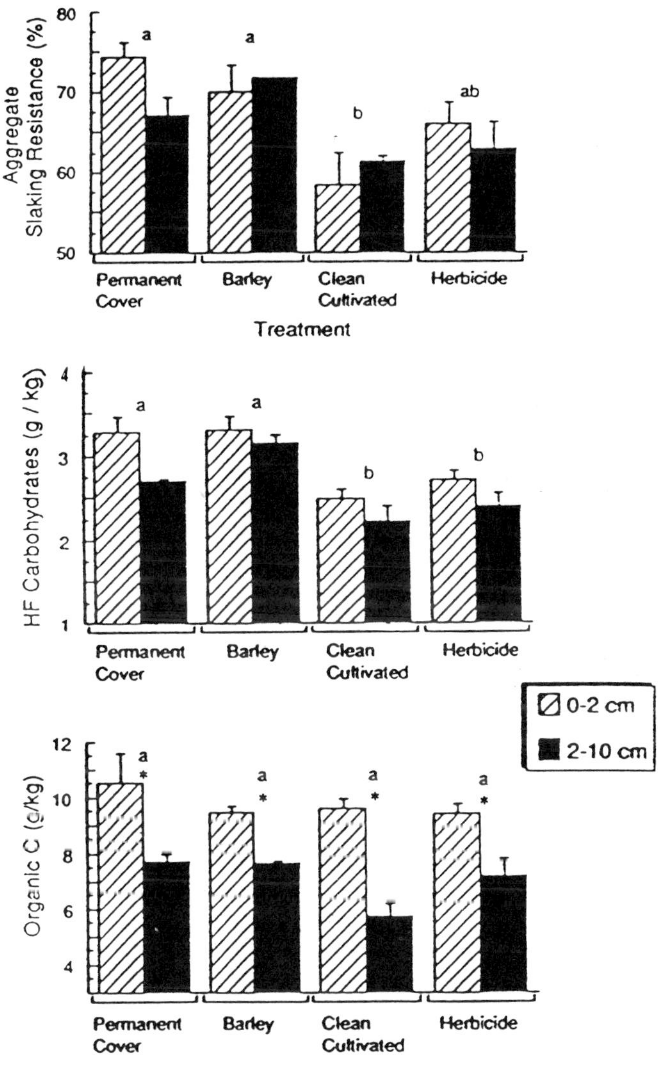

Fig. 4 Response of aggregate slaking resistance, heavy-fraction (HF) carbohydrates and organic C to various treatments. Treatments labeled with the same letter do not differ at the $p < 0.05$ significance level. An asterisk indicates depths differ at the $p < 0.05$ significance level. (From Roberson, Sarig and Firestone, 1991).

This result illustrates the two main effects of organic matter on soil erosion: (i) effect on aggregate stability through clay-organic matter interaction and binding action of polymers, and (ii) physical protection by mulching the soil surface (Singer and Blackard, 1977) and therefore encouraging infiltration.

The effect of organic matter varies in time because of differences in organic matter decomposition rates (Monnier, 1965) and because of seasonal or between-year climatic variations (Alberts et al., 1987). Perfect et al. (1990a) found that aggregate stability varied during a growing season in relation with water content and microbial biomass (Fig. 5). They (1990b) also observed long-term evolution of aggregate stability: aggregate stability increases with grass cultivation because of increases in root biomass. The same authors also observed an increase of aggregate stability over the winter months due to freezing and thawing. The same result was obtained by Lehrsch et al. (1991).

The microbial effect on aggregate stability is due to the presence of microorganisms (fungal hyphae, bacterial mucilage) and to their by-products, which have a chemical and physicochemical action. Among these products, polysaccharides are considered to be very efficient aggregate stabilizers (Molope et al.,

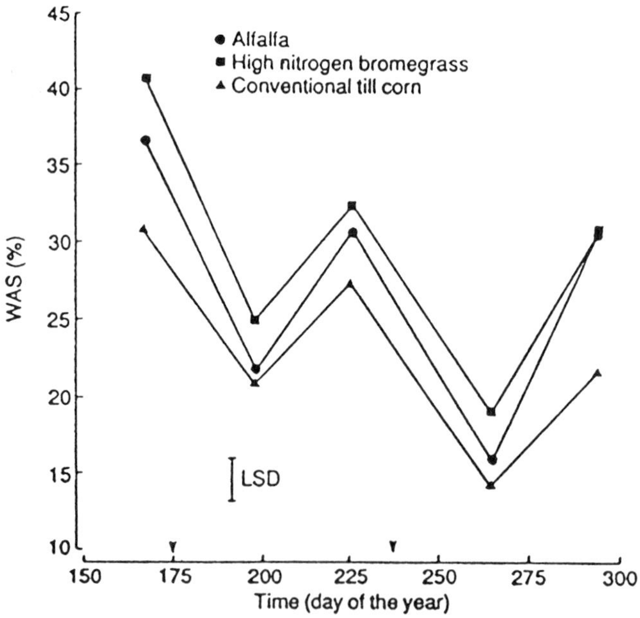

Fig. 5 Temporal variation in wet aggregate stability (WAS) under selected cropping treatments during the 1987 growing season. Arrows on the abscissa indicate cutting date for the forages. LSD is calculated across all times. (From Perfect et al., 1990a).

1985; Metzger et al., 1987; Levy et al., 1991; Roberson et al., 1991). The mechanism of action of polysaccharides and other natural or synthetic polymers was studied experimentally by several authors (Chenu, 1985; Shainberg et al., 1990; Le Souder et al., 1991). The molecular weight, the charge of the polymer and the electrolyte concentration influence the effect of polymers. Cationic polymers are effective in flocculating the clay and maintaining high infiltration rate even with distilled water, while anionic polymers are efficient only in solutions with higher electrolyte concentration (Theng, 1982).

In conclusion, most studies found organic matter reduces soil erodibility. Organic matter is a bonding agent between mineral soil particles. The relation is probably not linear and depends on the interactions between organic matter and other soil properties. For very high organic matter content, hydrophobicity may be a factor increasing runoff and erosion. Furthermore, the different organic matter fractions may have different effects: the litter has a physical effect (mulch, infiltration maintenance, reduction in detachment) but also increases biological activity when it becomes degraded; the microorganisms and their by-products increase aggregate stability by physicochemical and chemical actions.

D. Sodium and Other Cations

The nature and amount of exchangeable cations influence erosion through their effect on the clay dispersion/flocculation processes. The effects are closely related to clay mineralogy. Several authors have shown the effect of increasing exchangeable sodium percentage (ESP) on the decrease of the infiltration rate and subsequent erosion.

Physicochemical dispersion results from reduction of the attractive forces between colloidal particles while wetting (Sumner, 1992). Most studies found increased ESP caused more dispersion, crust formation and erosion, but the effect differed between soils. Some soils are affected at very low ESP, others are affected at high ESP only, and some are not affected at all. Levy and Van Der Watt (1988) found that dispersion of a kaolinitic soil was not significantly affected in the ESP range of 1% to 9%, while two other soils (i) with mixed kaolinite, illite and montmorillonite and (ii) with illite and interstratified minerals, were significantly affected at ESP 4.3. Further reduction of infiltration was noticed at ESP 11.3. Kazman et al. (1983) found the infiltration of a sandy loam already affected at ESP 2.2 under very high rainfall intensity (Fig. 6). Conversely, it was found that ESP did not affect the behavior of Hawaiian soils (El-Swaify and Dangler, 1977). The high iron oxide content of these soils is responsible for controlling their aggregation. When treated with citrate bicarbonate dithionite (CBD) to remove the free iron oxide, these soils became susceptible to dispersion by sodium.

Infiltration rate is more affected by ESP than soil hydraulic conductivity (Shainberg and Letey, 1984) because of the mechanical impact and the stirring

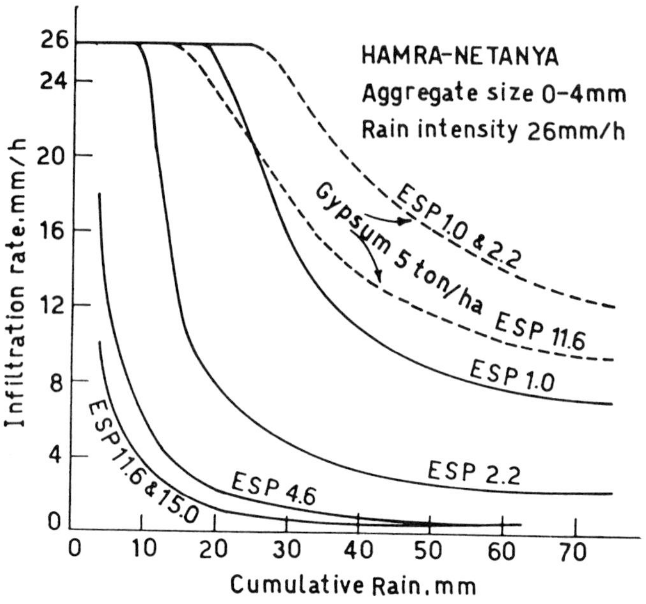

Fig. 6 The infiltration rate of the Netanya soil as a function of cumulative rain. Effect of soil ESP and phosphogypsum application. (From Kazman, Shainberg and Gal, 1983).

effect of raindrops which enhance chemical dispersion. Because of the bridging effect of cations, and in accordance with the DLVO theory (the stability of colloids depends on the interaction between the Van Der Waals attraction forces and the electrical double-layer repulsion), it is logical that stability or dispersion depends on the cation size and valence. Therefore the cations can be classified in the following order: $Ca^{2+} > Mg^{2+} > K^{+} > Na^{+}$. In fact, Levy and Van Der Watt (1990) observed an intermediate effect of K between Ca and Na. Mg was shown to be less efficient in aggregating than Ca (Shainberg et al., 1988; Keren, 1989) although much more efficient than Na (Levy et al., 1988).

The increase of electrolyte concentration reduces the dispersive effect of Na, and chemical dispersion may be prevented by spreading phosphogypsum or another readily available electrolyte source on the soil surface (Kazman et al., 1983) (Fig. 6).

In summary, increasing ESP increases soil dispersion, but this effect is influenced by clay amount and mineralogy. The effect of ESP is probably more important with swelling clay and when there is no other bonding agent. Dispersion is also affected by the electrolyte conductivity (EC) of the soil solution, the EC of applied water, and by raindrop impact. The effect of ESP was not observed in tropical soils with high levels of iron oxides, due to the cementing action of iron and the nonswelling nature of kaolinite. Furthermore, the nature of the cations influences the adsorption of polymers participating in the bridging of soil particles. Dispersion is one of the most effective processes of aggregate breakdown and crusting. It generally induces very fast crusting, low infiltrability and very high particle mobility in water due to the size of the dispersed particles. However, Abu-Sharar et al. (1987) showed that dispersion has to be associated with slaking to reduce hydraulic conductivity. Dispersed clay is mobile and not able to clog large conducting pores.

E. Fe and Al Oxides

Studies of structural and hydraulic parameters of tropical and western or southeastern U.S. soils have shown the importance of iron and aluminum in soil physical behavior. Most workers have found good positive correlation between sesquioxides and soil structural stability (El-Swaify and Dangler, 1977; Trott and Singer, 1983; Churchman and Tate, 1987; Farres, 1987; Panayiotopoulos and Kostopoulou, 1989). Beyond these empirical results, there are different theories explaining the mechanisms of sesquioxides effect: (i) Iron and aluminum in solution act as flocculants in lowering zeta potential of clays, preventing dispersion in about the same way that calcium does in other soils. Because of the pH of the soil solution, it is probable that this process concerns aluminum more than iron. (ii) Sesquioxides play a role in clay particle–organic polymer interactions. We assume in this case that iron and organic matter together create conditions for stable aggregation. Le Bissonnais and Singer (1993) showed that soils with more than 3% organic C and 2.5% CBD extractable Fe + Al do not induce infiltration decrease or erosion, which support the assumption (Fig. 7). (iii) Sesquioxides can precipitate as gel on clay surfaces. In many cases the coatings are difficult to observe even by electron microscopy. However, Robert et al. (1983) observed some modifications of clay properties due to iron and aluminum polycation precipitation.

Iron and aluminum can behave differently under different conditions such as pH, clay minerals, climate, organic matter and soil solution composition. There is a debate whether iron or aluminum is more important in aggregation of soil particles. Aluminum seems to be more efficient because of its higher solubility over a wide range of pH, but it is usually found in smaller amounts than iron in soil. A more important concern is the form of the soil sesquioxides. Two basic forms of iron are present in soil: (i) iron present in primary minerals and (ii) free iron, which includes crystalline and amorphous iron oxides and organically associated iron. The total free iron is extracted by CBD, which has been correlated with erodibility (Römkens et al., 1977; Trott and Singer, 1983). Other extracts of organic forms of iron and aluminum have also been correlated with

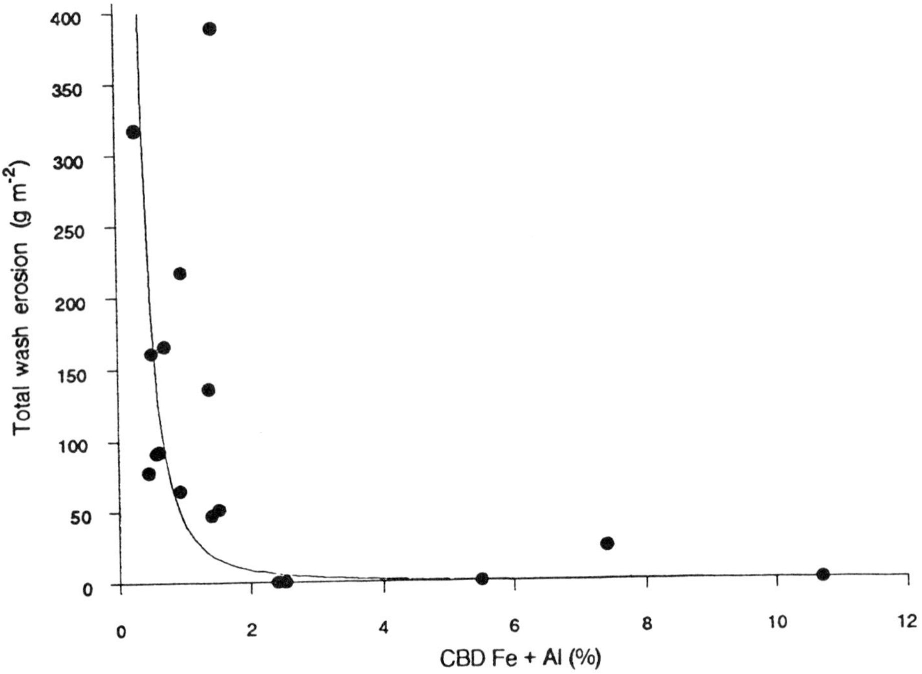

Fig. 7 Total wash erosion as a function of CBD extractable Fe + Al for 17 California soils subjected to simulated rainfall. (Le Bissonnais and Singer, 1993).

aggregate stability. So the various forms of free iron and aluminum oxides may perform variably as aggregating agents.

In summary, there is good evidence for the positive effect of sesquioxides on aggregate stability for tropical and lateritic soils. They may act in several ways: (i) as a flocculant like calcium, (ii) as links between clay and polymers, (iii) as a cement after precipitation as gel on clay surfaces. The effect of sesquioxides on aggregate stability is pH dependent.

F. $CaCO_3$

The effect of $CaCO_3$ on aggregate stability and infiltration or erosion has not been intensively studied. Ben-Hur et al. (1985), comparing calcareous and noncalcareous soils, observed no effect of $CaCO_3$ on infiltration rate under simulated rainfall (Fig. 8). However, Merzouk and Blake (1991) in a study of soils in Morocco found that, among other parameters, active $CaCO_3$ was a good indicator of soil erodibility. Castro and Logan (1991) found an ambivalent effect of

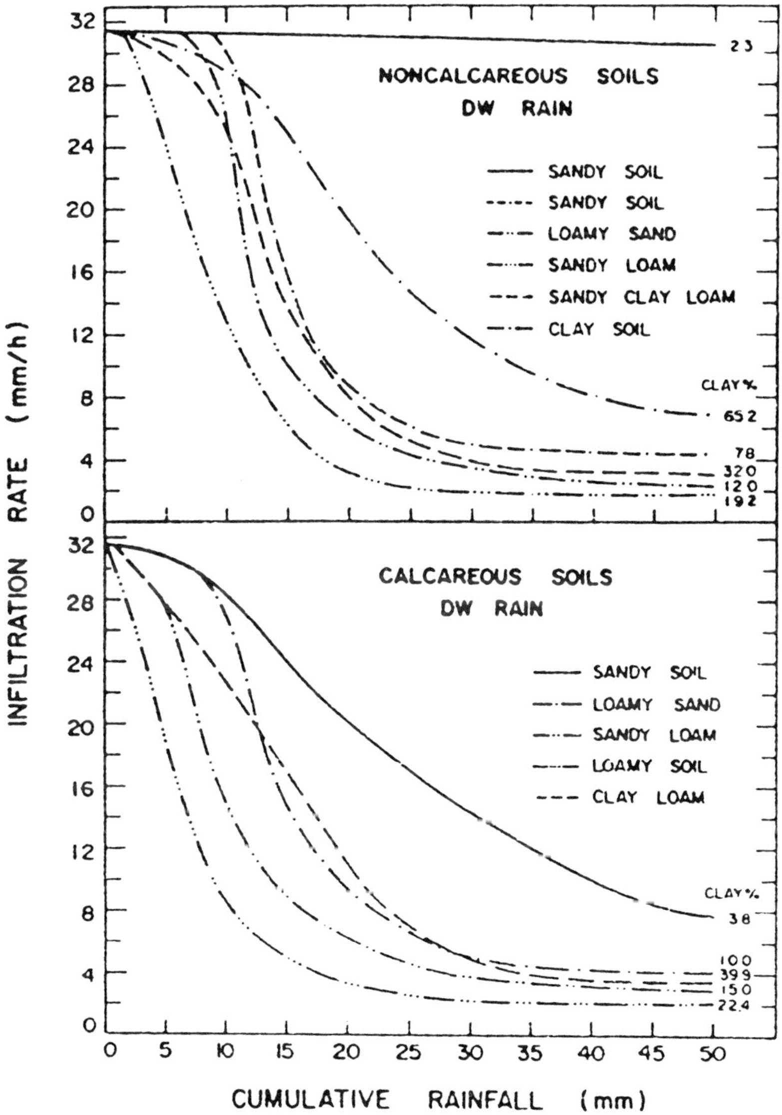

Fig. 8 The infiltration rate of calcareous and noncalcareous soils as a function of the cumulative rainfall. (From Ben-Hur et al., 1985).

liming, which is a classical soil improvement method. They observed a short-term negative effect on the soil structure and a long-term positive effect on the reduction of erosion.

From a chemical point of view, carbonates in soil should be favorable to aggregate stability and infiltration rate because of the Ca cations. However, the effect probably depends on the size distribution of $CaCO_3$ particles and on the clay content. When $CaCO_3$ particles are the size of silt and the clay content is low, the soil may behave as a silty soil (i.e., as a very unstable soil). For a beneficial aggregating effect of $CaCO_3$, sufficient clay must be present.

G. Conclusion

Although the effect of some of these single primary properties is well established in experimental studies, the relationships do not necessarily give a mechanistic explanation of the soil behavior. Furthermore, these relationships are generally difficult to apply to the whole range of soils and situations.

We must keep in mind that interactions and correlations between these properties often exist and that there may exist a threshold for some soil components above which this component determines the soil behavior whatever the level of other components (e.g., sodium or iron).

Another problem is that these primary properties affect secondary soil properties like aggregate stability or permeability, which are more closely related to erosion processes but also much more difficult to measure. The effects of these secondary soil properties depend on the methods used to measure them and therefore are difficult to compare. Their significance with regard to erosion needs further discussion.

III. RELATIONSHIP BETWEEN AGGREGATE STABILITY AND EROSION

As aggregate stability measurements are supposed to contribute to the assessment of the soil erodibility, the relationship between the two processes should be examined carefully. Water erosion consists of detachment and transport of soil particles by water. However, there are several problems of definition: what is the minimal distance of transport which is considered to be erosion? The answer will vary, depending on where and how we measure erosion. The main cause of water erosion is surface runoff. In some circumstances splash by drop impact may detach and displace soil particles without runoff; however this process is most often related to runoff (Le Bissonnais and Singer, 1993).

Soil erodibility may be defined as the inherent reaction of soil to water in (i) limiting water infiltration and decreasing soil surface roughness because of aggregate breakdown, and (ii) being detached and transported by the resulting runoff and drop impacts.

Considering these two processes, it is possible to distinguish between two main types of erosion situations:

Without or with little change in water infiltration capacity

With progressive decrease in infiltration capacity

For the first type, the soil structure does not considerably change during the rainfall. Runoff results from particular rainfall and soil conditions: (i) very high intensity exceeding infiltrability whatever the soil properties, (ii) very long rainfall events resulting in the saturation of the whole soil profile, and (iii) permanent low percolation capacity of the profile due to surface or subsoil physical conditions (compacted subsurface soil layers, bedrock, heavy clay, etc.).

This type of erosion may occur in heavily vegetated nonarable lands during long-duration rainstorms. Generally, the vegetation and land use greatly influence the erosion; therefore soil characteristics are less important. For this type of erosion it is clear that aggregate stability is not a good indicator of soil erodibility because the aggregates themselves are transported.

For the second type, the infiltration capacity decreases following rainfall events because of surface sealing and crusting (Le Bissonnais, 1990; Moore and Singer, 1990): infiltration may be reduced to 1 mm/h (Boiffin, 1984). This situation corresponds to most of the cultivated lands, which are generally medium textured. Silty and loamy soils are the most prone to sealing. Furthermore, these soils are more easily transported by runoff. However, this effect is complicated because crusting may increase soil shear strength and resistance to rill formation (Bradford et al., 1987).

Water erosion in cultivated areas is divided into two components: rill and interrill erosion. However, they are most often closely associated in the field. Sealing may affect both interrill and rill erosion. It increases runoff and therefore splash, transport and detachment capacity (i.e., rill initiation).

For the crusting-related erosion situations, aggregate stability is certainly closely related to erodibility. In fact, aggregate stability measures the soil susceptibility to aggregate breakdown, which is the main process responsible for crusting. However, as stressed by Farres (1987), the magnitude of erosion is not just controlled by when breakdown takes place. Equally important are the characteristics of the detached particles (Le Bissonnais et al., 1989; Loch, 1989). In fact, the sizes of the detached particles determine crust physical properties (infiltration) and transportability of particles.

IV. CONCLUSION AND PERSPECTIVES

According to several authors (De Ploey and Poesen, 1985; De Tapia, 1986; Bradford and Huang, 1992), we need to improve our knowledge of the basic processes involved in erosion in order to build deterministic models to relate primary soil characteristics to erosion. Progress should be made in the understanding of the interaction between these soil characteristics, since aggregate stability and erosion measurements often show contradictory results where single parameters are studied.

The problem of dealing with aggregate stability is that, unlike primary characteristics, there is no single standardized method to measure it. Several authors tried to compare various aggregate stability tests (Matkin and Smart, 1987; Chisci et al., 1989). Most often the results of the different tests are correlated despite the fact that not always the same processes were measured. The problem is the lack of relationships between these tests and the actual erodibility measured in the field. Usually the tests used do not reflect all the mechanisms involved, and they do not exactly correspond to the field processes (Bryan et al., 1989; Kreznor et al., 1992).

In order to analyze the effect of soil characteristics on erosion, we have to keep in mind what is measured and the physical conditions of the studied system. We also have to realize that the results are highly dependent on the soil types under study. Contradictory results in different studies probably indicate that the considered properties do not have the same influence in all conditions and for the whole range of soils.

However, a standardized method for the study of aggregate stability, including the size distribution of produced particles, would be of great interest for predicting the erodibility of cultivated soils.

ACKNOWLEDGMENTS

The author would like to thank Drs. M. Singer, M. Robert and M. Jamagne for their review and comments on the manuscript.

REFERENCES

Abu-Sharar, T. M., Bingham, F. T., and Rhoades, J. D. (1987). Reduction in hydraulic conductivity in relation to clay dispersion and disaggregation. *Soil Soc. Am. J. 51*: 342–346.

Alberts, E. E., Laflen, J. M., and Spomer, R. G. (1987). Between year variation in soil erodibility determined by rainfall simulation. *Trans. ASAE 30*: 982–987.

Ben-Hur, M. I., Shainberg, I., Bakker, D., and Keren, R. (1985). Effect of soil texture and $CaCO_3$ content on water infiltration in crusted soils as related to water salinity. *Irrig. Sci. 6*: 181–194.

Boiffin, J. (1984). La dégradation structurale des couches superficielles du sol sous l'action des pluies. Thèse de docteur-ingénieur, INA-PG.

Bradford, J. M., Ferris, J. E., and Remley, P. A. (1987). Interrill soil erosion processes: relationship of splash detachment to soil properties. *Soil Sci. Soc. Am. J. 51*: 1571–1575.

Bradford, J. M., and Huang, C. (1992). Mechanisms of crust formation: physical components. In *Advance in Soil Science. Soil Crusting: Physical and Chemical Processes* M. E. Sumner and B. A. Stewart, eds.). pp. 55–72.
Bryan, R. B., Govers, G., and Poesen, J. (1989). The concept of soil erodibility and some problems of assessment and application. *Catena 16*: 393–412.
Castro, C. F., and Logan, T. (1991). Liming effect on soil stability and erodibility. *Soil Sci. Soc. Am. J. 55*: 1407–1413.
Chenu, C. (1985). Etude Physico-chimique des interactions argiles-polysaccharides neutres. Thèse Univ. Paris VII.
Chisci, G., Bazzoffi, P., and Mbagwu, J. S. C. (1989). Comparison of aggregate stability indices for soil classification and assessment of soil management practices. *Soil Technol. 2*: 113–133.
Churchman, G. J., and Tate, K. R. (1987). Stability of aggregates of different size gradees in allophanic soils from volcanic ash in New Zealand. *J. Soil Sci. 38*: 19–27.
Collinet, J. (1988). Etude expérimentale de l'érosion hydrique de sols représentatifs de l'Afrique de l'Ouest. *Cahier Orstom Sér. Pédol. 24*: 235–254.
De Ploey, J., and Poesen, J. (1985). Aggregate stability, runoff generation and interrill erosion. In *Geomorphology and Soils* (K. S. Richards, R. R. Arnett, and S. Ellis, eds.), Allen and Unwin, pp. 99–120.
De Tapia, M. (1986). A mechanistic approach towards the evaluation of soil erodibility. In *Assessment of Soil Surface Sealing and Crusting* (F. Caillebaut et al., eds.), University of Ghent, pp. 121–129.
Ekwue, E. I. (1990). Effect of organic matter on splash detachment and the process involved. *Earth Surf. Process. Landforms 15*: 175–181.
Ellison, W. D. (1945). Some effects of raindrops and surface flow on soil erosion and infiltration. *Trans. Am. Geophys. Union 24*: 452–459.
El-Swaify, S. A., and Dangler, E. W. (1977). Erodibility of selected tropical soils in relation to structural and hydrologic parameters. In *Soil Erosion: Prediction and Control*, Spec. Pub. 21 Soil Conserv. Soc. Am., Ankeny, IA.
Emerson, W. W. (1964). The slaking of soil crumb as influenced by clay mineral composition. *Austr. J. Soil Res. 2*: 211–217
Farres, P. J. (1987). The dynamics of rainsplash erosion and the role of soil aggregate stability. *Catena 14*: 119–130.
Frenkel, H., Goertzen, J. O., and Rhoades, J. D. (1978). Effects of clay type and content, exchangeable sodium percentage and electrolyte concentration on clay dispersion and soil hydraulic conductivity. *Soil Sci. Soc. Am. J. 42*: 32–39.
Gollany, H. T., Schumacher, T. E., Evenson, P. D., Lindstrom, M. J., and Lemme, G. D. (1991). Aggregate stability of an eroded and desurfaced Typic Argiustoll. *Soil Sci. Soc. Am. J. 55*: 811–816.
Hénin, S. (1938). Etude physico-chimique de la stabilité structurale des terres. Thèse, Paris.
Kazman, Z., Shainberg, I., and Gal, M. (1983). Effect of low level of exchangeable sodium and applied phosphogypsum on the infiltration rate of various soils. *Soil Sci. 135*: 184–192.
Keren, R. (1989). Water-drop kinetic energy effect on water infiltration in calcium and magnesium soils. *Soil Sci. Soc. Am. J. 53*: 1624–1628.

Kreznor, W. R., Olson, K. R., and Johnson, D. L. (1992). Field evaluation of methods to estimate soil erosion. *Soil Sci. 153*: 69–81.

Le Bissonnais, Y. (1988). Comportement d'agrégats terreux soumis à l'action de l'eau: analyse des mécanismes de désagrégation. *Agronomie 8*: 87–96.

Le Bissonnais, Y. (1990). Experimental study and modeling of soil surface crusting processes. In *Catena Supplement 17*: *Soil Erosion-Experiments and Models* (R. B. Bryan, ed.), pp. 13–28.

Le Bissonnais, Y., Bruand, A., and Jamagne, M. (1989). Laboratory experimental study of soil crusting: relation between aggregates breakdown and crust structure. *Catena 16*: 377–392.

Le Bissonnais, Y., and Singer, M. J. (1993). Seal formation, runoff and interrill erosion from 17 California soils. *Soil Sci. Soc. Am. J. 57*: 224–229.

Lehrsch, G. A., Sojka, R. E., Carter, D. L., and Jolley, P. M. (1991). Freezing effects on aggregate stability affected by texture, mineralogy and organic matter. *Soil Sci. Soc. Am. J. 55*: 1401–1406.

Le Souder, C., Le Bissonnais, Y., and Robert, M. (1991). Influence of a mineral conditioner on the mechanisms of disaggregation and sealing of a soil surface. *Soil Sci. 152*: 395–402.

Levy, G., Ben-Hur, M., and Agassi, M. (1991). The effect of polyacrylamide on runoff, erosion, and cotton yield from fields irrigated with moving sprinkler systems. *Irrig. Sci. 12*.

Levy, G. J., and Van Der Watt, H. (1988). Effect of clay mineralogy and soil sodicity on soil infiltration rate. *S. Afric. J. Plant Soil 5*: 92–96.

Levy, G. J., and Van Der Watt, H. (1990). Effect of exchangeable potassium on the hydraulic conductivity and infiltration rate of some South African soils. *Soil Sci. 149*: 69–77.

Levy, G., Van Der Watt, H., du Plessis, H. M. (1988). Effect of sodium-magnesium and sodium-calcium systems on soil hydraulic conductivity and infiltration. *Soil Sci. 146*: 303–310.

Loch, R. J. (1989). Aggregate breakdown under rain: its measurement and interpretation. Ph.D. thesis, University of New England, Australia.

Matkin, E. A., and Smart, P. (1987). A comparison of tests of soil structural stability. *J. Soil Sci. 38*: 123–135.

Merzouk, A., and Blake, G. R. (1991). Indices for the estimation of interrill erodibility of Moroccan soils. *Catena 18*: 537–550.

Metzger, L., Levanon, D., and Mingelgrin, U. (1987). The effect of sewage sludge on soil structural stability: microbiological aspects. *Soil Sci. Soc. Am. J. 51*: 346–351.

Miller, W. P., and Baharuddin, M. K. (1986). Relationship of soil dispersibility to infiltration and erosion of southeastern soils. *Soil Sci. 142*: 235–240.

Molope, M. B., Grieve, I. C., and Page, E. R. (1985). Thixotropic changes in the stability of molded soil aggregates. *Soil Sci. Soc. Am. J. 49*: 979–983.

Monnier, G. (1965). Action des matières organiques sur la stabilité structurale des sols. Thèse, Paris.

Moore, D. C., and Singer, M. J. (1990). Crust formation effect on soil erosion processes. *Soil Sci. Soc. Am. J. 54*: 1117–1123.

Obi, M. E., Salako, F. K., and Lal, R. (1989). Relative susceptibility of some southeastern Nigeria soils to erosion. *Catena 16*: 215–225.

Panayiotopoulos, K. P., and Kostopoulou, S. (1989). Aggregate stability dependence on size, cultivation and various soil constituents in red mediterranean soils (alfisols). *Soil Technol. 2*: 79–89.

Perfect, E., Kay, B. D., van Loon, W. K. P., Sheard, R. W., and Pojasok, T. (1990a). Factors influencing soil structural stability within a growing season. *Soil Sci. Soc. Am. J. 54*: 173–179.

Perfect, E., Kay, B. D., van Loon, W. K. P., Sheard, R. W., and Pojasok, T. (1990b). Rate of change in structural stability under forages and corn. *Soil Sci. Soc. Am. J. 54*: 179–186.

Pierson, F. B., and Mulla, D. J. (1990). Aggregate stability in the Palouse region of Washington: effect of landscape position. *Soil Sci. Soc. Am. J. 54*: 1407–1412.

Poesen, J. (1986). Surface sealing as influenced by slope angle and position of simulated stones in the top layer of loose sediments. *Earth Surf. Process. Landforms 11*: 1–10.

Roberson, E. B., Sarig, S., and Firestone, M. K. (1991). Cover crop management of polysaccharides-mediated aggregation in an orchard soil. *Soil Sci. Soc. Am. J. 55*: 734–739.

Robert, M., Veneau, G., and Hervio, M. (1983). Influence des polycations du fer et de l'aluminium sur les propriétés des argiles. *Science Sol 3–4*: 235–251.

Römkens, M. J. M., Roth, C. B., and Nelson, D. W. (1977). Erodibility of selected clay subsoils in relation to physical and chemical properties. *Soil Sci. Soc. Am. J. 41*: 954–960.

Shainberg, I., Alperovitch, N., and Keren, R. (1988). Effect of magnesium on the hydraulic conductivity of sodic smectites-sand mixtures. *Clays Clay Miner. 36*: 432–438.

Shainberg, I., and Letey, J. (1984). Response of soils to sodic and saline conditions. *Hilgardia 52*: 1–57.

Shainberg, I., Warrington, D., and Rengasamy, P. (1990). Effect of PAM and gypsum application on rain infiltration and runoff. *Soil Sci. 149*: 301–307.

Singer, M. J., and Blackard, J. (1977). Effect of mulching on sediment in runoff from simulated rainfall. *Soil Sci. Soc. Am. J. 42*: 481–485.

Smith, H. J. C. (1990). The crusting of red soils as affected by parent material, rainfall, cultivation and sodicity. M. Sc. agricultural thesis, University of Pretoria.

Stern, R., Ben-Hur, M., and Shainberg, I. (1991). Clay mineralogy effect on rain infiltration, seal formation and soil losses. *Soil Sci. 152*: 455–462.

Sumner, M. E. (1992). The electrical double layer and clay dispersion. *Advance in Soil Science. Soil Crusting: Physical and Chemical Processes* (M. E. Sumner and B. A. Stewart, eds.). pp. 7–37.

Theng, B. K. G. (1982). Clay-polymer interactions: summary and perspectives, *Clays Clay Miner. 30*: 1–10.

Trott, K. E., and Singer, M. J. (1983). Relative erodibility of 20 California range and forest soils. *Soil Sci. Soc. Am. J. 47*: 753–749.

Van Der Watt, H., and Valentin, C. (1992). Soil crusting: the African view. In *Advance in Soil Science. Soil Crusting: Physical and Chemical Processes* (M. E. Sumner and B. A. Stewart, eds.). pp. 301–338.

Wischmeier, W. H., Johnson, C. B., and Cross, B. V. (1971). A soil erodibility nomograph for farmland and construction sites. *J. Soil Water Conserv. 20*: 150–152.
Wischmeier, W. H., and Mannering, L. V. (1969). Relation of soil properties to its erodibility. *Soil Sci. Soc. Am. Proc. 33*: 131–137.
Yair, A., and Lavee, H. (1976). Runoff generation process and runoff yield from arid talus mantled slopes. *Earth Surf. Process. Landforms 1*: 235–247.
Yoder, R. E. (1936). A direct method of aggregate analysis of soils and a study of the physical nature of erosion losses. *J. Am. Soc. Agron. 28*: 337–351.
Young, R. A., and Mutchler, C. K. (1977). Erodibility of some Minnesota soils. *J. Soil Water Conserv. 32*: 180–182.

4

Splash and Detachment by Waterdrops

Joe M. Bradford
*Subtropical Agricultural Research Laboratory,
Agricultural Research Service, U.S. Department of Agriculture,
Weslaco, Texas*

Chi-hua Huang
*National Soil Erosion Research Laboratory,
Agricultural Research Service, U.S. Department of Agriculture,
and Purdue University, West Lafayette, Indiana*

I. INTRODUCTION

Soil erosion by water occurs due to complex interactions of subprocesses of detachment and transport of soil materials by raindrop impact and overland flow and of deposition. The dominant subprocess varies according to whether the source area is rills or interrills. Interrill erosion occurs on an area where all detachment is due to the forces of raindrop impact and transport is primarily by overland flow. Slope length and slope steepness are sufficiently low that particle detachment by flow does not occur. On very steep and long slopes, interrill erosion can occur; however, for very short distances (centimeters). Surface flow depth is very low (millimeters) and uniformly distributed across the area. If flows concentrate and soil detachment by flow occurs, rills are initiated and develop at a rate dependent on flow shear stress and soil resistance against flow shear.

In this chapter we limit our discussion primarily to the mechanics and processes associated with soil detachment by raindrop impact. Factors controlling the process of detachment and the interrelationship of raindrop detachment with transport of detached particles by shallow flow in interrill erosion are discussed. Since soil detachment by raindrop impact plays a minor role in rill erosion processes, rill erosion will not be discussed.

II. SINGLE WATERDROPS AND SPLASH

Understanding of interrill erosion or multiple-raindrop splash detachment begins with empirical and theoretical studies of the behavior of single raindrops or waterdrops. Single drop studies have examined shapes of raindrops, forces of raindrop impact, and reaction of soil to the impact forces.

A. Methodology

Knowledge of the behavior of single waterdrops was advanced by study techniques developed by Al-Durrah and Bradford (1981). An improved and inexpensive raindrop splash collection device was designed so that a single drop after falling 9 m would hit a soil target area 16 mm in diameter and the airborne splash material was collected. Gantzer et al. (1985) developed an electronic discriminator consisting of two optical light source sensors to eliminate falling waterdrops that drifted away from the target area. By use of high-speed photography, Al-Durrah and Bradford (1982b) were able to study the interrelations of splash angle, splash weight, and soil strength.

B. Raindrop Shape and Size

Early studies on raindrop shape were conducted by Laws (1941), who photographed waterdrops falling in still air and found that the waterdrop is not spherical but oblate flattened on the bottom. Disrud et al. (1969) reported that if waterdrops fall in wind for sufficient time, they will have the shape of an oblate spheroid with a flattened area between the bottom and upwind side. The stable shape of raindrops is reached only when the terminal velocity is reached. Before a waterdrop reaches its terminal velocity, it oscillates between vertical and horizontal oblateness with a frequency that depends upon its size. McDonald (1954) reported that the factors affecting the equilibrium shape of waterdrops falling at terminal velocity are surface tension, hydrostatic pressure gradients, aerodynamic pressures, electronic charge, and internal circulations.

Drop shape affects the amount of soil splash (Riezebos and Epema, 1985). By changing the height of fall from 0.57 to 0.62 to 0.67 m, the amount of splash loss changed from 0.78 to 0.28 to 0.88 g per drop. The drop shape at 0.57- and 0.67-m fall height was prolate, and at 0.62-m fall height the drop shape was oblate.

Drop sizes are generally distributed from a fraction of a millimeter to an upper limit of about 5 mm in diameter; drops bigger than this break into smaller drops (Hudson, 1971). Median drop size is dependent upon rainfall intensity and ranges from 1.4 to 2.7 mm for rainfall intensities of 2.5 to 51 mm/h (Rogers et al., 1967). Hudson (1971) showed an increase in median drop diameter for intensities up to 100 mm/h and then a decrease at higher intensities. Current thinking is that median drop size remains essentially constant above 100 mm/h.

C. Force and Impact Stress

Waterdrop impact forces are highly dependent upon drop size and shape and fall height. Impact forces and stresses have been measured (1) by penetration of the drops through nylon meshes, (2) from analysis of splash rebound measurements from high-speed photography, and (3) from the output of miniature force transducers (Ghadiri and Payne, 1981; Nearing et al., 1986; Nearing and Bradford, 1987) and calculated by solutions to the Navier-Stokes equations (Huang et al., 1982, 1983). For 3.5- and 6.2-mm drops falling from 0.7 to 6.2 m, maximum stresses ranged between 2 to 6 MPa and for a duration of about 50 μs on the perimeter of a circle around the drop center (Ghadiri and Payne, 1981). Peak forces for 3.31- to 5.25-mm-diameter waterdrops falling from a height of 14.0 m onto piezoelectric transducers occurred within 13 to 21 μs of initial contact, ranged from 1.0 to 3.8 N, and decreased to 0.5 N after approximately 100 μs (Nearing et al., 1986).

Average pressure curves increased and decreased much more quickly than did the force curves because of the increase of contact area with impact time. Average pressures of over 500 kPa were present for about 7 μs for 3.31-mm-diameter drops and for about 17 μs for 4.51-mm diameter drops. Mean peak force of waterdrop impact may be described as a linear function of either kinetic energy or momentum for any given drop height, but the function is not the same for each height (Nearing and Bradford, 1987). In other words, impact stress does not correlate with kinetic energy if impact studies are conducted at several heights (Ghadiri and Payne, 1977).

D. Erosivity

The most-used erosivity parameters for single waterdrops are kinetic energy ((1/2) mass × velocity squared) and momentum (mass × velocity) (Al-Durrah and Bradford, 1982a; Riezebos and Epema, 1985). Based on the experiments of Al-Durrah and Bradford (1982a), there is no reason to select kinetic energy over momentum as the waterdrop erosivity parameter. Sharma and Gupta (1989) presented evidence for the existence of a threshold kinetic energy or momentum before the detachment process can be initiated by raindrop impact. The threshold erosivity concept of Sharma and Gupta assumes that a minimum energy is needed to overcome the inherent strength of soil.

E. Soil Resisting Forces

Numerous attempts have been made to measure or model the reaction of the soil to waterdrop impact forces or to relate soil variables to waterdrop splash

(Rose, 1960, 1961; Mazurak and Mosher, 1968; Bubenzer and Jones, 1971; Yamamoto and Anderson, 1973; Sloneker et al., 1976; Al-Durrah and Bradford, 1982a; Nearing and Bradford, 1985; Bradford et al., 1986). Soil variables identified as affecting splash are soil texture, organic matter, bulk density, water potential, exchangeable cations, aggregate size and stability, and shear strength. Of all the soil variables, soil shear strength appears to be most highly correlated with splash due to raindrop impact. The idea of using shear strength as an index of soil erodibility (Chorley, 1959) was advanced by Al-Durrah and Bradford (1981). They proposed a simple relationship between splash weight and a linear function of the ratio of waterdrop kinetic energy to soil shear strength as measured by the Swedish fall-cone device. A high correlation coefficient ($r^2 = 0.97$) was found for an Ida silt loam (mesic Typic Udorthents) at three bulk densities, four water potentials, and three drop diameters (Fig. 1). A constant drop height of 8.9 m was maintained. Eight additional soils having a wide range of particle sizes were later tested (Al-Durrah and Bradford, 1982a). The coefficient of de-

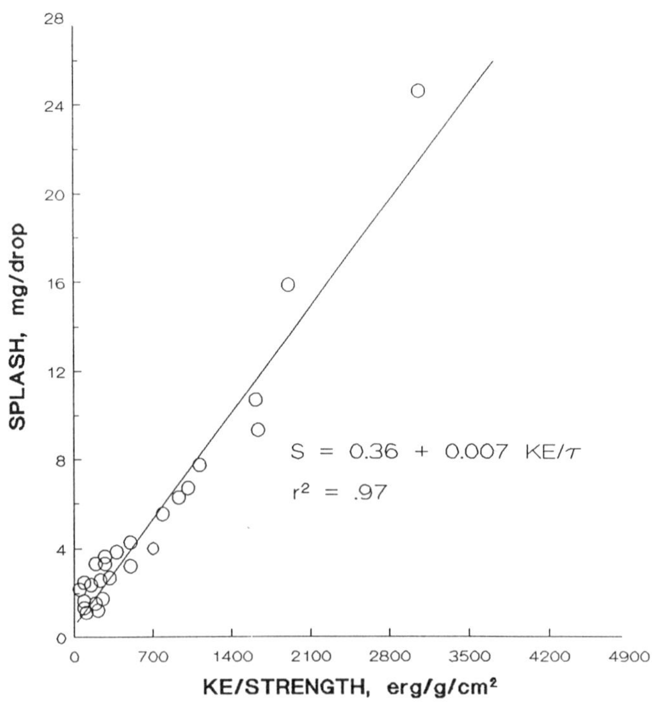

Fig. 1 The relation between splash weights (S) and the ratio of kinetic energy (KE) of a raindrop to the soil shear strength (τ). (From Al-Durrah and Bradford, 1981.)

termination values between splash weight and the force-resistance ratio for individual soils were high (r^2 = 0.88 to 0.97); however, the coefficient of determination among all soils was lower (r^2 = 0.81). Possible reasons given for the lower fit among soils were (1) improper calibration of the fall-cone device, (2) differences in fall velocities between waterdrops (7 to 9 m/s) and the fall-cone (0.35 m/s), and (3) differences in soil failure processes between the fall-cone and the impacting waterdrop. Nearing and Bradford (1985) found that variation in the relationship among soils could be reduced by reducing the fall-cone strength as a function of the total stress friction angle for each soil. The authors assumed that the fall-cone strength adequately estimates soil resistance to the initial splash phase of cavity formation upon raindrop impact, but overestimated soil resistance to lateral jetting forces. The model was later extended to soil surface crusts (Bradford et al., 1986). Splash was accurately predicted for silt loam or finer textures, but not for coarse-textured soils. The fall-cone test was designed and calibrated for use in fine-textured soils and for saturated soils with failure under undrained conditions. Thus the fall-cone device has limited usefulness in sandy soils for predicting splash.

III. MULTIPLE WATERDROPS/RAINDROPS AND SPLASH

Understanding of the mechanics of the soil splash process was advanced by single-drop experiments, and some of the relationships derived with single drops were upheld in multiple-drop experiments. On the other hand, because of differences in scale between single- and multiple-drop studies, other factors became important in estimating splash in multiple-drop experiments. Some of these factors will be discussed.

A. Methodology

Several techniques have been used to measure splash under natural conditions or multiple-raindrop simulators. In the field, splash is often measured with splash cups by determining splash loss from cups filled with soil material or by measuring the mass of sediment collected in an empty cup (Poesen and Torri, 1988). Using plot pans in the laboratory, splash is measured either as the amount of splash falling outside the pan (Bradford et al., 1987a) or as the material collected within the pan boundaries in a splash collector (Bradford and Huang, 1993). Splash expressed as a unit of ground area cannot be accurately estimated by simply determining the mass of soil leaving a plot pan. Proper consideration of cup dimensions and rim effects is necessary (Kinnell, 1974; Poesen and Torri, 1988; Torri and Poesen, 1988). For example, splash was determined by Bradford and Huang (1993) for two pan designs: (1) a 320-mm-wide by 450-mm-long plot pan (0.14 m^2) with a 540-mm-high splash collection shield surrounding the box (Bradford et al., 1987a); (2) a newly designed pan similar to the pan of Lattanzi et al. (1974). Features of the new design were (a) a 530-mm-wide by 610-mm-long test area (0.32 m^2) surrounded by a 300-mm-wide soil buffer, (b) two 35-mm-wide by 610-mm-long troughs located along both sides of the test area for splash collection, (c) a slot along the lower end of the test area for collection of runoff and wash, and (d) drainage outlets in the bottom of each compartment for water percolation and suction control. The unique feature of the pan was the ability to measure separately within the pan both splash and sediment yield. Table 1 shows that splash per unit of soil area is much less for the 0.14-m^2 pan compared to the 0.32-m^2 pan with the soil buffer area. The calculation for the 0.32-m^2 pan assumes that the amount of sediment collected in the collection slot times 1.33 equals the mass of sediment detached at the soil surface with an identical area as the collection slot (Poesen and Torri, 1988). The factor 1.33 assumes that the 35-mm-wide collector collects about 75% of the maximum possible splashed soil (Bradford and Huang, 1993). For the rectangular-type pans, as the area increases, splash per unit soil area decreases because the central part of the plot box contributes less sediment than parts closer to the edge.

Caution must be exercised in interpreting splash data from different pan designs. Splash collected from the rectangular 0.14-m^2 pan with shield can be interpreted only on a unit-pan-area basis. This splash value can be used only as an index because the actual area conversion is unknown. Splash collected within the slots bounding the 0.32-m^2 pan, which is expressed in unit collector area, can be interpreted as the amount of raindrop-splashed material available for transport.

B. Rainfall Intensity and Energy

Amount of soil splash increases as both rainfall intensity and rainfall energy increases (Free, 1960; Truman, 1990; Agassi et al., 1994). The rate of increase, however, depends on antecedent soil moisture, mechanisms of aggregate break-

Table 1 Comparison of Total Splash Detachment during 1-h Rainstorm at Rainfall Intensity of 64 mm/h for a 0.14-m^2 Rectangular Erosion Pan and a 0.32-m^2 Pan with 0.30-m-Wide Surrounding Soil Border

Soil[a] series	Total soil splash ($kg/m^2/h$)	
	0.14-m^2 pan	0.32-m^2 pan
Vicksburg	0.68	1.27
Brooksville	1.28	4.66

[a]Soil < 20 mm in diameter was placed dry in pans at a 9% slope.

down, and soil properties, such as clay mineralogy, texture, organic matter, and exchangeable sodium content (Agassi et al., 1994).

C. Soil Characteristics

Erosion is a process of detachment and transport of soil materials. Those soil properties that affect soil detachment by raindrop impact presumably are different from those soil properties that control transport of soil. Numerous studies have been conducted to identify those soil properties that affect soil erosion or erodibility, but few studies have quantified the effect of soil properties on splash detachment. Soil variables identified as affecting splash were soil texture (Rose, 1961), percent clay (Bubenzer and Jones, 1971), soil structure (McIntyre, 1958), aggregate size (Mazurak and Mosher, 1970), and soil sodicity (Agassi et al., 1994).

D. Soil Strength

As with the case for splash detachment by single waterdrops, rainsplash under multiple-waterdrop impact is highly correlated with rainfall kinetic energy divided by the fall-cone shear strength term (Bradford et al., 1987b). A low correlation was found with the torvane and pocket penetrometer strengths as the shear strength term (Bradford et al., 1992). A high correlation, using the fall-cone, exists because the mechanism of failure of both fall-cone and raindrop impact on soil is a process of compression and shear (Al-Durrah and Bradford, 1982a; Huang et al., 1983). Low correlations between splash and strength exist when other factors such as surface roughness, depth of surface water ponding, and size and density of splashed material vary.

E. Antecedent Moisture

Antecedent soil water plays an important role in controlling splash detachment through its effects on soil resistance to raindrop impact and the magnitude and mechanisms of aggregate breakdown on the soil surface (le Bissonnais, 1990; Truman et al., 1990). For single-drop splash, Al-Durrah and Bradford (1982a) found that weight of soil splash decreased as matric water potential decreased from −0.5 to −6.2 kPa. This decrease in splash was due to an increase in soil strength as water potential decreased. For multiple-raindrop impact, rate of wetting and aggregate breakdown exert a greater control on splash than initial soil strength. Highly aggregated soils resist aggregate breakdown upon wetting and splash detachment is lower. Table 2 indicates lower splash detachment for an initially wet, as compared to an initially dry, surface for five soils ranging in texture from sandy loam to clay (Truman and Bradford, 1990). Splash was determined as the amount of soil splashed from a 320-mm-wide by 450-mm-

Table 2 Soil Splash from a 320-mm-Wide by 450-mm-Long Erosion Pan during an Initial and 5-Min Sampling of a Simulated Rainstorm at an Intensity of 63 mm/h

Soil series	Water content	Splash ($kg/m^2/h$) 0–5 min	55–60 min
Cecil	air-dry	2.69	4.70
	prewetted	2.57	1.97
Miami	air-dry	1.43	1.18
	prewetted	0.75	1.09
Heiden	air-dry	0.27	5.65
	prewetted	0.16	1.66
Pierre	air-dry	1.07	5.66
	prewetted	0.46	2.83
Broughton	air-dry	0.77	3.58
	prewetted	0.16	2.48

Source: Truman and Bradford (1990).

long plot pan expressed on a per-unit soil area basis. The tendency for prewetted soils to have lower splash detachment than air-dry soils is due to the reduction in slaking forces with slower wetting. In initially dry soil, both slaking forces and mechanical forces of raindrop impact occur simultaneously and aggregate breakdown is rapid.

F. Water Ponding Depth

Depth of water ponding or flow depth on the soil surface influences the amount of splashed material (Palmer, 1963). Liu (1991) determined average flow depths of 0.8, 0.6, and 0.5 mm for 5%, 20%, and 40% slopes, respectively. These values were determined for <4 mm soil placed in laboratory erosion pans under a rainfall intensity of 70 mm/h. Airborne detachment appears to be most intense at zero water depth and is greatly reduced or inhibited at water depths greater than 2 to 5 mm (Kirkby and Kirkby, 1974; Moss and Green, 1983), depending on drop size. Under conditions of shallow flow, raindrop impact effects on particle detachment cannot easily be separated from effects on sediment transport. The importance of raindrop impact in enhancing transport of raindrop detached sediment in shallow, interrill flow is the subject of several papers by Kinnell (Kinnell, 1988, 1990, 1991). Although the effect of water ponding on splash detachment is recognized, its quantification for soil surfaces under rainfall is difficult due to complex interactions between microtopography and the evolving surface sealing and erosion processes.

G. Surface Sealing

The extent to which a soil seals during a rainstorm greatly influences the amount of splash. Sealing depends upon many variables, including rainfall characteristics, antecedent soil moisture, and aggregate breakdown, but soil texture seems to be one of the most important soil variables influencing surface sealing and splash detachment (Bradford and Huang, 1992). Figure 2 compares splash into collection slots associated with the 0.32-m^2 soil pan with 0.30-m-wide surrounding soil border for a well-aggregated Brooksville silty clay and a poorly aggregated Vicksburg silt loam soil. The Vicksburg soil sealed rapidly and formed a seal of high strength (40.2 kPa) and low infiltration rate (3.1 mm/h). Splash detachment was low. For soils that are not highly susceptible to surface sealing, such as Brooksville soil, splash detachment is higher and remains as a high rate throughout a storm.

H. Slope Steepness

As slope steepness increases, the number of drop impacts per unit surface area and the normal component of drop impact both decrease, thereby decreasing splash detachment (Poesen, 1984). Conversely, as slope steepness increases,

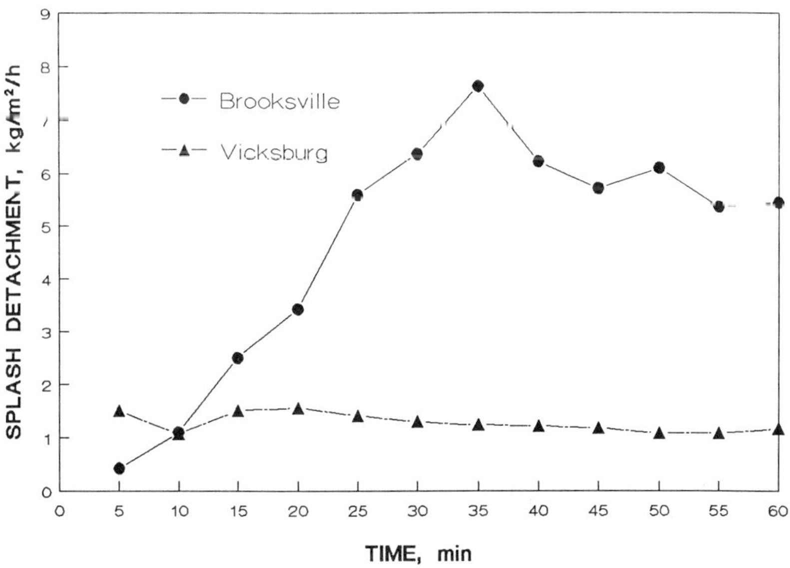

Fig. 2 Splash detachment vs. time into collection troughs for a 0.32-m^2 soil pan. Soil was initially dry, slope was 9%, and rainfall intensity was 64 mm/h.

degree of surface sealing decreases and rate of increase in soil resistance or strength decreases, thereby increasing splash detachment (Poesen, 1984). Table 3 reports total splash detachment values for collection slots within the 0.32-m^2 plot pans. Rainfall intensity was 65 mm/h for 1 h. As slope steepness increased for the three silt loam soils, splash detachment values for a 20% slope were about 1.3 times splash values for a 9% slope, indicating less sealing on the 20% slope. For the clay loam and clay soils, splash values at a 20% slope were less than at 9%. A significant interaction between soil properties (aggregate stability, soil strength, and surface sealing) and slope steepness is shown.

I. Surface Drying and Crusting

Drying of a rain-impacted surface seal forms a hard crust in some soils. Upon rewetting of the crust by raindrop impact, crust breakdown and the resulting splash detachment depend upon the mechanism of crust breakdown, strength of the crust, and soil properties such as texture, organic matter content, clay mineralogy, exchangeable ions, and electrolyte concentration in the soil solution (le Bissonnais et al., 1989; le Bissonnais, 1990; le Bissonnais and Singer, 1992). Table 4 shows the effect of a single drying cycle on total splash detachment for a 1-h rainstorm with an intensity of 65 mm/h. Rainfall was first applied on a

Table 3 Total Splash Detachment during 1-h Rainstorm of 65 mm/h Using a 0.32-m^2 Plot Pan with Soil Buffer

Soil[a] series	Slope steepness (%)	Splash ($kg/m^2/h$)	Ratio 20% to 9%	Sediment yield ($kg/m^2/h$)	Ratio 20% to 9%	Proportion of splash transported by overland flow
Vicksburg	9	1.27	1.27	0.50	4.46	0.39
silt loam	20	1.62		2.23		1.38
Russell	9	2.65	1.29	0.87	2.13	0.33
silt loam	20	3.43		1.85		0.54
Miami	9	3.18	1.24	0.69	2.01	0.22
silt loam	20	3.93		1.39		0.35
Sharpsburg	9	3.34	0.95	1.27	1.36	0.38
clay loam	20	3.18		1.73		0.54
Heiden	9	4.39	0.52	0.99	0.55	0.23
clay	20	2.32		0.55		0.24

[a]Air-dry soil < 20 mm.

Table 4 Total Splash and Sediment Yield during 1-h Rainstorm of Intensity 65 mm/h as Influenced by Number of Drying Cycles

Soil series	No. of drying cycles	Splash ($kg/m^2/h$)	Ratio of 1 to 0 cycle	Sediment yield ($kg/m^2/h$)	Ratio of 1 to 0 cycle	Proportion of splash transported by overland flow
Vicksburg	0	1.27	0.68	0.51	0.78	0.40
silt loam	1	0.87		0.40		0.46
Russell	0	2.25	0.83	0.87	0.76	0.39
silt loam	1	1.87		0.66		0.35
Miami	0	3.18	0.86	0.69	1.24	0.22
silt loam	1	2.76		0.86		0.31
Sharpsburg	0	3.34	1.13	1.27	1.20	0.38
clay loam	1	3.79		1.52		0.40
Heiden	0	4.39	1.27	0.99	1.75	0.22
clay	1	5.59		1.74		0.31

For both runs the soil was initially air dry. Plot pan was a 0.32-m^2 pan with soil buffer at a 9% slope steepness.

dry soil surface placed in the 0.32-m^2 plot pan. Following rainfall, the soil was dried for one week at room temperature, allowing the sealed soil to dry and form a crust. A second rainfall was applied on the air-dried crust for an additional hour at the same intensity. Splash was collected in splash troughs located within the pan (Bradford and Huang, 1993). For the three silt loam soils, total splash detachment was less for the dried crust. For the clay loam and clay soils, splash from the dried crust was greater. The greater splash values from the clay loam and clay crusted soils result from greater slaking and differential swelling upon wetting for the higher clay soils (Cernuda et al., 1954). Without a knowledge of soil properties and an understanding of crust breakdown processes, the effect of drying and rewetting on splash detachment cannot be predicted.

IV. INTERRILL EROSION

Soil detachment by raindrop impact is the principal erosion process controlling interrill soil erosion, even though sufficient surface flow must be available for transport of the detached particles. The detachment capacity of interrill flow is negligible compared to that of raindrop splash (Young and Wiersma, 1973) due to the low shear stresses of the thin sheet flow. Young and Wiersma (1973) reported that by decreasing rainfall impact energy by 89%, without reducing rainfall amount, interrill soil losses were decreased by 90% or more. Even though soil detachment by raindrop impact is the primary factor controlling interrill erosion, the amount of soil splash does not necessarily correlate with interrill soil erosion (Bradford and Huang, 1993). Other processes and factors are critical to sediment yield from interrill areas.

A. Methodology

Since the origination of the concept of separating soil erosion into interrill and rill subprocesses (Meyer et al., 1975), numerous researchers have measured interrill erosion both in the laboratory and in the field. Interrill field plot size varies from about 1 m wide by 1 m long (Meyer and Harmon, 1979) to 1 m wide by 2 m long. Laboratory plot boxes come in a variety of designs and dimensions. Bryan and de Ploey (1983) noted, however, that little attention has been paid to the comparability of data acquired using different simulators (or plot pans) or to the way in which the results obtained may differ from the situation under natural rainfall. Bradford and Huang (1993) compared a frequently used design, a 320-mm-wide by 450-mm-long (0.14 m^2) rectangular erosion pan (Bradford et al., 1987a), with the 0.32-m^2 pan surrounded with a soil buffer. Data from the pan compare favorably well with data collected in the field with rainfall simulators (Bradford and Huang, 1993; Truman, 1990). Data collected with the 0.14-m^2 pan compared neither with the 0.32-m^2 pan nor with the field results.

B. Splash and Sediment Yield

As stated above, sediment yield from interrill areas is a function of detachment by raindrop impact and transport by shallow flow. As shown in Tables 3 and 4, the percent of collected splash material transported by overland flow was, in general, considerably less than 100%. As slope steepness increased from 9% to 20% (Table 3), the percentage of splash transported as sediment yield increased. The increase was shown to be a function of soil properties, such as texture, strength, and aggregate stability. The increase in percentage of splash transported in overland flow due to slope steepness was greatest for the highly erodible Vicksburg silt loam (1.38/0.39 = 3.54) and slight for the less erodible Heiden clay (0.24/0.23 = 1.04). For the Vicksburg silt loam at 20% slope, the sediment yield value (2.23 kg/m^2/h) was greater than the splash value (1.62 kg/m^2/h) due to the formation of rills in the erosion pan. Thus the primarily detachment process has changed from detachment by raindrop impact to detachment by concentrated surface flow. Table 4 shows the effect of drying cycles on splash detachment, sediment transport, and percentage of splash transported. As shown, a greater percentage of splash is transported as sediment yield during a simulated rainstorm on a dried, crusted soil surface (1) than on condition representing a freshly cultivated condition (0). Presumably, the reason for the greater percentage of material transported following a drying cycle was greater aggregate breakdown and a smaller mean sediment size as compared to the initial rainfall condition. For example, mean weight diameter of sediment transported for the initial condition for the Vicksburg soil was 0.14 mm and for the second rainfall was 0.04 mm. Transport of detached soil particles is influenced also by slope steepness (flow velocity) and soil surface roughness (tortuosity of flow paths).

V. CONCLUSIONS

In this chapter, we have not attempted to provide a thorough literature review of splash detachment experiments but have reviewed the studies of Bradford, Huang, and colleagues studying at the National Soil Erosion Research Laboratory. We have attempted to clarify concepts and processes of detachment by raindrop impact and clarify the relative importance of splash to sediment yield in interrill areas. Much research has yet to be done. Future plans include using a large database on splash detachment, interrill sediment transport, soil and sediment properties, and rainfall characteristics to develop a processed-based, interrill erosion model. Such a model would allow for the development of interrill erodibility parameters based on detachment and transport.

REFERENCES

Agassi, M., Bloem, D., and Ben-Hur, M. (1994). Effect of drop energy and soil and water chemistry, on infiltration and erosion. *Water Resources Res. 30*(4): 1187–1193.

Al-Durrah, M., and Bradford, J. M. (1981). New methods of studying soil detachment due to waterdrop impact. *Soil Sci. Soc. Am. J. 45*: 949–953

Al-Durrah, M. M., and Bradford, J. M. (1982a). Parameters for describing soil detachment due to single waterdrop impact. *Soil Sci. Soc. Am. J. 46*: 836–840.

Al-Durrah, M. M., and Bradford, J. M. (1982b). The mechanism of raindrop splash on soil surfaces. *Soil Sci. Soc. Am. J. 46*: 1086–1090.

Bradford, J. M., Ferris, J. E., and Remley, P. A. (1987a). Interrill soil erosion processes. I: Effect of surface sealing on infiltration, runoff, and soil splash detachment. *Soil Sci. Soc. Am. J. 51*: 1566–1571.

Bradford, J. M., Ferris, J. E., and Remley, P. A. (1987b). Interrill soil erosion processes. II: Relationship of splash detachment to soil properties. *Soil Sci. Soc. Am. J. 51*: 1571–1575.

Bradford, J. M., Remley, P. A., Ferris, J. E., and Santini, J. B. (1986). Effect of soil surface sealing on splash from a single waterdrop. *Soil Sci. Soc. Am. J. 50*: 1547–1552.

Bradford, J. M., and Huang, C. (1992). Mechanisms of crust formation: Physical components. In *Soil Crusting—Chemical and Physical Processes* (M. E. Sumner and B. A. Stewart, eds.). Advances in Soil Science, Lewis Pub., Ann Arbor, MI, pp. 55–72.

Bradford, J. M., and Huang, C. (1993). Comparison of interrill soil loss for laboratory and field procedures. *Soil Tech. 6*: 145–156.

Bradford, J. M., Truman, C. C., and Huang, C. (1992). Comparison of three measures of resistance of soil surface seals to raindrop splash. *Soil Tech. 5*: 47–56.

Bryan, R. B., and de Ploy, J. (1983). Comparability of soil erosion measurements with different laboratory rainfall simulators. In *Rainfall Simulation, Runoff and Soil Erosion* (J. de Ploey, ed.). Catena Suppl. 4, Braunschweig, pp. 4–56.

Bubenzer, G. D., and Jones, B. A. (1971). Drop size and impact velocity effects on the detachment of soils under simulated rainfall. *Trans. ASAE 14*: 625–628.

Cernuda, C. F., Smith, R. M., and Vincente-Chandler, J. (1954). Influence of initial soil moisture condition on resistance of macroaggregates to slaking and to water-drop impact. *Soil Sci. 77*: 19–27.

Chorley, R. J. (1959). The geomorphic significance of some Oxford soils. *Am. J. Sci. 257*: 503–515.

Disrud, L. A., Lyles, L., and Skidmore, E. K. (1969). How wind affects the size and shape of raindrops. *Agr. Engr. 50*: 617.

Free, G. R. (1960). Erosion characteristics of rainfall. *Agric. Eng. 41*: 447–449, 445.

Gantzer, C. J., Alberts, E. E., and Bennett, W. H. (1985). An electronic discriminator to eliminate the problem of horizontal raindrop drift. *Soil Sci. Soc. Am. J. 49*: 211–215.

Ghadiri, H., and Payne, D. (1977). Raindrop impact stress and the breakdown of soil crumbs. *Soil Sci. 28*: 247–258.

Ghadiri, H., and Payne, D. (1981). Raindrop impact stress. *J. Soil Sci. 32*: 41–49.

Huang, C., Bradford, J. M., and Cushman, J. H. (1982). A numerical study of raindrop impact phenomena: The rigid case. *Soil Sci. Soc. Am. J. 46*: 14–19.

Huang, C., Bradford, J. M., and Cushman, J. H. (1983). A numerical study of raindrop impact phenomena: The elastic deformation case. *Soil Sci. Soc. Am. J. 47*: 855–861.

Hudson, N. (1971). *Soil Conservation*. Cornell University Press, Ithaca, NY.

Kinnell, P. I. A. (1974). Splash erosion: Some observations on the splash-cup technique. *Soil Sci. Soc. Am. J. 38*: 657–660.

Kinnell, P. I. A. (1988). The influence of flow discharge on sediment concentrations in raindrop induced flow transport. *Aust. J. Soil Res. 26*: 575–582.

Kinnell, P. I. A. (1990). The mechanics of raindrop-induced flow transport. *Aust. J. Soil Res. 28*: 497–516.

Kinnell, P. I. A. (1991). The effect of flow depth on sediment transport induced by raindrops impacting shallow flows. *Trans. ASAE 34*: 161–168.

Kirkby, A., and Kirkby, M. J. (1974). Surface wash and the semi-arid break in slopes. *Z. Geomorphol. Suppl. 21*: 151–176.

Lattanzi, A. R., Meyer, L. D., and Baumgardener, M. F. (1974). Influences of mulch rate and slope steepness on interrill erosion. *Soil Sci. Soc. Am. Proc. 8*: 946–950.

Laws, J. O. (1941). Measurements of fall-velocity of water drops and raindrops. *Trans. Am. Geophys. Union. 22*: 709–721.

le Bissonnais, Y. (1990). Experimental study and modelling of soil surface crusting processes. In *Soil Erosion—Experiments and Models* (R. B. Bryan, ed.). Catena Suppl. 17, Catena Verlag, Cremlingen-Destedt, Germany, pp. 13–28.

le Bissonnais, Y., Bruand, A., and Jamagne, M. (1989). Laboratory experimental study of soil crusting: Relations between aggregate breakdown mechanisms and crust structure. *Catena. 16*: 377–392.

le Bissonnais, Y., and Singer, M. J. (1992). Crusting, runoff, and erosion response to soil water content and successive rainfalls. *Soil Sci. Soc. Am. J. 56*: 1898–1903.

Liu, B. (1991). Interrill erosion processes as affected by slope steepness. M.S. thesis, Purdue University.

Mazurak, A. P., and Mosher, P. M. (1970). Detachment of soil aggregates by simulated rainfall. *Soil Sci. Soc. Am. Proc. 34*: 798–800.

McDonald, J. E. (1954). The shape and aerodynamics of large raindrops. *J. Meteor. 11*: 478–494.

McIntyre, D. S. (1958). Soil splash and the formation of surface crusts by raindrop impact. *Soil Sci. 85*: 261–266.

Meyer, L. D., Foster, G. R., and Romkens, M. J. M. (1975). Source of soil eroded by water from upland slopes. In *Present and Prospective Technology for Predicting Sediment Yields and Sources*. Proc. Sediment Yield Workshop, SUDA Sediment. Lab., Oxford, MS, 28–30 Nov. 1972, USDA-ARS, ARS-S-40, U.S. Department of Agriculture, Washington, DC, pp. 177–189.

Meyer, L. D., and Harmon, W. C. (1979). Multiple-intensity rainfall simulator for erosion research on row sideslopes. *Trans. ASAE 22*: 100–103.

Moss, A. J., and Green, P. (1983). Movement of solids in air and water by raindrop impact. Effects of drop-size and water-depth variations. *Aust. J. Soil Res. 21*: 257–269.

Nearing, M. A., and Bradford, J. M. (1985). Single waterdrop impact detachment and mechanical properties of soils. *Soil Sci. Soc. Am. J. 49*: 457–552.

Nearing, M. A., and Bradford, J. M. (1987). Relationships between waterdrop properties and forces of impact. *Soil Sci. Soc. Am. J. 51*: 425–430.

Nearing, M. A., Bradford, J. M., and Holtz, R. D. (1986). Measurements of force vs. time relations for waterdrop impact. *Soil Sci. Soc. Am. J. 50*: 1532–1536.

Palmer, R. S. (1963). The influence of a thin water layer on waterdrop impact forces. *Int. Assoc. Sci. Hydrol. 65*: 141–148.

Poesen, J. (1984). The influence of slope angle on infiltration rate and Hortonian overland flow volume. *Z. Geom. Supp. Bd. 49*: 1117–1143.

Poesen, J., and Torri, D. (1988). The effect of cup size on splash detachment and transport measurements. I: Field measurements. *Catena Suppl. 12*: 113–126.

Riezebos, H. Th., and Epema, G. F. (1985). Drop shape and erosivity. II: Splash detachment, transport and erosivity indices. *Earth Surf. Process. Landforms 10*: 69–74.

Rogers, J. S., Johnson, L. C., Jones, D. M. A., and Jones, B. A. (1967). Sources of error in calculating the kinetic energy of rainfall. *J. Soil Water Conserv. 22*: 101–104.

Rose, C. W. (1960). Soil detachment caused by rainfall. *Soil Sci. 89*: 28–35.

Rose, C. W. (1961). Rainfall and soil structure. *Soil Sci. 91*: 49–54.

Sharma, P. P., and Gupta, S. C. (1989). Sand detachment by single raindrops of varying kinetic energy and momentum. *Soil Sci. Soc. Am. J. 53*: 1005–1010.

Sloneker, L. L., Olson, T. C., and Moldenhauer, W. C. (1976). Effect of pore water pressure on sand splash. *Soil Sci. Soc. Am. J. 40*: 948–951.

Torri, D., and Poesen, J. (1988). The effect of cup size on splash detachment and transport measurements. II: Theoretical approach. *Catena Suppl. 12*: 127–137.

Truman, C. C. (1990). Effect of antecedent moisture content, rainfall intensity, aggregate stability, and plot configuration on interrill erosion processes. Ph.D. thesis, Purdue University.

Truman, C. C., and Bradford, J. M. (1990). Effect of antecedent soil moisture on splash detachment under simulated rainfall. *Soil Sci. 150*: 787–798.

Truman, C. C., Bradford, J. M., and Ferris, J. E. (1990). Antecedent water content and rainfall energy influence on soil aggregate breakdown. *Soil Sci. Soc. Am. J. 54*: 1385–1392.

Yamamoto, T., and Anderson, H. W. (1973). Splash erosion related to soil erodibility indexes and other forest soil properties in Hawaii. *Water Resour. Res. 9*: 336–345.

Young, R. A., and Wiersma, J. L. (1973). The role of rainfall impact in soil detachment and transport. *Water Resour. Res. 9*: 1629–1636.

5

Slope, Aspect and Surface Storage

Dino Torri
Soil Genesis, Classification, and Cartography Research Center, National Research Council, Florence, Italy

I. INTRODUCTION

Slope affects soil erosion rate through its morphological characteristics and aspect. Two of these morphological characteristics, namely, gradient and slope length, were first introduced in quantitative relationships estimating soil loss (Zingg, 1940; Wischmeier and Smith, 1978; see also Morgan, 1986, or Mitchell and Bubenzer, 1980, for a general survey). Aspect is still scarcely considered in soil loss modeling, despite its acknowledged importance.

A quick glance at the slope parametrization of one of the best-known equations for soil loss estimation (i.e., the universal soil loss equation, proposed by Wischmeier and Smith, 1978) will help to introduce this chapter. Here length and gradient are described as follows:

$$L_F = \left(\frac{x}{22.13}\right)^m \tag{1}$$

where

L_F = length effect
x = slope length (m)
m = 0.5 if 5% $\leq$ slope, m = 0.4 if 3% $\leq$ slope < 5%, m = 0.3 if 1% $\leq$ slope < 3%, m = 0.2 if slope < 1%

and

$$S_F = 0.065 + 4.53 \tan \alpha + 65 \tan^2 \alpha \qquad (2)$$

where S_F = slope factor and α = slope angle.

The interaction between gradient and length was further investigated by Mutchler and Greer (1980), Murphree and Mutchler (1981), and Mutchler and Murphree (1985), resulting in the following equations:

$$m = 1.2 (\sin \alpha)^{0.33} \qquad (3)$$

where m is the exponent of the length factor formula, and

$$S'_F = C_r S_F \qquad (4)$$

where S'_F = corrected slope factor and C_r = correction factor, given by

$$C_r = 1 - 0.67e^{-160 \sin \alpha} \qquad (5)$$

These relations were further examined and rearranged by McCool et al. (1987). They proposed the following set of equations:

$$S_F = 16.8 \sin \alpha - 0.5 \quad \text{if } \tan \alpha > 0.09 \text{ and slope length} > 4 \text{ m} \qquad (6)$$

$$S_F = 10.8 \sin \alpha + 0.03 \quad \text{if } \tan \alpha < 0.09 \text{ and slope length} > 4 \text{ m} \qquad (7)$$

$$S_F = 3.0 (\sin \alpha)^{0.8} + 0.56 \quad \text{if slope length} < 4 \text{ m} \qquad (8)$$

The last equation was proposed for unrilled surfaces as they are generally shorter than 4 m.

McIsaac et al. (1987), while examining soil loss data for disturbed land (reclaimed mine lands, construction sites, etc.), found a satisfying correspondence with Eqs. (7) and (8) but proposed the following relation instead of Eq. (6):

$$S_F = (12 \pm 7) \sin \alpha + B \qquad (9)$$

The constant B is chosen so that Eq. (9) gives 1 at $\tan \alpha = 0.09$; i.e., B depends on which value of the sine coefficient is chosen.

Equations (3)–(9) have been proposed to improve the original equation of Wischmeier and Smith (1978). More recently, an equation valid for interrill areas (hence approximately equivalent to Eq. (8) was developed by Liebenow et al. (1990):

$$S_F = 1.05 - 0.85e^{-4 \sin \alpha} \qquad (10)$$

Equation (10) gives $S_F = 1$ at $\alpha = 45°$; hence it should be divided by its value at $\tan \alpha = 0.09$ to be comparable with the other equations. It must be observed that part of the differences between Eqs. (8) and (10) may be due to different definitions of rain erosivity (respectively, rain kinetic energy times maximum rain intensity in 0.5 h and rain intensity squared).

The foregoing equations show two main trends, one for rilled surfaces (Fig. 1, curves 1, 2a and 3) and one for interrill areas and short slopes (Fig. 1, curves 2b and 4). This exemplifies the differences in effects due to different combinations of erosion agents. Many researchers have proposed explanations for gradient and length effects using a combination of processes (e.g., Moore and Burch, 1986). Others have found confirmations of the general trends combining field observations and theoretical considerations (e.g., Govers, 1991a). Sometimes, opposite trends have been observed (Bryan, 1979), depending on soil characteristics and antecedent surface and moisture conditions. More generally, the effects of slope gradient and length depend on the locally dominant soil erosion processes (Kirkby, 1969, 1971). In the same way that processes are influenced by many factors, such as climate, vegetation, animals and men, so are the effects of slope (gradient, length and aspect). In this chapter we will try to examine the pertinent processes and their interrelations in order to clarify how slope gradient, length and aspect built up their effects on soil loss. The relations and models that will be used in this chapter have almost no predictive power, but the way of putting them together, looking for interactions and changes in trends, represents a useful tool either for programming research or conservation measures.

II. GRADIENT AND LENGTH EFFECTS

On a macroscopic scale the forces binding soil particles and aggregates to the soil mass are generally measured in terms of apparent cohesion. The shape and weight of grains, their degree of packing and water tension determine another term of the macroscopic resistance force, i.e., the friction term of Coulomb's equation for shear resistance. Soil shear strength is usually expressed in terms of pressure as follows:

$$\tau = \frac{c + w \cos \alpha \tan \Phi}{A} \tag{11}$$

where

τ = soil shear strength
c = cohesion force
w = grain weight (often, submerged weight)
α = slope angle
Φ = angle of friction
A = surface of contact between grain and soil mass

Soil shear resistance is not a constant for a given soil. It varies with soil moisture, rapidity of wetting and drying, freezing and thawing, time span during which the soil is wet, soil density, development of a sealed (or crusted) layer,

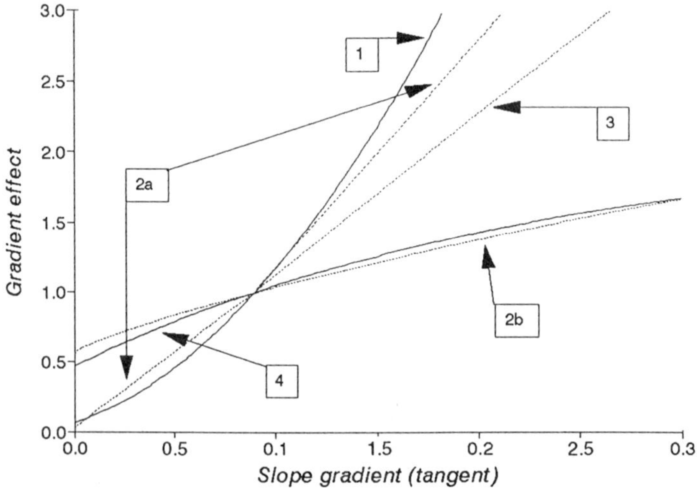

Fig. 1 Steepness effect on erosion as modeled by different authors: (1) Wischmeier and Smith (1978); (2a,b) McCool et al. (1987); (3) McIsaac, Mitchell and Hirschi (1987); (4) Liebenow et al. (1990).

soil compaction, etc. (Coote et al., 1988; Dexter, 1988). At the moment when rain begins, the soil is characterized by a certain arrangement of grains, a certain moisture content and a soil shear resistance. Depending on the initial moisture content, wetting may be a very disruptive agent. One of the main effects is the decrease of all the resistive forces to a minimum which depends on the speed of wetting and initial moisture content. In some cases deflocculation may reduce the resistive force to the friction term rewritten for hydrated single particles.

As water penetrates into the aggregates, the air trapped inside is confined in progressively smaller volumes. This can result in aggregate slaking by excess pressure of the trapped air (R. E. Yoder, 1936, as quoted in Bryan, 1969). Such a mechanical slaking is sometime accompanied by deflocculation which takes place when certain conditions, often characterized by the electrical conductivity of the solution and exchangeable sodium percentage, are reached (see Agassi et al., 1981; Gal et al., 1984; Ben-Hur et al., 1985; Gerits et al., 1987; Rengasamy and Olsson, 1991; Shainberg et al., 1992). Slaking causes grain detachment but not transport. It is more efficient on dry soils because it depends on the force with which the soil attracts water (matrix and osmotic forces). While suction forces decrease as moisture increases, slaking decreases with time (it generally lasts 15–20 min, at least, at the soil surface where erosion is taking place). It is independent from slope length and gradient. Deflocculation also causes only detachment and is independent from slope gradient and length. It influences soil transport because sediment load may be mainly composed of primary particles.

A. Raindrop Impact

While wetting decreases soil strength, rain drops impact the soil surface detaching grains. The impinging drop generates a crown of lateral jets of water whose velocities depend on that of the drop (Engels, 1955; Harlow and Shannon, 1967). The jets detach soil grains and eject them all around (Huang et al., 1983). If the drop trajectory is vertical and the soil surface is locally horizontal, then the material is uniformly ejected all around. If the soil surface is inclined, then the distribution of the ejected mass is asymmetrical with a predominance of downslope ejection. Such an effect has been investigated by many authors (e.g., Ekern and Muckenhirn, 1947). The most complete description of splash detachment and transport is the one first presented by De Ploey and Savat (1968), subsequently improved by Poesen and Savat (1981) until its present form (Poesen, 1985, 1986). This model explicitly introduces a positive effect of slope on detachment in conformity with empirical evidence. The splash detachment model where slope effect is more clearly explained is the one derived by Torri and Poesen (1992). Here the slope effect is attributed to the distribution of the drop mass during impact, which depends on the relative orientation of the drop trajectory and the soil surface (Fig. 2a,b). The slope effect, for vertically falling drops, may be written as follows:

$$S_d = \frac{1}{\pi} \int_0^{2\pi} \frac{1 - (\mu - \sin \mu)/2\pi}{\cos \alpha \tan \Phi - \sin \alpha \cos w'} dw \tag{12}$$

where

S_d = mass of soil detached per unit of rain detaching power (which may be estimated through rain kinetic energy, momentum, etc.)
$w' = \arccos[\cos w/\sqrt{\cos^2 \alpha \sin^2 w + \cos^2 w}]$
w = auxiliary angle used for solving splash detachment for a three-dimensional drop
$\mu = 2 \arccos\{\sin \alpha \cos w/\sqrt{\cos^2 \alpha \sin^2 w + \cos^2 w})/2\}$

Equation (12) is relatively complicated and can be calculated using numerical methods for the integral (e.g., Simpson's rule). In Fig. 3 the effect of slope on detachment rate is shown. It can be seen that slope effect depends on the angle of fricton of the soil grains (Φ), i.e., on soil erodibility.

Splash transport is characterized by an exponential decrease of the ejected soil mass from the point where the drop has impacted (Mosley, 1973; Poesen

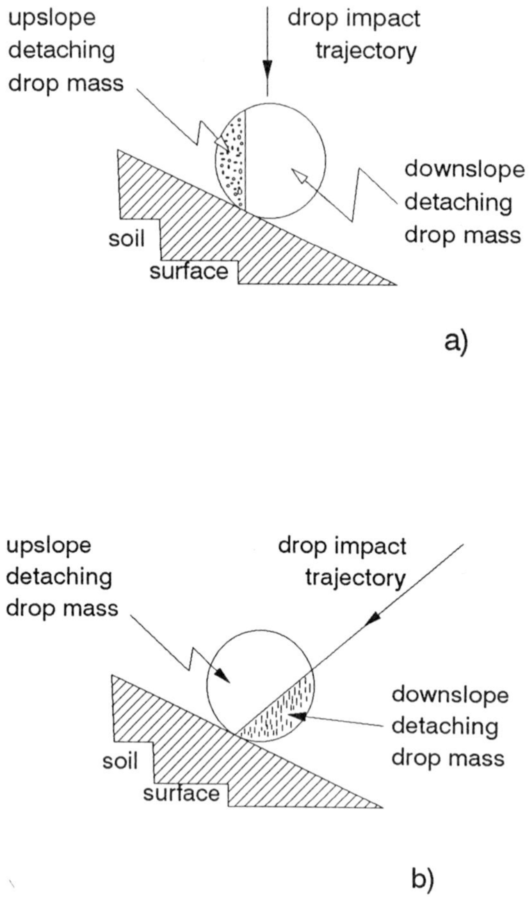

Fig. 2 Distribution of drop mass during impact as affected by the angle of incidence of the drop and by the soil surface slope angle.

and Savat, 1981; Riezebos and Epema, 1985; Torri et al., 1987). In one meter distance, virtually all the ejected grains are deposited. Here too, downslope splash has a longer jump length for the obvious reasons sketched in Fig. 4. If there is some wind, the drop trajectory is inclined and the trajectory of droplets and ejected soil grains are disturbed and far from a parabola. This makes it difficult to guess any jump length. In a field scale, splash transport is generally considered negligible because the large majority of eroded sediments are transported by overland flow.

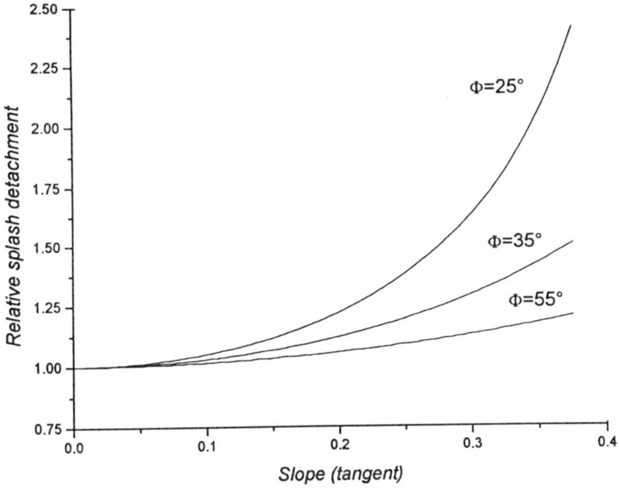

Fig. 3 Soil detachment due to drop impact increase with gradient (modified from Torri and Poesen, 1992).

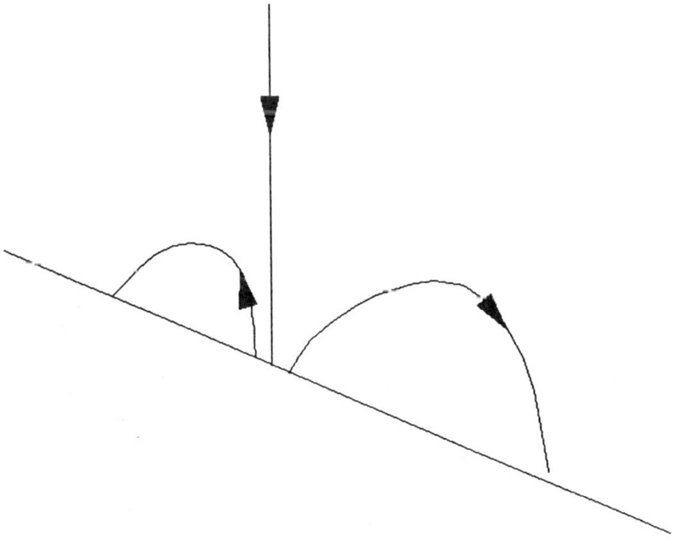

Fig. 4 Upward and downward ejection lengths of soil grains as affected by soil surface slope angle.

B. Overland Flow

Once runoff is generated detachment and transport forces of the flow appear. Detachment of grains is mainly due to runoff shear stress which is usually expressed as follows:

$$\tau_r = \delta_w g R s \tag{13}$$

where

τ_r = fluid shear stress
δ_w = fluid density
g = acceleration of gravity
R = hydraulic radius
s = slope (tangent or, more correctly, sine)

Sediment load is the result of the balance between sediment intake into the flow and sedimentation. Grain and flow characteristics determine both the amount of sediment load and the way in which sediment is transported (bed load and suspended load: see Yalin, 1977; or Raudkivi, 1976; for insight into the matter). At present there are many empirical equations and algorithms dealing with it. Many papers have been written about it recently (e.g., Govers, 1990, and Torri and Borselli, 1991 for overland flow; Celik and Rodi, 1988, and Leng, 1989, for river and channel flow), and every model must deal with it (Foster et al., 1989; Styczen and Nielsen, 1990; Woolhiser et al., 1990; Morgan et al., 1992). In order to avoid excessive complication, one of the simplest transport capacity equations will be used here. This equation (Kirkby, 1980) is based on stream power (per unit of width) and simply states what follows:

$$TC \approx (qs)^a s \tag{14}$$

where TC is the sediment rate at flow transport capacity, and a is an exponent depending on the characteristics of the soil grains.

Following the suggestion given by Kirkby (1980), in this study a will usually be set equal to 2. It must be noted that other simple empirical relationships exist (e.g., Govers, 1992; Slattery and Bryan, 1992), and there is no experimental or theoretical reason for expressing any preference.

Now both detachment force and transport capacity of overland flow are expressed in terms of hydraulic radius, unit discharge and slope gradient. The dependence of transport capacity and flow shear stress on slope is not limited to the one explicitly expressed in Eqs. (13) and (14) because R, q and s are interdependent. If R is used as the dependent variable, then the relationship is as follows:

$$R = (nq)^b s^{-c} \tag{15}$$

where

s = slope gradient (sinus or tangent)
b,c = exponents
n = hydraulic roughness

The exponent c shows little variation between laminar and turbulent flow (0.33 and 0.3 respectively), while b varies from 0.33 to 0.6 and hydraulic roughness values vary with the type of flow.

C. Combining Detachment and Transport Processes

The processes sketched in the previous paragraphs will now be merged in a simple model that should give us a better understanding of gradient and length effect. In order to keep the model simple we will limit ourselves to the examination of sediment transport rate in a rill, at a given instant during a rain (no temporal changes allowed). Moreover, the model will be deterministic. This will produce sharp changes from one erosion regime to another, contrary to what happens in reality where processes are characterized by noticeable stochastic components (Kirchener et al., 1990; Torri et al., 1990; Nearing, 1991).

The rill receives sediment produced in interrill areas where the dominant processes are drop impact, slaking and deflocculation. As already mentioned, detachment by slaking and deflocculation does not change with gradient or length, hence they will be represented by a constant. The same is valid for splash detachment and length because when interrill erosion is dominant, splash detachment will decrease with increasing runoff depth, i.e., with slope length. As our example deals with an established rill, this will not be the case. Slope effect on drop detachment will be represented by Eq. (12). The sum of the three components (drop impact, slaking and deflocculation) will be considered as interrill contribution. Rill contribution to sediment rate will be represented with the following set of equations (based on trends found by Nearing et al., 1991, and Ciampalini, 1992):

$$S_R = a\tau_d - b \quad \text{if } \tau_a > b \tag{16$_1$}$$

$$S_R = a'\tau_d \quad \text{if } \tau_d \leq b \tag{16$_2$}$$

where τ_d = runoff shear stress, and a,b,a' = constants depending on soil characteristics.

The coefficients a and a' vary inversely with soil shear resistance while the ratio b/a represent a shear stress value below which flow erosion is selective. Generally $a' < a$. Both slaking and deflocculation break some of the bonds between particles and aggregates. This causes soil resistance to decrease and, hence, a and a' to increase and b/a to decrease. This is taken into account into the model and a soil prone to slaking and deflocculation is consequently characterized by larger values of a and a' and smaller values of b/a. Deflocculation is also allowed to influence transport capacity because it tends to decrease the size of the transported grains (less aggregates). Kirkby (1980) observed that the exponent of Eq. (14) increases with the decrease in grain size of sediment. Hence the exponent changes from 2 to 3 for the deflocculation-prone soil.

The results of the simulation are shown in Fig. 5 where the bold line indicates the gradient effect on the sediment rate at a given cross section along the rill. Transport capacity exceeds sediment detachment at point A, and soil loss changes from a sediment rate which equals transport capacity to a rate equal to detachment. Point B indicates where rill flow turns from selectivity to unselectivity (i.e., where runoff shear stress exceeds b/a). The step between A and B in the sediment rate curve of Fig. 5a can be shorter or longer or even disappear, depending on the balance among the three erosion components, namely, interrill contribution, runoff detachment rate, and transport capacity. This is exemplified in Fig. 5b where A and B almost coincide in their abscissa. The effect of an increase of the overall soil erodibility pushes point A toward larger gradients. If transport capacity is also modified by the processes, A can be pushed back at lower gradient values (Fig. 5c) also if sediment rate is larger.

In order to compare the different trends, sediment rates at 9% slope were made equal to unity (Fig. 6). The overall steepness of the trend increases with decreasing erodibility. This is due to the fact that the more a soil is resistant to erosion (e.g., no slaking or deflocculation, which do not depend on slope) the more important are runoff detachment and transport (which depend on slope). Moreover, the curves diverge significantly, indicating that general relationships, such as Eq. (1), depend on prevailing local conditions.

Relations a, b and c of Fig. 6 have been calculated for soil with the same rill density. This is not generally true. For example, on agricultural soils, different crops require different spacing among rows (e.g., corn and winter wheat) which affects the number of rills per unit of length across the slope. Hence, a fourth curve has been drawn (d) in order to introduce the effect of a larger rill density. Curve d has been calculated as for the soil prone to deflocculation (c), simply adjusting runoff and interrill data for a double rill density. Curve d has the inflection point (corresponding to A of Fig. 5) at a higher slope than its equivalent, represented by curve c. This is due to the decrease in runoff drained by each rill (d-rills drain areas half the size of c-rills). A similar result would have been obtained simply decreasing runoff rate, without any rill density change. Hence, as runoff rate depends on rain characteristics then gradient effect on erosion depends on storm characteristics.

Since slaking takes place when a dry soil is being wetted, the effect of the initial moisture content is partially represented by the differences between curves a (moist) and b (dry) where the only different process is slaking. A similar

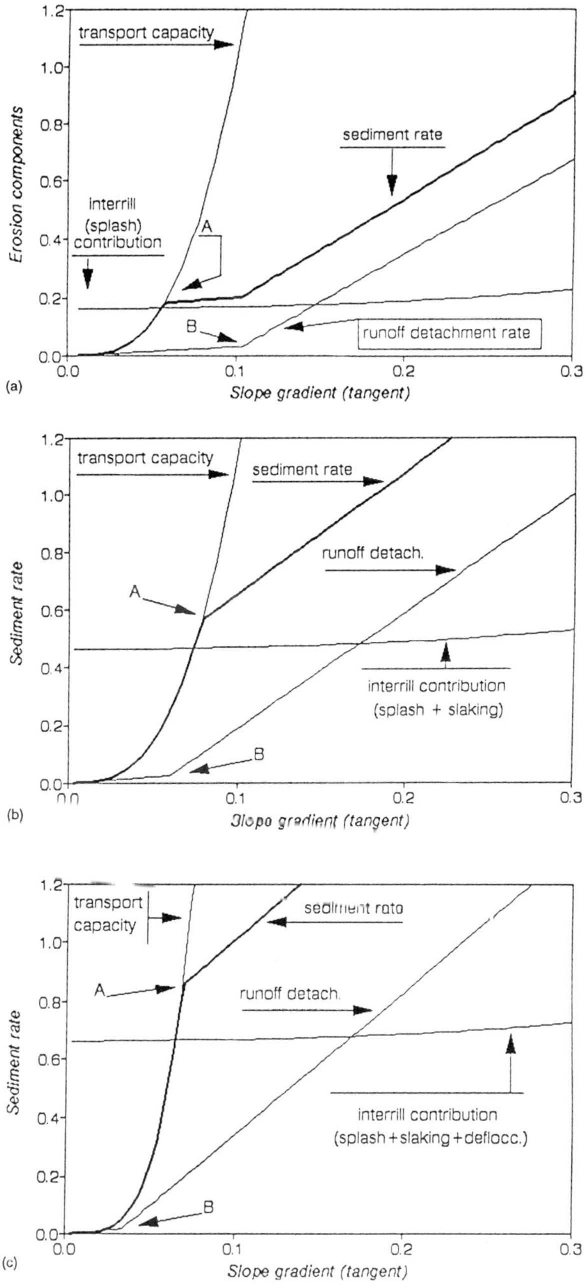

Fig. 5 Effect of slope gradient on different erosion processes: a, b and c differ because of changes in interrill processes and in the overall erodibility of soil.

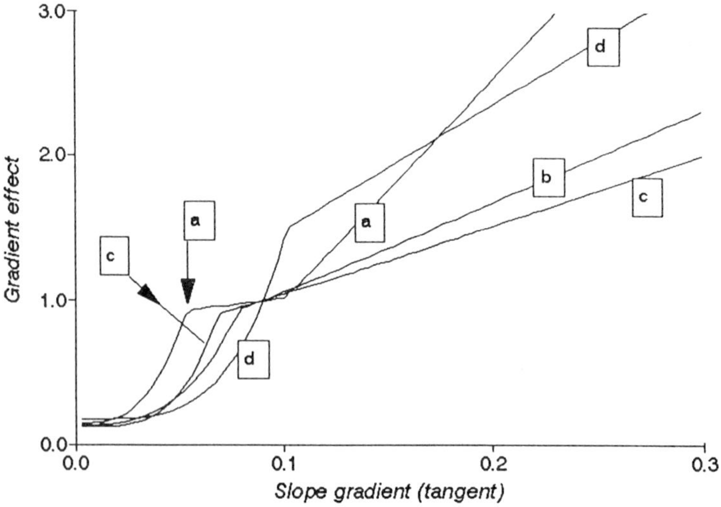

Fig. 6 Steepness effect on erosion (when gradient = 9% then gradient effect = 1): curves a, b, c correspond to cases a, b and c of Fig. 5. Case d is the equivalent of c but with a doubled rill density.

variation should be observed for curves c and d (both valid for dry conditions) if slaking is subtracted.

Let us now examine some further examples in order to discuss the relations between soil loss and slope length. Some assumptions must be made on the dependence of unit discharge on slope length. One of the simplest and most common assumptions is that runoff accumulates proportionally to the distance x from the upslope divide as given here (e.g., Kirkby, 1980; Govers, 1992):

$$q = \sigma x \tag{17}$$

where σ = runoff intensity (e.g., mm/h).

Actually, such an assumption is not always a good approximation of reality. Data examined by Lal (1983), Agassi et al. (1985), Poesen and Bryan (1990) and Ben-Hur (1991) indicate that runoff intensity often varies inversely to slope length. There are many reasons for such nonlinear behavior. The competition between seal formation and erosion (more erosion, less seal and more infiltration) is one of the causes. Runoff collected at a given section along a slope depends on the real area contributing runoff to that part. Sometimes, the upper part of the slope does not contribute because the storm duration is too short for that part of runoff to reach the control section. This is particularly true in arid and semiarid environments (Yair and Lavee, 1985) or during the dry season in mediterranean climates. Hence, the following variation of overland flow intensity σ will also be used:

$$\sigma = 0.01e^{-0.001x} \quad (18)$$

In order to depict a length effect on erosion, interrill contribution per unit of slope length has been considered constant. It was added to a rill detachment rate (calculated using Eq. (16)) on the same slope segment. This value was then compared with the deficit between actual sediment rate and transport capacity in that segment. The lower of the two values was added to sediment rate calculated in the nearby upslope segment. The value obtained was then expressed as erosion per unit of surface. Soil loss at 22-m length was made equal to unity. Results of the simulations are shown in Figs. 7 and 8.

Let us examine Fig. 7. The four curves (two almost completely overlap) have been calculated for the same interrill contribution rate, for two different susceptibilities to runoff detachment (two different sets of a,b and a' in Eq. (16), and two different transportabilities (the exponent a of Eq. (14) was made equal to 2 and to 1.5). All the other pertinent characteristics were kept constant while Eq. 17 was used for runoff calculation along the slope. The curves corresponding

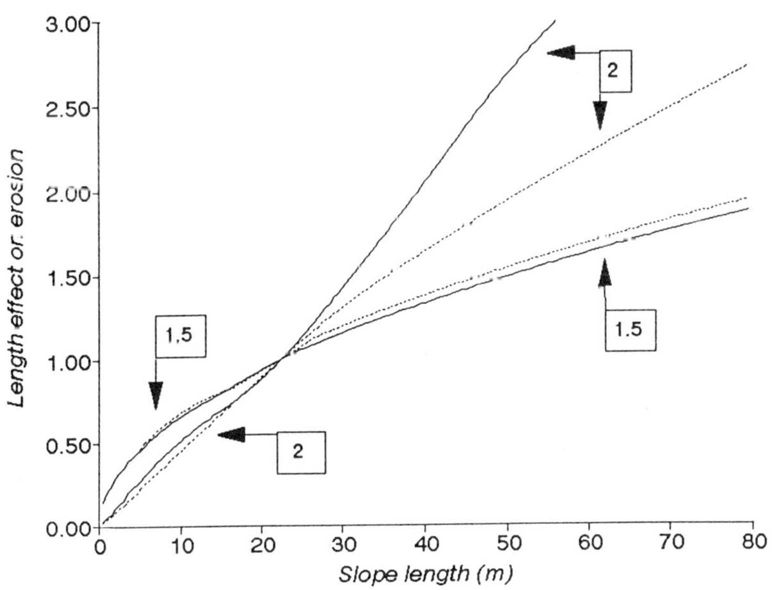

Fig. 7 Length effect equals unity when slope length is 22 m. Numbers indicate the value of the exponent of the transport capacity (Eq. (14)). The steepest curve of each pair corresponds to the lower soil erodibility.

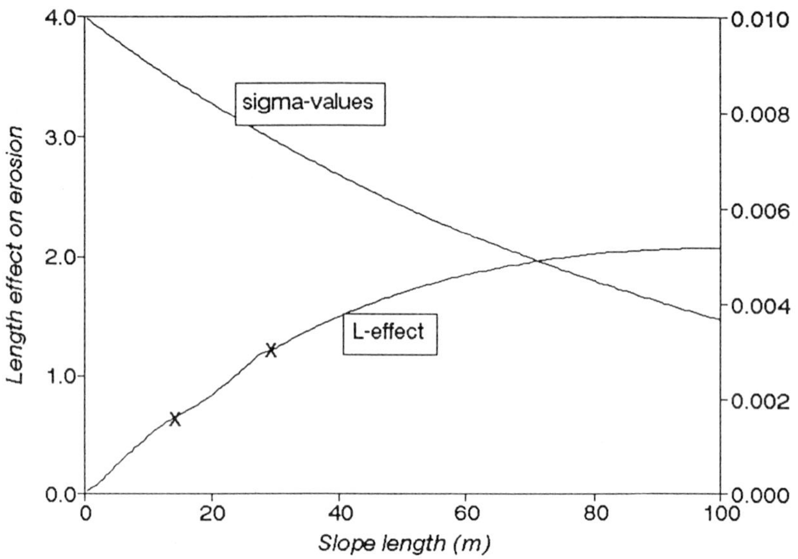

Fig. 8 Length effect as modified by a nonlinear dependence of runoff intensity (sigma) on slope length.

to the lower exponent of the transport formula coincide mainly because they are almost always transport capacity limited; i.e., detachment rate is generally larger than the amount that can be carried by the flow. Hence the curves are very close to the one obtainable by introducing Eq. (17) in Eq. (14) and then dividing by the slope length. The situation is different when the other two are examined. Here the two curves diverge because the erosion is detachment limited. So, the different erodibility of the two soils causes differences in the length effect. The curves present some point of inflection. They are determined by changes in sediment pickup regime (from Eq. (16_2) to Eq. (16_1)) and changes from detachment to transport-limited erosion, and vice versa. This is better shown in Fig. 8, where the segment of detachment limited erosion is delimited by two X's.

Figure 8 depicts a length effect due to the adoption of Eq. (18) instead of Eq. (17). All the other parameters were kept at the same values as those used for drawing the lowest of the curves with transport capacity exponent equal to 2 in Fig. 7. Hence the differences between the two curves are only due to differences in runoff accumulation along the slope. Following Bryan and Poesen (1989) or Poesen (1984), erosion rate, particularly rill incision, determines a decrease in sealing effectiveness. This causes a lower reduction in infiltration rate with respect to a case with the same soil but lower erosion rate or rill density. As the number of rills often increases with downslope distance, infiltration also increases and runoff decreases. In other words, it is again the soil susceptiblity to erosion that causes such changes in the slope length effect.

We have already seen that the length factor of the universal soil loss equation and the gradient factor are not independent (see Eqs. (1) and (3). This is a clear result of our simulation models. If we use the same parameter values already used for the lower of curves 2 (Fig. 7) and we calculate the length effect again changing slope gradient (from 10% to 20% and 5%), it is clear that the processes of detachment and transport do not produce the same result everytime (Fig. 9). Length is more important in generating erosion at higher gradient values in conformity with the trend depicted in Eq. (3).

III. SURFACE STORAGE

The simulations discussed above were run on perfectly identical soil surfaces and only some parameters, such as erodibility and slope gradient and length, were modified. This is not the normal situation for natural soil surfaces, and it is even less true for agricultural soil were crop management systems strongly modify the morphology of the plowed horizon. The main effects of differences in soil surface morphology can be summarized as follows:

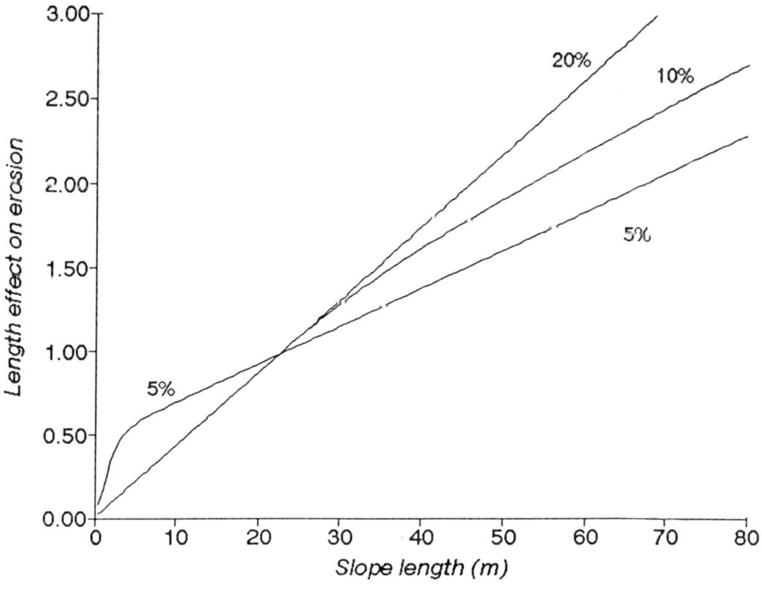

Fig. 9 Simulated length effect on slopes differing in gradient only.

1. Changes in surface depression: water and sediment are trapped in, modifying rates of processes and their balances.
2. Changes in hydraulic roughness (hence in overland flow detachment and transport capacity).
3. Changes in macropore frequency and in sealing pattern and intensity: infiltration is influenced—hence runoff and erosion.

Let us now examine surface depression in terms of storage capacity and of trap efficiency. Surface storage is due to depressions that, retaining water, subtract it from overland flow. A general model for storage (Ponce, 1989) is as follows:

$$S = S_f(1 - e^{-r/S_f}) \tag{19}$$

where

S = amount of water stored in depressions at time t
t = time since beginning of rain
r = cumulative precipitation excess at time t
S_f = total surface storage

In reality, rain excess is responsible for the largest part of the mass needed to fill the storage, but it is accompanied by trapped sediment. The latter component is important because, contrary to captured water which eventually infiltrates or evaporates, the trapped sediment decreases storage permanently and reduces infiltration rate. Splash erosion decreases the relative elevations of the rim defining a given depression, while microrills and protochannels are generally unstable and ''valley'' capture is common. Hence, erosion decreases storage directly acting on the elevation. In the meantime, it locally modifies size and shape of the basin draining into a given depression. The processes responsible for such dynamic situations, well exemplified by the experiments conducted by Moore and Larson (1979) and Wang (1988), are those described in the previous paragraph. Hence splash detachment, interrill flow transport capacity (flow detachment is negligible), slaking and deflocculation determine the amount of sediment flowing into each depression. Generally, the actual sediment load of the incoming flow is close to transport capacity, particularly in sparsely vegetated or bare soils. The sediment carried by the flow is totally deposited inside the depression until its storage is filled up with water and sediment. Then only part of the incoming sediment is trapped, particularly the largest and heaviest particles and aggregates. In other words, trap efficiency for flow transported sediment decreases during the storm.

Contrary to this is what happens with direct splash contribution. At the beginning of the storm, there is a splash directed toward the depression and splash directed out of it. The intensity of the former is always larger because of the gradient effect on splash detachment. Actually the splash into is mainly due to a downslope splash (where the slopes to be considered are those encircling the depression, e.g., those of the surrounding clods); hence it is larger then its counterpart, which is mainly due to upslope particle ejection (see literature already quoted when introducing Eq. (12). This is also the reason why the rim encircling the depression decreases its elevation in time: here particle ejection outnumbers sedimentation.

Such a net splash flow toward the depression is further enhanced by the water that is accumulating in the depression. Actually water acts as a cushion between the impinging drop and the soil surface, dissipating much of the drop impact force (Palmer, 1963; Torri et al., 1987; Hairsine and Rose, 1991; Profitt and Rose, 1991). It has been shown that splash detachment generally decreases exponentially with increasing depth of the water (flowing or ponding) covering soil grains. Following Torri et al. (1987), this mechanism can be expressed as

$$D_s(h) = D_s(0)e^{-h/p} \tag{20}$$

where

D_s = splash detachment rate
h = flow depth
p = experimental constant depending on drop and grain sizes

Such a relation has thus far been ignored in this Chapter because the examples were given on an established rill network. While runoff is changing in interrill areas or ponding occurs, then the processes described by Eq. (20) play an important role (Torri and Sfalanga, 1986; Profitt et al., 1991). In particular, the surface inside the depression is protected by the accumulated water and detachment and ejection of soil particles decrease. The consequence is an increase of the net flow of splashed sediment into the depressions.

All these erosion processes are more intense if the gradient is larger. Consequently, depressions will be filled more rapidly on a steep than on a gentle slope. This implies that the depression storage on a steep slope will generally be found lower than the depression storage on a gentle slope. Moreover, tillage practices and conservation measures based on depression storage will show a longer-lasting effect on gentle slopes.

The negative effect of gradient on depression storage is further enhanced by the fact that a given depression volume in a steep slope contains less water than an equal volume in a gentle slope. This can be deduced from intuitive geometrical considerations. Actually, a depression on a plane surface is characterized by a well-defined volume. It coincides with the potential storage only on a horizontal surface. As soon as the surface is inclined, the depression storage decreases while the volume remains unchanged, similar to, e.g., slowly tilting a spoon full of water: the relationship between spoon storage and its slope is similar to those found by Huang and Bradford (1990). The reality of such a negative trend is supported by observations and field data collected in several studies (e.g., Onstad, 1984; Linden et al., 1988; Ponce, 1989, p. 38; Huang and Bradford, 1990).

Every soil surface is made of concavities and convexities that make it more or less rough. Such roughness causes water accumulation, and water diversion, and delays in water flow. When a flow is described using characteristics measured on too small a scale (e.g., gradient is often estimated with the slope angle, while locally water may run almost transversal to the main downslope direction), hydraulic roughness must reflect what water flow senses and we ignore, i.e., the sequence of concavities and convexities that make surface roughness. Obviously as surface roughness decreases also hydraulic roughness decreases (Gilley and Finkner, 1990). This fact has important consequences on flow transport capacity as well as on detachment rate. In fact, the forces generated by the overland flow are partly dissipated by obstacles. These obstacles are soil aggregates and particles, clods, grass, mulch elements, rock fragments, etc. They (and especially those larger in size) contribute to surface roughness but only particles and aggregates can be removed in normal rainstorms. Hence the force that the overland flow dissipates on immovable obstacles is lost to erosion (e.g., Kirkby, 1980; Govers and Rauws, 1986). Generally, the larger the surface roughness the smaller the fraction of runoff forces that are used for detaching and transporting. This will naturally modify the effects of slope gradient and length on erosion. Hence other curves should be added to those shown in Figs. 6–8.

As already mentioned, roughness and surface depression are continuously smoothed by rain (Zobeck and Onstad, 1987). Such a trend should be more rapid on steep slopes where erosion is more effective. After a while, depression storage reaches its minimum. When this happens, surface roughness may be dominated by erosion features and increase again (Huang and Bradford, 1992).

A. Surface Roughness and Depression Storage Evaluation

The evaluation of surface depressional storage is not a simple one. The more straightforward methods use techniques such as stereophotographs for building a digital model of the soil surface (e.g., Ullah and Dickinson, 1979). Other techniques are based on measurements of elevations of soil microtopography. These measurements compelled the researchers to introduce the concepts of random and ordered roughness. The latter is the one caused by furrows, dike-furrows, some tillage operations, etc. It has the property of repeating itself regularly several times along a field. On the contrary, the former includes the effects of clods, aggregates and particles. Techniques of data filtering were introduced in order to separate the two components (e.g., Allmaras et al., 1966). Ordered roughness has a storage capacity that can be estimated more easily than the one generated by random roughness. Hence, most of the studies have been devoted to the latter, producing many indices for random roughness characterization. Examples are the standard deviation of the elevation values, the logarithm of the standard deviation, the ratio between the actual length of a transect and its projection, the mean of the absolute difference between elevation of consecutive points, the product of the peak frequency and a microrelief index, defined as the area per unit length between the measured surface profile and the least-square line interpolating the measured elevations (Kuipers, 1957; Burwell et al., 1963; Allmaras et al., Holt, 1966; Dexter, 1977; Boiffin, 1984; Linden and Van Doren, 1986; Römkens and Wang, 1986). Obviously, some of these indices are correlated (Lehrsch et al., 1988; Bertuzzi et al., 1990b), while the spacing between consecutive elevation values affects the values of random roughness indices. Additional data are given by Imeson (1983), who used the storage parameter proposed by Seginer (1971).

Those indices have been used for characterizing the evolution of soil surface. Particularly, the effect of rainfall on roughness reductions allowed Zobeck and Onstad (1987) to determine some relationships between those indices and precipitation. The indices proposed by Linden and Van Doren (1986) were successfully related to surface storage by Linden et al. (1988).

Relatively recent application of noncontact laser profilometers (Huang et al., 1988; Römkens et al., 1988; Destain et al., 1989; Bertuzzi et al., 1990a) has allowed the collection of accurate elevation measurements. This has made possible some original approach for estimating surface storage. Chadoeuf et al. (1989) proposed a model where a rough soil surface (random roughness only) is generated using a procedure based on Boolean functions.

A different approach is the one proposed by Huang and Bradford (1990, 1992) whose data sets indicate that soil surface roughness can be well described by a combination of fractional Brownian motion and Markov-Gaussian processes. Both approaches allow a three-dimensional reconstruction of soil surfaces. This makes deductions on rivulets meandering (i.e., protorills), surface storage, extent of sealing and crusting, etc., possibilities that indices do not allow.

IV. ASPECT

It has been shown that the antecedent soil moisture content influences the subsequent erosion response of the soil. Consequently all the factors affecting soil moisture content influence erosion. From an examination of the best known evapotranspiration equations the most common factors are direct radiation and wind speed. Both have a given dominant orientation; i.e., the direct radiation received by a given slope depends on slope aspect (e.g., N- and S-oriented slopes) and gradient. A model describing slope evolution in semiarid environments was presented by Kirkby et al. (1990). The authors, examining the effect due to differences in incident radiation, found that much more runoff was produced in the S-facing slope (in the northern hemisphere: from now on we will always refer to the northern one).

Generally, there is a predominant wind direction which causes a larger evaporation in the windward slope, similar to radiation on a S-facing slope. Hence, the chances for the south (or the windward) slope to get dry are greater. This means that slaking will take place more often on this slope. Also soil resistance to detachment will generally be lower because of the greater disorganization of the soil particles due to more rapid drying processes.

Wind, when combined with rain, also has other effects. It causes rain to fall with an oblique trajectory (hence, a sloping-surface facing wind will collect more precipitation than a horizontal surface), and drops to have a greater speed and, generally, a smaller size (e.g., Hudson, 1964; Lyles, 1976, who suggested a direct wind action in particle detachment and aggregate disintegration; Torri, 1979; Sharon, 1980; see also elsewhere in this book). From 1980 to 1985, Poesen (1986) experimentally measured an upslope net splash transport on an experimental field in Belgium. The effect was due to the predominant wind direction during rainfall which was generally toward the instrumented slope.

Such an effect is produced by the combination of two subprocesses that can be explained using the hypothesis proposed by Torri and Poesen (1992). Detachment was assumed proportional to the fraction of drop mass compelled to splash in that direction (Fig. 2). In a bidimensional approximation the straight line passing through the point of first contact between the drop and the soil surface and parallel to the drop trajectory, splits the drop mass in two fractions. Particularly when the drop trajectory is almost normal to the slope surface, the fractions become equal. When the drop trajectory is sufficiently inclined, the upslope detaching mass becomes larger than the downslope one. Also the trajectories of the ejected grains are influenced. The component moving opposite to wind is counteracted by a strong friction and the jump length is consequently reduced. Obviously an opposite effect is experienced by the grains ejected in the wind direction. The global effect is that splash transport rate in the wind direction is enhanced.

On the leeward slope, wind effect on distribution of detached material and jump length will be lower than the one described on the windward slope. At the same time the differences in splash detachment will be further increased by differences in rain aggressiveness due to differences in drop impact velocities and in rain amounts. Lyles (1977) found a threefold increase in soil detachment if rain was driven by a wind blowing at 11.2 m/s. The overall effect will be an increase in interrill erosion, in sealing and in the depletion rate of roughness and surface storage (with consequent decrease in infiltration, increase in runoff and its detachment and transport capacity) in the windward slope. If the effect of splash erosion is coupled with the effect due to slaking and to more rapid soil drying it is obvious that the windward slope should generally be more eroded. On the other hand, runoff is more likely to occur on the leeward slope because of the generally higher antecedent soil moisture content.

Let us now examine the overall effect of wind on soil loss, using the data collected by researchers working in Israel (center of the coastal plain, near Morasha: Agassi and Ben-Hur, 1991; northern Negev, near Kibbutz Bet Qama; Agassi et al., 1990) and in Italy (Tuscany, Mugello valley, near Scarperia and the Chianti region, near Volterra; Panicucci, 1983).

The Israeli experiments were conducted using data collected in field plots. The Morasha plots were located both on a W-aspect slope and on a N-aspect slope with dominant winds blowing from the West. Bet Qama plots were located on a western (windward) slope and on a eastern (leeward) slope.

The Italian experiments were conducted using remolded soil settled in containers 1 m long, 0.4 m wide and 0.3 m deep. Both sites were equipped with two windward exposed containers and two with leeward exposure. Prevalent wind during rainfall was from the southwest.

The data collected are presented in Table 1. The values of precipitation, runoff and erosion given in the table are the ratios between leeward slope values and windward ones. The values relative to Morasha are the ratio between the northward and the westward (windward) slope.

The values relative to Bet Qama, 8.7% slope, do not follow the general trend for erosion. Here the most eroded slope is the leeward one. This probably de-

Table 1 Effect of Wind Bearing and Slope Aspect Interaction on Erosion

Locality	Rain Amount	Runoff	Erosion	Slope (%)	Soil clay content (%)
Scarperia[a]	0.95	1.07	0.78	40	39
Volterra[a]	0.85	1.05	0.63	40	46
Morasha interrill[b]	0.98	0.95	0.66	48	13
Morasha rill[b]	0.98	0.91	0.72	48	13
Bet Qama[c]	1.01	1.01	1.62	8.7	48
Bet Qama[c]	0.72	0.50	0.16	57.5	48

[a]Panicucci (1983).
[b]Agassi and Ben-Hur (1991).
[c]Agassi, Shainberg and Morin (1990).
Values of rain amount, runoff and erosion are given as ratios between the leeward slope values and the corresponding windward values.

pends on local variations which may hide differences due to aspect simply because they are still imperceptible (slope = 8.7%).

Let us now examine the trends for the other steeper slopes. Generally the rain amount is lower on the leeward slope. This points to a lower erosion potential. In any case, the decrease in precipitation cannot explain the larger decrease in soil loss. The drop erosive potential is larger on the windward slope because the drop impact velocity is increased due to wind drag, causing a larger soil detachment rate in the windward slope. Runoff decrease in the leeward slope is responsible for a lower transport capacity and rilling potential. This interpretation seems to be supported by the striking differences in erosion found on the Bet Qama, 57.5% slope. This could also explain the differences observed in Morasha rill plots.

Differences in the Italian values cannot be attributed to runoff (which increases slightly leeward) because of the small plot sizes. Here, it is splash which is more effective on the windward slope, as already mentioned. The same can be said of the Morasha interrill plots. Also, slaking should have contributed as the windward slopes dry more quickly.

Let us now return to the gradient and length effect on erosion, and let us try to move from one curve to another in Figs. 5–8 while comparing slopes with different aspects. If the N- (or leeward) slope (P-slope, P for "protected") is characterized by a slope effect as the one described in Fig. 5a, then the other slope (U-slope, for "unprotected") has a larger slaking component and a larger runoff detachment capacity. Figure 5b represents a better approximation of its gradient effect. Also slope length effect is represented by the steepest of curves 2 (Fig. 7) in the P-slope and by the lower of curves 2 for the U-slope. In area where runoff accumulation along the slope is close to the passage from the situation described in Eq. (17) (constant σ-value), which was used to calculate curves 2, to the one described in Eq. (18), a lower overland flow intensity may determine the selection of the latter equation. In such a case the length effect is better described by the curve of Fig. 8.

Aspect has also long-term effects on soil erosion due to microclimatic differences that may lead the slope development as exemplified by many authors (e.g., Yair et al., 1980). A recent example of the interaction aspect–natural vegetation–erosion is given in Marqués and Mora (1992). They measured soil loss and runoff in plots located in a N- and in a S-facing slope in Montserrat (Catalunya, Spain) where vegetation was burnt in 1986 during a forest fire. They always observed more intense erosion on the S-slope. This was attributed to the overall effect of the two slope aspects, which included more rapid recover of vegetation on the N-slope.

Another extreme effect of slope aspect on soil loss is given by a typical erosion form of some Italian badlands, known as *biancana* and reproduced in Fig. 10 (a bibliography on those hummock-shaped hills, generated on pliocenic marine clay or silty-clay deposits, can be found in Alexander, 1982; Mazzanti and Rodolfi, 1988; Torri et al., 1993). The biancana of Fig. 10 has the vegetated slope NNE-oriented, which is typical of biancana developed in Tuscany, Italy (the northmost area where such badlands are located).

Let us examine the situation that can generate such a striking difference between the two opposite slopes. Following the experiments conducted by Torri et al. (1993), soil loss on the eroded slope, under initially dry conditions, was found to be 1.2 times greater than the amount eroded under initially moderately moist conditions. As the biancana slopes are fairly steep (around 40°), the difference between the direct solar radiation received by the NNE- and the SSW-facing slopes is very large (Fig. 11). The balance between the erosion rate and the weathering rate on the SSW-slope reaches the equilibrium value of some centimeters of weathered regolith (5–10 cm in one year as measured by the author in 1990–1991). The lower soil loss values in the wetter NNE-slope, together with a greater water penetration depth (due to a much lower evaporation intensity), allows a deeper weathered layer to develop. In such a situation leaching of salts (in which the sediment is rich) should be more efficient, and eventually vegetation takes over.

Fig. 10 The dome-shaped relief is a typical *biancana* (Val d'Orcia, Tuscany, Italy). The vegetated slope is exposed to NNE.

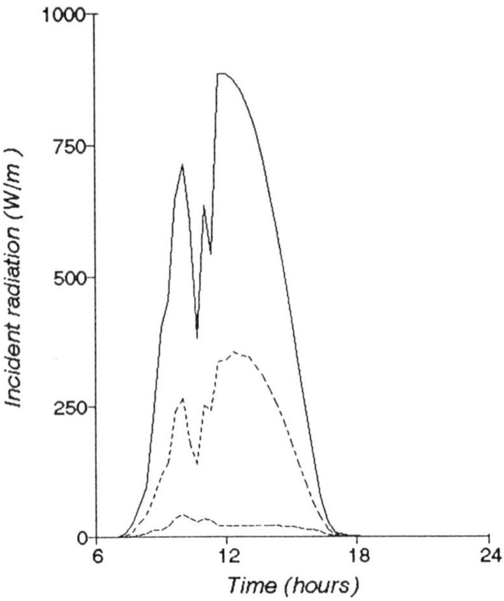

Fig. 11 Direct radiation received on the S-facing slope of a *biancana* (bold line), on a horizontal surface (intermediate broken line) and on the N-facing *biancana* slope (lower broken line).

V. CONCLUSIONS

Erosion rates are modified by the main slope characteristics such as length, gradient and aspect. The way in which the three characteristics interact among themselves and with soil loss depends on the way in which basic erosion processes merge. In other words, gradient, length, aspect, surface conditions, storm characteristics (other than just kinetic energy and amount), etc., build up the slope effects on erosion. One of the consequences is that there is no unique relationship between soil loss and length or steepness, unless long-term trends are of interest. Moreover, conservation measures or management techniques can be more safely applied if their interactions with the processes are thoroughly analyzed.

The examples used in this chapter are static representations. In other words, they are the equivalent of instantaneous snapshots. This was done to avoid excessive complications. The reality is more complex, and many of the parameters we used change during a single storm. This was recently exemplified by Govers (1991b) for overland flow erosion, by Borselli (1993) for interrill erosion, and by Rose et al. (1990) for soil shear strength. As pointed out by Borselli (1993), such changes cause gradient effect to be perceived differently, depending on storm duration. Also storm pattern influences amounts and rates of overland flow and erosion (Flanagan et al., 1988).

Consequently, locations where storms have different characteristics will probably show different relationships between slope characteristics and soil loss. The difficulties caused by the lack of universal, all-solving, relationships could be tackled with personal knowledge and experience, coupled with a clever use of simulation models.

ACKNOWLEDGMENTS

The author gratefully acknowledges R. B. Bryan for reviewing this chapter and M. Agassi for his many suggestions.

REFERENCES

Agassi, M., and Ben-Hur, M. (1991). Effect of slope, length, aspect and phosphogypsum on runoff and erosion from steep slopes. *Aust. J. Soil Res. 29*: 197–207.

Agassi, M., Shainberg, I., and Morin, J. (1981). Effect of the electrolyte concentration and soil sodicity on infiltration rate and crust formation. *Soil Sci. Soc. Am. J. 45*: 848–851.

Agassi, M., Shainberg, I., and Morin, J. (1985). Infiltration and runoff in wheat fields in the semi-arid region of Israel. *Geoderma 36*: 263–276.

Agassi, M., Shainberg, I., and Morin, J. (1990). Slope, aspect and phosphogypsum effects on runoff and erosion. *Soil Sci. Soc. Am. J. 54*(4): 1102–1106.

Alexander, D. (1982). Differences between ''calanchi'' and ''biancane'' badlands in Italy. In *Badland: Geomorphology and Piping* (R. B. Bryan and A. Yair, eds.), Geobooks, Norwich, pp. 71–88.

Allmaras, R. R., Burwell, R. E., Larson, W. E., and Holt, R. F. (1966). Total porosity and random roughness of the interrow zone as influenced by tillage. USDA Conserv. Res. Rep. No. 7.

Ben-Hur, M. (1991). The effect of dispersants, stabilizer and slope length on runoff and water-harvesting farming. *Aust. J. Soil Res 29*: 553–63.

Ben-Hur, M., Shainberg, I., Keren, R., and Gal, M. (1985). Effect of water quality and drying on soil crust properties. *Soil Sci. Soc. Am. J. 49*: 191–196.

Bertuzzi, P., Caussignac, J. M., Stengel, P., Morel, G., Lorendeau, J. Y., and Pelloux, G. (1990a). An automated non contact laser profile meter for measuring soil roughness in situ. *Soil Sci. 149*: 169–178.

Bertuzzi, P., Rauws, G., and Corault, D. (1990b). Testing roughness indices to estimate soil surface roughness changes due to simulated rainfall. *Soil Tillage Res. 17*:87–99.

Boiffin, J. (1984). La dégradation structurale des couches superficielles sous l'action des pluies. ThΦse de Docteur-Ingénieur, Inst. Nat. Agron., Paris.

Borselli, L. (1993). *Temporal Changes in Soil Erodibility*. Quaderni di Scienza del Suolo, Firenze, 4, v, 23–46.

Bryan, R. B. (1969). The relative erodibility of soils developed in the Peak District of Derbyshire. *Geograf. Ann. 51*: 145–159.

Bryan, R. B. (1979). The influence of slope angle on soil entrainment by sheetwash and rainsplash. *Earth Surf. Process. Landforms 4*: 43–58.

Bryan, R. B., and Poesen, J. (1989). Laboratory experiments on the influence of slope length on runoff, percolation and rill development. *Earth Surf. Process. Landforms 14*: 211–231.

Burwell, R. E., Allmaras, R. R., and Amemiya, M. (1963). A field measurement of total porosity and surface microreliefs of soils. *Soil Sci. Soc. Am. Proc. 27*: 697–900.

Celik, I., and Rodi, W. (1988). Modeling suspended sediment transport in nonequilibrium situations. *J. Hydraul. Eng. 114*(10): 1157–1191.

Chadoeuf, J., Monestiez, P., Bertuzzi, P., and Stengel, P. (1989). Parameter estimation in a boolean model of rough surface. Application to soil surfaces. In *Proceedings of the 5th European Congress for Stereology* (M. Kalisnik and O. Leder, eds.), Freiburg im Breisgau, 3–8 Sept. 1989, Acta Stereol., 2, pp. 635–640.

Ciampalini, R. (1992). Erosione del suolo: distacco e trasporto dei sedimenti ad opera delle acque di deflusso superficiale. Unpublished thesis, Department of Earth Sciences, University of Florence.

Coote, D. R., Malcolm-McGovern, C. A., Wall, C. A., Dickinson, W. T., and Rudra, R. P. (1988). Seasonal variation of erodibility indeces based on shear strength and aggregate stability in some Ontario soils. *Can. J. Soil Sci. 68*: 405–416.

De Ploey, J., and Savat, J. (1968). Contribution à l'étude de l'erosion par le splash. *Z. Geomorph. 2*: 174–193.

Destain, M. F., Descornet, G., Roisin, C., and Frankinet, M. (1989). Investigation of soil degradation by means of opto-electronic microreliefmeter. *Soil Tillage Res. 13*: 299–315.

Dexter, A. R. (1977). Effect of rainfall on the surface micro-relief of a tilled soil. *Terramechanics 14*: 11–22.

Dexter, A. R. (1988). Strength of soil aggregates and of aggregate beds. *Catena Suppl. 11*: 35–52.

Ekern, P. C., and Muckenhirn, R. J. (1947). Water drop impact as a force in transporting sand. *Soil Sci. Soc. Am. Proc. 12*: 441–444.

Engels, O. G. (1955). Waterdrop collisions with solid surfaces. *J. Res. Nat. Bur. Stand. 54*(5): 281–298.

Flanagan, D. C., Foster, G. R., and Moldenhauer, W. C. (1988). Storm pattern effect on infiltration, runoff, and erosion. *Trans. ASAE 31*(2): 414–420.

Foster, G. R., Lane, L. J., Nearing, M. A., Finkner, S. C., and Flanagan, D. C. (1989). Erosion component. In *USDA—Water Erosion Prediction Project: Hillslope Profile Model Documentation* (L. J. Lane and M. A. Nearing, eds.), NERSL Report No. 2, USDA-ARS National Soil Erosion Research Laboratory, West Lafayette, IN.

Gal, M., Arcan, L., Shainberg, I., and Keren, R. (1984). Effect of exchangeable sodium and phosphogypsum on crust structure—Scanning electron microscope observations. *Soil Sci. Soc. Am. J. 48*: 872–878.

Gerits, J., Imeson, A. C., Verstraten, J. M., and Bryan, R. B. (1987). Rill development and badland regolith properties. *Catena Suppl.* 8: 141–160.

Gilley, J. E., and Finkner, S. C. (1991). Hydraulic roughness coefficients as affected by random roughness. *Trans. ASAE 34*(3): 897–903.

Govers, G. (1990). Empirical relationships on the transporting capacity of overland flow: a laboratory study. Proceedings of the Jerusalem Workshop, IAHS Spec. Publ. No. 189, pp. 45–63.

Govers, G. (1991a). Rill erosion on arable land in central Belgium: rates, controls and predictability. *Catena 18*: 133–155.

Govers, G. (1991b). Time-dependency of runoff velocity and erosion: the effect of the initial soil moisture profile. *Earth Surf. Process. Landforms 16*: 713–730.

Govers, G. (1992). Relationship between discharge, velocity and flow area for rills eroding loose, non-layered material. *Earth Surf. Process. Landforms 15*: 515–528.

Govers, G., and Rauws, G. (1986). Transporting capacity of overland flow on plane and irregular beds. *Earth Surf. Process. Landforms 11*: 515–524.

Hairsine, P. B., and Rose, C. W. (1991). Rainfall detachment and deposition: sediment transport in the absence of flow-driven processes. *Soil Sci. Soc. Am. J. 55*: 320–324.

Harlow, F. H., and Shannon, J. P. (1967). The splash of a liquid drop. *J. Appl. Phys. 38*(10): 3855–3866.

Huang, C., and Bradford, J. M. (1990). Depressional storage for Markov-Gaussian surfaces. *Water Resources Res. 26*(9): 2235–2242.

Huang, C., and Bradford, J. M. (1992). Application of a laser scanner to quantify soil microtopography. *Soil Sci. Soc. Am. J. 56*: 14–21.

Huang, C., Bradford, J. M., and Cushman, J. H. (1983). A numerical study of raindrop impact phenomena: the elastic deformation case. *Soil Sci. Soc. Am. J. 47*: 855–861.

Huang, C., White, I., Thwaite, E. G., and Bendeli, A. (1988). A non-contact laser system for measuring soil surface topography. *Soil Sci. Soc. Am. 52*: 350–355.

Hudson, N. W. (1964). Bearing and incidence of sub-tropical convective rainfall. *Q. J. R. Meteorol. Soc. 90*(385): 323–328.

Imeson, A. C. (1983). Studies of erosion thresholds in semi-arid areas: field measurements of soil loss and infiltration in northern Morocco. *Catena Suppl. 4*: 79–89.

Kirchner, J. W., Dietrich, W. E., Iseya, F., and Ikeda, H. (1990). The variability of critical shear stress, friction angle, and grain protusion in water-worked sediments. *Sedimentology 37*: 647–672.

Kirkby, M. J. (1969). Erosion by water on hillslopes. In *Water, Earth and Man* (R. J. Chorley, ed.), Methuen, pp. 229–38.

Kirkby, M. J. (1971). Hillslope process-response models based on continuity equation. In *Slopes: Form and Process* (D. Brunsden, ed.), Inst. Br. Geogr. Spec. Pub. 3, pp. 15–30.

Kirkby, M. J. (1980). Modelling water erosion processes. In *Soil Erosion* (M. J. Kirkby and R. P. C. Morgan, eds.), Wiley, Chichester, pp. 183–216.

Kirkby, M. J., Atkinson, K., and Lockwood, J. (1990). Aspect, vegetation cover and erosion on semi-arid hillslopes. In *Vegetation and Erosion* (J. Thornes, ed.), Wiley, New York, pp. 25–39.

Kuipers, H. (1957). A reliefmeter for soil cultivation studies. *Neth. J. Agric. Sci. 5*: 255–262.

Lal, R. (1983). Effect of slope length on runoff from Alfisols in western Nigeria. *Geoderma 31*: 185–193.
Lehrsch, G. A., Whisler, F. D., and Römkens, M. J. M. (1988). Spatial variation of parameters describing soil surface roughness. *Soil Sci. Soc. Am. J. 52*: 311–319.1.
Liebenow, A. M., Elliot, W. J., Laflen, J. M., and Kohl, K. D. (1990). Interrill erodibility: collection and analysis of data from cropland soil. *Trans. ASAE 33*: 1882–1888.
Linden, D. R., and Van Doren, D. M. (1986). Parameters for characterizing tillage-induced soil surface roughness. *Soil Sci. Soc. Am. J. 50*: 1550–1565.
Linden, D. R., Van Doren, D. M., Allmaras, J. R., and Allmaras, R. R. (1988). A model of the effects of tillage-induced soil surface roughness on erosion. Int. Soil Tillage Res. Org., Proc. 11th Int. Conf., Vol. 1, pp. 373–378.
Low, H. S. (1989). Effect of sediment density on bed-load transport. *J. Hydraul. Eng. 115*(1): 124–138.
Lyles, L. (1977). Soil detachment and aggregate disintegration by wind-driven rain. In *Soil Erosion, Prediction and Control*, Soil Cons. Soc. Am., Ankeny, IA, pp. 152–159.
Marqués, M. A., and Mora, E. (1992). The influence of aspect on runoff and soil loss in a Mediterranean burnt forest (Spain). *Catena 19*: 333–344.
Mazzanti, R., and Rodolfi, G. (1988). Evoluzione del rilievo nei sedimenti argillosi e sabbiosi dei cicli neogenici e quaternari italiani. In *La Gestione delle Aree Franose* (P. Canuti and E. Pranzini, eds.), Edizioni delle Autonomie, Roma, pp. 13–60.
McCool, D. K., Brown, L. C., Foster, G. R., Mutchler, C. K., and Meyer, L. D. (1987). Revised slope steepness factor for the universal soil loss equation. *Trans. ASAE 30*: 1005–1013.
McIsaac, G. F., Mitchell, J. K., and Hirschi, M. C. (1987). Slope steepness effects on soil loss from disturbed lands. *Trans. ASAE 30*: 1005–1013.
Moore, I. D., and Burch, G. J. (1986). Physical basis of the length-slope factor in the universal soil loss equation. *Soil Sci. Soc. Am. J.* 1294–1298.
Moore, I. D., and Larson, C. L. (1979). Estimating micro-relief surface storage from point data. *Trans. ASAE 22*: 1073–1077.
Morgan, R. P. C. (1986). *Soil Erosion and Conservation*. Longman, p. 298.
Morgan, R. P. C., Quinton, J. N., and Rickson, R. J. (1992). *Eurosem: Documentation Manual*. Version 1, Silsoe College, Silsoe, UK.
Mitchell, J. K., and Bubenzer, G. D. (1980). Soil loss estimation. In *Soil Erosion* (M. J. Kirkby and R. P. C. Morgan, eds.), Wiley, Chichester, pp. 17–62.
Mosley, M. P. (1973). Rainsplash and the convexity of badland divides. *Z. Geomorph. N.F., Suppl. 18*: 10–25.
Murphree, C. E., and Mutchler, C. K. (1981). Verication of the slope factor in the USLE for low slopes. *J. Soil Water Cons. 38*(5): 135–142.
Mutchler, C. K., and Greer, J. D. (1980). Effect of slope length on erosion from low slopes. *Trans. ASAE 23*(4): 866–876.
Mutchler, C. K., and Murphree, C. E., Jr. (1985). Experimentally derived modification of the USLE. In *Soil Erosion and Conservation* (S. A. El-Swaify, W. C. Moldenhauer, and A. Lo, eds.), Soil Conservation Society of America, Ankeny, IA, pp. 523–527.
Nearing, M. A. (1991). A probabilistic model of soil detachment by shallow turbulent flow. *Trans. ASAE 34*(1): 81–85.

Nearing, M. A., Bradford, J. M., and Parker, S. C. (1991). Soil detachment by shallow flow at low slopes. *Soil Sci. Soc. Am. J. 55*: 339–344.

Onstad, C. A. (1984). Depressional strage on tilled soil surfaces. *Trans. ASAE 27*: 729–732.

Palmer, R. S. (1963). The influence of a thin water layer on waterdrop impact forces. *I.A.H.S. Publ. 65*: 141–148.

Panicucci, M. (1983). Erodibilità del suolo in relazione all'orientamento dei versanti: nota I—rilievi su piccoli lisimetri. Vol. XIV, pp. 109–126, Annali Istituto Sperimentale Studio Difesa Suolo, Firenze, Italia.

Poesen, J. (1984). The influence of slope angle on infiltration rate and Hortonian overland flow volume. *Z. Geom. Suppl. 49*: 117–131.

Poesen, J. (1985). An improved splash transport model. *Z. Geomorph. 29*(2): 193–211.

Poesen, J. (1986). Field measurements of splash erosion to validate a splash transport model. *Z. Geomorph. N.F., Suppl. 58*: 81–91.

Poesen, J., and Bryan, R. B. (1990). Influence de la longueur de la pente sur le ruissellement: rôle du rigoles et de croûtes de sédimentation. *Cah. ORSTOM, Sér Pédol. 25*: 1–2, 71–80.

Poesen, J., and Savat, J. (1981). Detachment and transportation of loose sediments by raindrop splash. II: Detachability and transportability measurements. *Catena 8*: 19–41.

Ponce, V. M. (1989). *Engineering Hydrology—Principles and Practices.* Prentice Hall, Englewood Cliifs, NJ.

Profitt, A. P. B., and Rose, C. W. (1991). Soil erosion processes. I: The relative importance of rainfall detachment and runoff entrainment. *Aust. J. Soil Res. 29*: 671–683.

Profitt, A. P. B., Rose, C. W., and Hairsine, P. B. (1991). Rainfall detachment and deposition: experiments with low slopes and significant water depths. *Soil Sci. Soc. Am. J. 55*: 325–332.

Raudkivi, A. J. (1976). *Loose Boundary Hydraulics.* Pergamon Press, Oxford.

Rengasamy, P., and Olsson, K. A. (1991). Sodicity and soil structure. *Aust. J. Soil Res. 29*: 935–952.

Riezebos, H. Th., and Epema, G. F. (1985). Drop shape and erosivity. II: Splash detachment, transport and erosivity indices. *Earth Surf. Process. Landforms 10*: 69–74.

Römkens, M. J. M., and Wang, J. Y. (1986). Effect of tillage on surface roughness. *ASAE 29*(2): 429–433.

Römkens, M. J. M., Wang, J. Y., and Darden, R. W. (1988). A laser microreliefmeter. *Trans. ASAE 31*: 408–413.

Rose, C. W., Hairsine, P. B., Proffitt, A. P. B., and Misra, R. K. (1990). Interpreting the role of soil shear strength in erosion processes. *Catena Suppl. 17*: 153–166.

Seginer, J. (1971). A model for surface drainage of cultivated soils. *J. Hydrol. 13*: 139–151.

Shainberg, I., Warrington, D., and Laflen, J. M. (1992). Soil dispersibility, rain properties, and slope interaction in rill formation and erosion. *Soil Sci. Soc. Am. J. 56*: 278–283.

Sharon, D. (1980). The distribution of hydrologically effective rainfall incident on sloping ground. *J. Hydrol. 46*: 165–188.

Slattery, M. C., and Bryan, R. B. (1992). Hydraulic conditions for rill incision under simulated rainfall: a laboratory experiment. *Earth Surf. Process. Landforms 17*: 127–146.

Styczen, M., and Nielsen, S. A. (1989). *A view of soil erosion theory, process-research and model building: possible interactions and future developments.* Quaderni di Scienza del Suolo, II, Firenze, pp. 27–45.

Torri, D. (1979). *Modello per la stima degli afflussi su un versante mediante misure standard di pioggia, direzione e velocità del vento.* Annali Istituto Sperimentale Studio Difesa Suolo, Firenze, Italia, Vol. X, pp. 225–233.

Torri, D., Biancalani, R., and Poesen, J. (1990). Initiation of motion of gravels in concentrated overland flow: cohesive forces and probability of entrainment. *Catena Suppl. 17*: 79–90.

Torri, D., and Borselli, L. (1991). Overland flow and soil erosion: some processes and their interactions. *Catena Suppl. 19*: 129–137.

Torri, D., Colica, A., and Rockwell, D. (1994). Preliminary study of the erosion mechanisms in a biancana badland (Tuscany, Italy). *Catena. 23*: 281–294.

Torri, D., and Poesen, J. (1992). The effect of soil surface slope on raindrop detachment. *Catena, 19*: 561–578.

Torri, D., and Sfalanga, M. (1986). *Some Aspects of Soil Erosion Modelling.* Development in Environmental Modelling, Vol. 10, Elsevier, Amsterdam.

Torri, D., Sfalanga, M., and Del Sette, M. (1987). Splash detachment: runoff depth and soil cohesion. *Catena 14*: 149–155.

Ullah, W., and Dickinson, W. T. (1979). Quantitative description of depression storage using a digital surface model. *J. Hydrol. 42*: 63–76.

Wang, Y. (1988). Effects of slope steepness, bulk density and surface roughness on interrill erosion. M.S. thesis, Faculty of the Graduate School of the University of Minnesota.

Wischmeier, W. H., and Smith, D. D. (1978). *Predicting Rainfall Erosion Losses.* Agriculture Handbook No. 537, USDA, Washington, DC.

Woolhiser, D. A., Smith, R. E., and Goodrich, D. C. (1990). *KINEROS, A Kinematic Runoff and Erosion Model: Documentation and User Manual.* USDA-ARS, Ars-77.

Yair, A., Bryan, R. B., Lavee, H., and Adar, E. (1980). Runoff and erosion processes and rates in the Zin Valley badlands, northen Negev, Israel. *Earth Surf. Process. Landforms 5*: 205–225.

Yair, A., and Lavee, H. (1985). Runoff generation in arid and semi-arid zones. In *Hydrological Forecasting* (M. G. Anderson and T. P. Burt, eds.), Wiley, New York, pp. 183–220.

Yalin, M. S. (1977). *Sediment Transport.* Pergamon Press, Oxford.

Zingg, A. W. (1940). Degree and length of land slope as it affects soil loss in runoff. *Agric. Eng. 21*: 59–64.

Zobeck, T. M., and Onstad, C. A. (1987). Tillage and rainfall effects on random roughness. A review. *Soil Tillage Res. 9*: 1–20.

6

The Effect of Surface Cover on Infiltration and Soil Erosion

James E. Box, Jr. and R. Russell Bruce
Southern Piedmont Conservation Research Center, Agricultural Research Service, U.S. Department of Agriculture, Watkinsville, Georgia

I. INTRODUCTION

Soil surface covers of crop canopies and permeable mulches are important in reducing soil erosion by water. These covers dissipate raindrop impact energy, reduce the area of erodible surface causing flow energy to be dissipated on nonerodible cover in contact with the surface, increase infiltration by reducing surface sealing, and reduce the velocity of runoff flow. A complete cover of the soil surface fully protects the soil from raindrop impact. In addition, crop management practices that enhance soil biological activity and aggregate stability affect interrill erosion. Surface cover effects can be divided into three classes: (a) canopy effects, (b) ground-cover effects, and (c) within-soil effects (Wischmeier, 1975). Surface cover may consist of crop canopy and stems, plant residue and accumulated organic matter in various stages of decomposition, and rock fragments. The most important erosive agent on interrill areas is raindrop impact. Raindrops range in size from about 0.2 to 6.0 mm, and impact at about 5 to 9 $m\ s^{-1}$ (Bubenzer, 1979). Raindrop impact on bare soil creates intense shear forces (Al-Durrah and Bradford, 1982), which can detach large quantities of sediment. Many of the same soil and management factors affecting detachment by raindrop impact also affect detachment by surface flow, but in a different way. Two soils equally susceptible to interrill erosion differ greatly in susceptibility to rill erosion (Meyer et al., 1975). Tillage greatly increases the susceptibility of some soils to rill erosion. Rill erosion from a range of discharge rates was 3 to 15 times more on a freshly tilled soil than on the same soil not tilled for a year (Foster, 1982). Similarly, the critical shear stress of a typical soil that has not been tilled for some time is about 10 to 15 times more than on a freshly tilled soil (Foster et al., 1980). However, the inherent erodibility of the soil is associated with the soil particles smaller than about 2 mm (sand, silt, and clay). Normally a soil increases in erodibility with an increase in silt fraction regardless of whether the corresponding decrease is in the sand or clay fraction (Wischmeier et al., 1971). Organic matter ranks next to particle size distribution as an indicator of erodibility. Soil erodibility can vary during the year (Mutchler and Carter, 1983). They found that Loring (Typic Fraguidalfs) and Lexington (Typic Paleudalfs) soils near Holly Springs, MS, varied from a high of 169% to a low of 31% of its annual average. Similar results for a Barnes loam (fine-loamy mixed Udic Haploboroll) near Morris, MN, were reported by them. Erodibility is also a function of complex interactions of a substantial number of a soil's physical properties and may vary within a standard textural class. Data from 17 soils on 50 sites in the southeastern United States indicated that soil structure and depth of the A soil horizon have a significant effect on soil erodibility (Barnett and Rogers, 1966). Rainfall simulation experiments were conducted on a Barnes loam near Morris, MN, at planting, midseason and harvest to determine the effect of bare soil surface, crop canopies of maize (*Zea mays* L.) and soybeans [*Glycine max* (L.) Merr.], and crop residue ground cover on steady-state infiltration rate (Rawls et al., 1993). The steady-state infiltration rate of the bare soil surface decreased and stabilized over the season. Canopy and residue maintained a higher steady-state infiltration rate than that of the bare soil surface. Increases in canopy increased the steady-state infiltration rate. The rapid drop in infiltration rate of soils during rainstorms is mainly due to crust formation on the soil surface (Duley, 1939; Morin and Benyamini, 1977). Crust formation in soils exposed to rain results from two mechanisms (Agassi et al., 1981); (a) a physical disintegration of the soil aggregates and their compaction, caused by raindrop impact action, and (b) a chemical dispersion of the clay particles and their movement into a region of decreased porosity, where they precipitate and clog the conducting pores producing a "washed-in" layer as discussed by McIntyre (1958). The relative importance and interaction of these two mechanisms depends on the impact energy of the raindrops, the electrolyte concentration of the applied water, and the exchangeable sodium percentage of the soil (Agassi et al., 1985).

II. PARTICLE DETACHMENT

Energy for particle detachment is derived from raindrops and canopy drip that strike the soil surface and from runoff. Thus, to reduce detachment, the energy must be dissipated by other means, such as surface cover. Hudson (1981) gave an example of the potential of reducing soil loss by dissipating the energy of raindrops before they strike the soil surface. Two 1.5-m-wide by 27.5-m-long plots were kept free of vegetation by hand weeding. The bare soil surface remained exposed on one plot. For the other, a double layer of fine-mesh wire gauze was suspended over the surface. Total soil losses for a 10-year period were 9.4 Mg ha^{-1} for the plot covered by gauze and 1265.7 Mg ha^{-1} for the bare plot. In 5 of the 10 years, soil loss was negligible on the covered plot, and the maximum one-year loss was 4.5 Mg ha^{-1}. The minimum yearly loss was 49.5 Mg ha^{-1} on the bare plot; it was from 121.4 to 204.5 Mg ha^{-1} in other years.

The erosion abatement effects of mulch on and incorporated in the soil surface and crop canopy are related to particle detachment and transport. These effects are described in the process-based erosion model used in the USDA Water Erosion Prediction Project (WEPP) described by Nearing et al. (1989) and discussed in a subsequent chapter of this text. The WEPP model uses the concept introduced by Meyer et al. (1975) of rill versus interrill sources of sediment. In this model, mulch diminishes the fluid forces by either dissipating raindrop impact energy or reducing the hydraulic shear forces associated with fluid flow in the interrill and rill areas.

Fluid forces on a bare soil surface, from the impact of a single raindrop, increase asymptotically toward a maximum as drop diameter increases. The presence of a surface layer of water affects the magnitude of these forces. As water depth increases from zero to about 0.3 drop diameter, the forces increase (Mutchler and Hansen, 1970). Rainfall simulator studies show that interrill soil erosion varies with the square of rainfall intensity (Meyer, 1981).

III. CANOPY COVER

A plant canopy intercepts raindrops before they hit the soil. If the canopy's drop distance is sufficiently large, such as from tall trees, the erosivity of the drops can be greater than rainfall (Chapman, 1948). Drops falling from plant surfaces are usually larger than natural raindrops. However, the canopy of most agricultural crops is close to the ground and their overall effect is to reduce the erosivity of rainfall. Even though interception increases raindrop diameter, fall height is too short for drops to reach a velocity sufficient to produce impact energy greater than that of natural rainfall. Maize and grain sorghum (*Sorghum bicolor* [L.] Moench) were observed to have high stem flow (Bui and Box, 1992). However, runoff (Fig. 1) and erosion (Fig. 2) due to stem flow appeared to be negligible compared to that caused by throughfall. Detached sediment, determined by using a rainfall simulator, under a 100% corn canopy cover was about one-half that

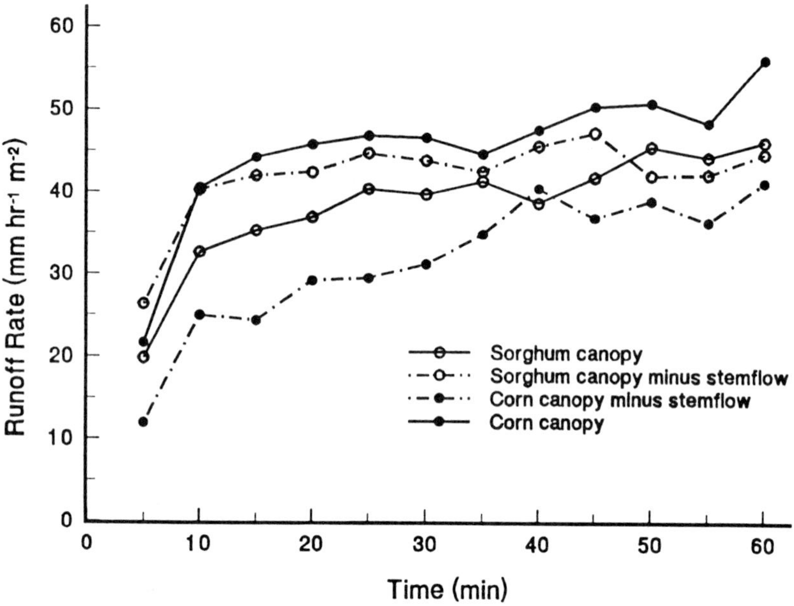

Fig. 1 The interaction of canopy and stemflow effects on runoff. (From Bui and Box, 1992.)

from fallow plots (Bui and Box, 1993) (Fig. 3). Runoff under the corn canopy was 80% of that from fallow plots (Fig. 4).

IV. CROP RESIDUE MULCH EFFECTS

Soil detachment is reduced by crop residue, stones, and live vegetation in direct contact with the soil surface and incorporated in the plow layer. There may also be a biological effect associated with crop residue. Examples are improved infiltration associated with mesofauna-induced soil macropores, improved soil aggregation associated with microbiological activity and earthworm castings, and root exudates functioning as aggregate binders. Runoff and soil loss were greatly reduced from a southern Piedmont U.S. conservation watershed when crop residues were left on the soil surface (Mills et al., 1986, 1988). The watershed had been in continuous no-tillage rotations of annual winter cover crops and summer annual crops for 10 years. All crop residues were left to accumulate as either undecomposed or decomposed organic matter on the soil surface. Barley (*Hordeum vulgare* L.), winter wheat (*Triticun aestivium* L.) or crimson clover (*Trifolium incarnatum* L.) was the winter cover, and soybean or grain sorghum

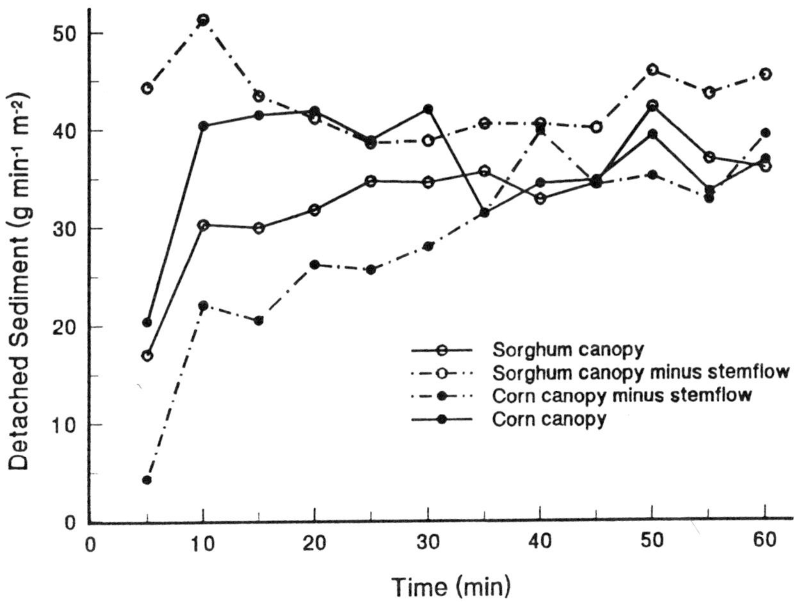

Fig. 2 The interaction of canopy and stemflow effects on sediment detachment. (From Bui and Box, 1992.)

was the summer crop. Similar results were observed for conservation tillage of cotton (*Gossypium hirsutum* L.) in Alabama (Yoo and Touchton, 1988). Most workers have attributed low soil loss rates from conservation tillage systems to increased amounts of crop residue at the surface, which protects the surface from raindrop impact and reduces the transport capacity of surface flow (Laflen et al., 1978; Foster et al., 1985; Meyer, 1985). Previously, the physical role of crop residues on the soil surface involving dissipation of raindrop energy, retardation of runoff, and consequent impedance to soil particle detachment, suspension, and transport has been emphasized (Foster et al., 1985).

Organic matter that accumulated on the soil surface from several cropping seasons of decomposing crop residue modifies soil surface properties (Hargrove et al., 1982; Kladivko et al., 1986; Bruce et al., 1987), which reduces runoff and soil erosion in addition to the physical effects of intercepting raindrop energy and impairing surface flow. The effectiveness of no-tillage systems in controlling erosion increases with the length of time the system is in use (Van Doren et al., 1984). Modification of surface soil properties associated with no-till cropping systems is an important factor influencing runoff and soil loss reductions. West et al. (1991) presented soil loss and runoff data for conventional tillage

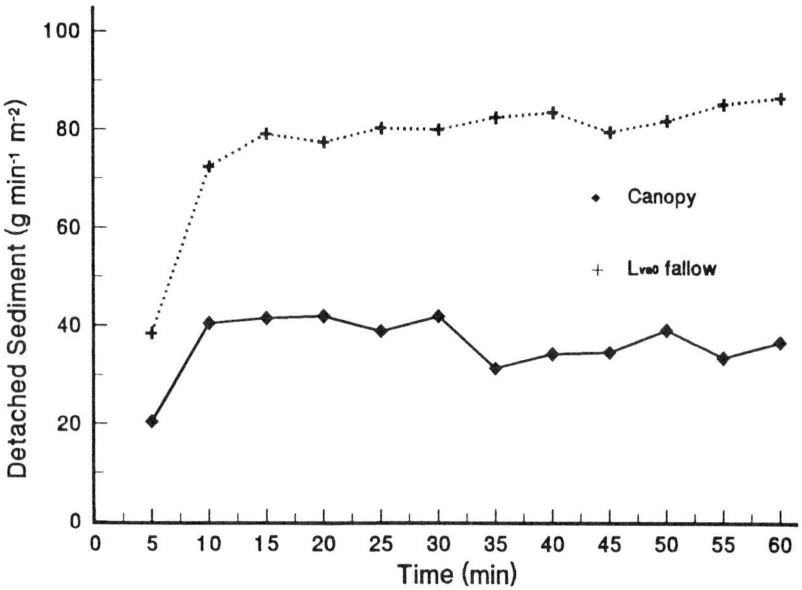

Fig. 3 Sediment detachment rate under a corn canopy as compared to freshly plowed fallow soil conditions. (From Bui and Box, 1993.)

and no tillage. Three sites were selected at Watkinsville, GA, that represented slightly eroded, moderately eroded, and severely eroded soils. The severely eroded site was a Pacolet and the others were Cecil soils (clayey, kaolinitic, thermic Typic Kanhapludults). Crops included were soybean, grain sorghum, and crimson clover. The cropping treatments on each site were (a) conventional tilled (soybean into disk-harrowed seedbed from winter fallow), CTS, (b) conventional tilled (grain sorghum into disk-harrowed seedbed from winter fallow), CTG, and (c) grain sorghum no-till planted into crimson clover, NTG. Removal of residue from soils in a consolidated state increased interrill soil loss by about 80% (Table 1). Lower soil loss from residue-covered plots than from bare plots appears to be related to runoff (Table 2). For the residue-covered plots, however, a relationship among amount of residue cover, runoff, and soil loss was not defined by the data. The no-till cropping system had the greatest residue coverage and the lowest amount of runoff and soil loss. The CTG treatment, compared to the CTS treatment, had about two times more residue cover, significantly less soil loss, but similar amount of runoff. Results from the West et al. (1991) study indicate that modification of soil surface properties induced by no-till cropping systems is an important component of the reduction in interrill soil loss realized for no tillage. For consolidated soil with residue left on the surface,

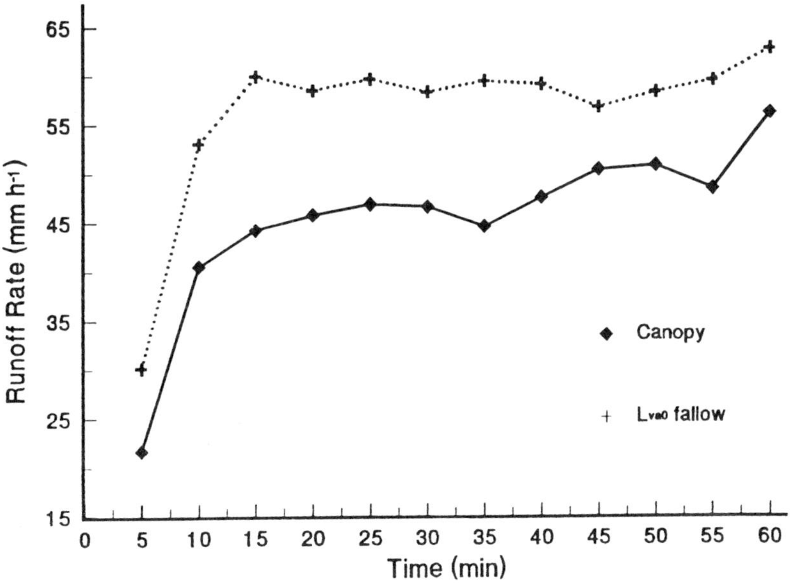

Fig. 4 Runoff rate under a corn canopy as compared to freshly plowed fallow soil conditions. (From Bui and Box, 1993.)

runoff and soil loss decreased as surface horizon clay content, organic C, and water stable aggregates increased among the sites. Tillage of the conventional sites increased soil loss by 44%, compared with the consolidated state.

West et al. (1991) reported that differences in interrill soil erodibility, K_i, among cropping systems and sites closely parallel the differences in total soil loss (Table 3). The K_i values were derived from the equation

$$D_i = K_i I^2$$

where

D_i = interrill soil-loss rate (kg m^{-2} s^{-1}) (soil loss rate taken as the mean rate over the last 15 min of rainfall simulation)
I = rainfall intensity (m s^{-1})
K_i = interrill soil erodibility (kg s mm^{-4}) (Foster, 1982).

For consolidated plots, the K_i for the no-till system was less than one-third of that for the conventionally tilled plots. No difference was observed in K_i between the two conventionally tilled systems or among the erosion classes. The expected relationship between erosion class and K_i was apparently altered by cropping

Table 1 Total Soil Loss for 1 h of Rainfall for Consolidated Plots with and without Residue and for Fresh-Tilled Plots on Three Erosion Classes

	Soil loss (kg m^{-2})			
Cropping system[a]	Slightly eroded	Moderately eroded	Severely eroded	Cropping system mean
Consolidated with residue				
CTS	0.520	0.325	0.173	0.337
CTG	0.347	0.207	0.130	0.213
NTG	0.074	0.034	0.015	0.041
Site mean	0.280	0.189	0.098	
LSD (0.05 site = 0.085[b]				
LSD (0.05) cropping system = 0.107[b]				
Consolidated without residue				
CTS	0.522	0.459	0.383	0.464
CTG	0.460	0.557	0.333	0.450
NTG	0.142	0.186	0.070	0.133
Site mean	0.375	0.401	0.247	
LSD (0.05) site = 0.182				
LSD (0.05) cropping system = 0.139				
Fresh tilled without residue				
CTS	0.686	0.685	0.553	0.659
CTG	0.717	0.791	0.481	0.660
Site mean	0.700	0.738	0.505	
LSD (0.05) site = 0.353				

[a]CT = conventionally tilled; NT = no-tilled; S = soybean; G = grain sorghum.
[b]LSD values are applicable to site and cropping system means as specified.
Source: West et al. (1991).

system. Tillage of consolidated conventionally tilled systems increased K_i by about 65%.

Bruce et al. (1992) studied infiltration rates on the sites which were evaluated for soil loss by West et al. (1991), and for which Langdale et al. (1990) reported the crop production aspects of this study. In Table 4, the infiltration rate after 1 h of simulated rainfall at 60 mm h^{-1} reinforces the observed effects of treatment on soil aggregate stability (Figs. 5 and 6). With residues removed the infiltration rate on NTG was 100% greater than CTS and CTG, which were not different. These data indicate a modification of the soil characteristics at the surface and a stability that sustains the infiltration rate. With residues on the

Table 2 Total Runoff for 1 h of Rainfall for Consolidated Plots with and without Residue and for Fresh-Tilled Plots on Three Erosion Classes

	Runoff[a] (mm)			
Cropping system[b]	Slightly eroded	Moderately eroded	Severely eroded	Cropping system mean
	Consolidated with residue			
CTS	17	13	5	12
CTG	13	8	8	10
NTG	1	1	1	1
Site mean	10	7	4	
LSD (0.05) site = 4[c]				
LSD (0.05) cropping system = 6[c]				
	Consolidated without residue			
CTS	13	21	16	17
CTG	16	23	17	19
NTG	3	3	6	5
Site mean	11	15	13	
LSD (0.05) site = 6				
LSD (0.05) cropping system = 5				
	Fresh tilled without residues			
CTS	17	19	12	16
CTG	18	23	16	19
Site mean	18	21	14	
LSD (0.05) site = 7				

[a]Simulated rainfall was applied at the rate of 60 mm h^{-1}.
[b]CT = conventionally tilled; NT = no-till; S = soybean; G = grain sorghum.
[c]LSD values are applicable to site and cropping system means.
Source: West et al. (1991).

surface, the infiltration rate for NTG was about 47% greater than CTS and CTG, compared with 100% greater with residues removed. The 23% reduction in infiltration rate for CTS and CTG, compared with 7% for NTG, on residue removal indicates the strong physical effect of residues on soil surfaces with low water stability of aggregates, e.g., CTS and CTG. The data in Table 4 suggest that there were no differences among infiltration rates for the erosion classes when residues were in place. Removal of the residues apparently reduced the infiltration on moderately and severely eroded sites to a greater degree than on slightly eroded sites, particularly on CTS and CTG.

Table 3 Interrill Soil Erodibility (K_i) for Consolidated Plots without Residue and for Fresh-Tilled Plots on Three Erosion Classes

	$K_i \times 10^{-3}$ (kg s m^{-4})			
Cropping system[a]	Slightly eroded	Moderately eroded	Severely eroded	Cropping system mean
	Consolidated without residue			
CTS	743	614	722	689
CTG	625	638	574	612
NTG	240	301	265	269
Site mean	536	518	495	
LSD (0.05) site = 16[b]				
LSD (0.05) cropping system = 122				
	Fresh tilled without residue			
CTS	960	1203	1041	1074
CTG	968	1243	975	1068
Site mean	964	1224	997	
LSD (0.05) site = 427				

[a]CT = conventionally tilled; NT = no-till; S = soybean; G = grain sorghum.
[b]LSD values are applicable only to site and cropping system means.
Source: West et al. (1991).

V. ROCK FRAGMENT MULCH EFFECTS

Several studies have shown that coarse fragments on the soil surface reduce soil erosion by water and affect infiltration or runoff. Infiltration into the soil surface is regulated in part by raindrop impact sealing. Koon (1968) theorized that when cover particles were impervious, infiltration could not occur on the covered area but would occur along the available perimeter of the cover particles. Water would then move in both vertical and lateral directions. Koon et al. (1970) reported that impervious particles on the soil surface affected infiltration in relation to the total perimeter of the particle and the lateral distance that water moves after entering the soil along the available particle perimeter. Historically, rock fragment covers have been viewed as protecting soil surfaces against erosion (Shaw, 1929; Cooke, 1970). Shaw (1929) reported that these coarse particles check the flow of runoff of the surface water and thus reduce its power to erode soils. Lamb and Chapman (1943) compared the effects of stone mulches, straw mulch and bare fallow soil on erosion, percolation and runoff. The stone mulches consisted of the naturally stony soil with 18% of its surface covered

Table 4 Infiltration Rate after 1 h of Simulated Rainfall for Three Crop Cultures with and without Surface Residue on Three Erosion Classes in March 1988

Culture[a]	Slightly eroded	Moderately eroded	Severely eroded	Mean[b]
		With residue (mm h^{-1})		
CTS	26.7	32.9	35.9	32.0B
CTG	33.0	38.6	34.5	35.7B
NTG	51.3	51.8	46.0	49.7A
Mean[b]	39.0a	41.1a	39.2a	
		Residue removed (mm h^{-1})		
CTS	29.5	20.5	20.2	23.8B
CTG	25.7	20.1	21.0	22.3B
NTG	50.7	46.4	41.6	46.3A
Mean[b]	35.3a	29.0b	28.5b	

[a]CTS = soybean into disk-harrowed seedbed from winter fallow; CTG = grain sorghum into disk-harrowed seedbed from winter fallow; NTG = grain sorghum no-till planted into crimson clover.
[b]Means in a given row or column having a common letter are not significantly different at 0.05 probability level. Lowercase letters apply to rows and uppercase to columns.
Source: Bruce et al. (1992).

by stones and a surface artificially covered to the extent of 65% by stones. The stones were mostly 5–15 cm long. Removal of the natural surface stones, larger than 5 cm long, doubled the runoff and increased soil loss by erosion six times. The 65% stone cover caused a large reduction in soil loss and increased infiltration when compared to the normal 18% cover. Removal of rocks larger than about 38 mm (Epstein and Grant, 1966) reduced soil surface cover from about 31% to 18% and increased erosion about 28%. In another study of unconventional mulches to control erosion on critical construction sites (Meyer et al., 1972), addition of gravel and crushed stone mulches on 20% slopes greatly reduced erosion for partial-cover rates and essentially eliminated erosion when the soil was entirely covered (Table 5). For three soils located in Stanley County, NC, Badin (clayey, mixed, thermic, Typic Kanhapludult), Georgeville (clayey, kaolinitic, thermic, Typic Kanhaplundult), and Goldston (loamy-skeletal, siliceous, thermic, shallow Ruptic-Ultic Dystrochrepts) containing large amounts of natural coarse fragments, Box (1981) showed that such fragments significantly reduced erosion from these soils. For soils with more than 50% rock fragment cover, removal of all fragments larger than 6 mm by sieving resulted in several times more erosion. Where cover was 15% to 25%, a 30% to 50% increase in erosion resulted from removal of the fragments. However, doubling

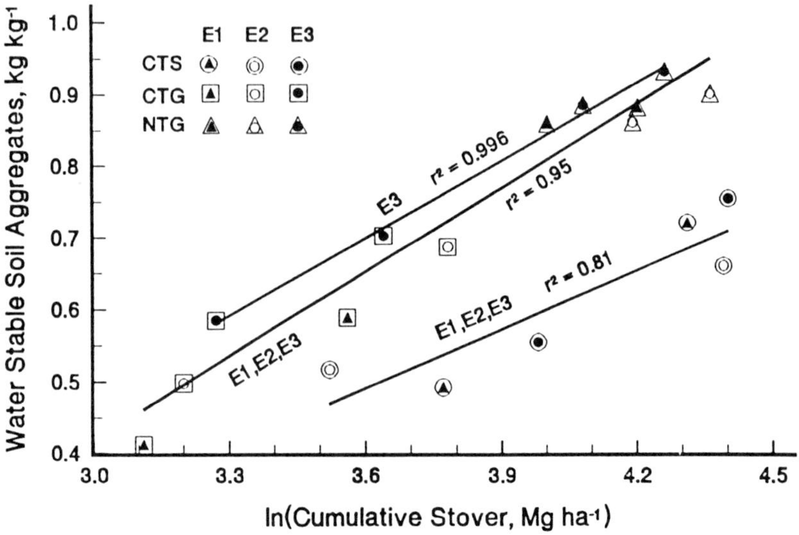

Fig. 5 Water stability of soil aggregates at 0 to 15 mm in March 1988, in relation to cumulative stover from each cultural treatment on three erosion classes. CTS = soybean into disk-harrowed seedbed from winter fallow; CTG = grain sorghum into disk-harrowed seedbed from winter fallow; NTG = grain sorghum no-till planted into crimson clover. E1, E2, and E3 = slightly, moderately and severely eroded sites, respectively. (From Bruce, 1992.)

on the soil surface the naturally occurring rock fragments did not decrease runoff, which was approximately 48 mm h^{-1}, when the rainulator application rate was 63.5 mm h^{-1} from the Badin and Georgeville soils but reduced to near-zero runoff from the Goldston soil. The naturally occurring surface covers for the Badin, Georgeville, and Goldston were approximately 70%, 60%, and 20%, respectively. Box and Meyer (1984) demonstrated that for these soils rock fragments on their surface reduced soil erosion the same as crop residue cover (Wischmeier and Smith, 1978). Agassi and Levy (1991) found that stone cover on a loamy soil (Calcic Haploxeralf) located in Israel had significant effect on both the infiltration rate and soil erosion. The levels of stone cover were 0%, 25%, and 50%.

The presence of rock fragments in the soil surface may affect erosion differently from those fragments on the soil surface. Investigations that have specifically addressed the relation between infiltration rates and rock fragments conclude that infiltration is both positively and negatively affected. The position of

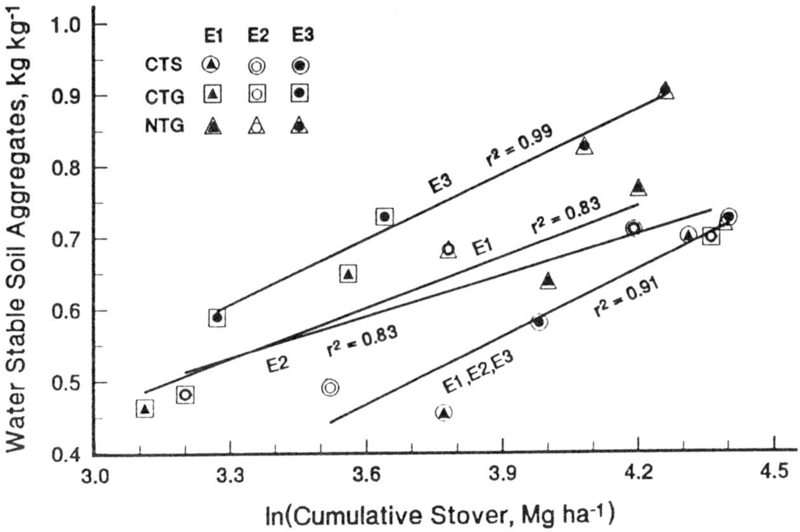

Fig. 6 Water-stable soil aggregates at 30 to 80 mm in March 1988, in relation to cumulative stover from each cultural treatment on all erosion classes. CTS = soybean into disk-harrowed seedbed from winter fallow; CTG = grain sorghum into disk-harrowed seedbed from winter fallow; NTG = grain sorghum no-till planted into crimson clover. E1, E2, and E3 = slightly, moderately, and severely eroded sites, respectively. (From Bruce, 1992.)

Table 5 Effect of Crushed Stone and Gravel Mulches on Soil Loss from 20% Construction Slopes

Mulch	Surface cover (%)	Soil loss ($t\ ha^{-1}$)	Soil loss ratio[b]
No mulch	3[a]	125.4	—
35 $t\ ha^{-1}$ stone	16	107.5	0.86
135 $t\ ha^{-1}$ stone	62	17.9	0.14
300 $t\ ha^{-1}$ stone	90	5.8	−0.05
540 $t\ ha^{-1}$ stone	100	<4.5	<0.04
155 $t\ ha^{-1}$ gravel	62	29.1	0.23

[a]Natural gravel of subsoil.
[b]Soil loss ratio is the loss during about 130 mm of moderately intense rain from the mulched treatment divided by the loss from the unmulched treatment.
Source: Meyer, Johnson and Foster (1972).

the fragments relative to the soil surface has been found to control the nature of the relation between infiltration rates and rock fragment cover (Poesen et al., 1990). They found in laboratory studies that infiltration rates increased when rock fragments were on the soil surface and decreased when they were embedded in the soil surface. Simanton et al. (1984) concluded that rock fragment cover can dominate the surface conditions of semiarid rangeland to such an extent that an erosion pavement is formed. They found that erosion rates, from simulator plots with natural and artificial rock fragment cover, decreased exponentially with increasing percent cover. Dadkhah and Gifford (1980) investigated the interactive effects of rock fragments with animal trampling and vegetation on infiltration. They concluded that on uncompacted soil, infiltration rates increased as the percentage of rock fragment cover increased. However, with trampling, the relation between infiltration and rock fragment cover was destroyed, resulting in no significant differences in infiltration rates for various amounts of rock fragment cover.

The influences of rock fragments on soil erosion as they interact with soil physical and chemical properties are complex. Soil porosity, aggregation, aggregate stability, consolidation and rock fragment incorporation into the soil surface affect particle detachment, surface sealing, flow velocity in rill and interrill areas, infiltration and runoff.

VI. SUMMARY

Surface cover reduces soil erosion by water because it absorbs raindrop impact energy, reduces the area of erodible surface causing flow energy to be dissipated on nonerodible cover in contact with the surface, increases infiltration by reducing surface sealing, and slows the velocity of runoff. The inherent erodibility of a soil is associated with soil particles smaller than about 2 mm. Organic matter ranks next to particle size distribution as an indicator of erodibility. Rill erosion from a range of discharge rates is 3 to 15 times greater on a freshly tilled soil than on the same soil not tilled for a year. A complete dissipation of raindrop energy requires complete surface cover. Surface cover of agricultural soils includes plant canopies, crop residue mulches and naturally occurring rock mulches. Cover on or above the soil surface is most effective in reducing interrill erosion, but may also reduce rill erosion. Generally, rill erosion is reduced most when the mulch is incorporated into the soil's surface. Biological effects associated with crop residue that reduce erosion include improved infiltration associated with mesofauna-induced macropores, improved aggregation associated with microbiological activity and earthworm castings, and root exudates functioning as aggregate binders. Rock fragments on the soil surface reduce soil erosion, but when incorporated into the soil surface may increase runoff and possibly erosion.

REFERENCES

Agassi, M., and Levy, G. J. (1991). Stone-cover and rain intensity: Effects on infiltration, erosion and water splash. *Aust. J. Soil Res. 29*: 565–575.

Agassi, M., Morin, J., and Shainberg, I. (1985). Effect of raindrop impact energy and water salinity on infiltration rates of sodic soils. *Soil Sci. Soc. Am. J. 49*: 186–190.

Agassi, M., Shainberg, I., and Morin, J. (1981). Effect of electrolyte concentration and soil sodicity on infiltration rate and crust formation. *Soil Sci. Soc. Am. J. 45*: 848–851.

Al-Durrah, M. M., and Bradford, J. M. (1982). The mechanism of raindrop splash on soil surfaces. *Soil Sci. Soc. Am. J. 46*: 1086–1090.

Barnett, A. P., and Rogers, J. S. (1966). Soil physical properties related to runoff and erosion from artificial rainfall. *Trans. ASAE 9*: 123–128.

Box, J. E., Jr. (1981). The effects of surface slaty fragments on soil erosion by water. *Soil Sci. Soc. Am. J. 45*: 111–116.

Box, J. E., Jr., and Meyer, L. D. (1984). Adjustment of the Universal Soil Loss Equation for cropland soils containing coarse fragments. In *Erosion and Productivity of Soils Containing Rock Fragments* (J. D. Nichols, P. L. Brown, and W. J. Grant, eds.), *Soil Sci. Soc. Am. Spec. Pub. No. 13*: 83–90.

Bruce, R. R., Langdale, G. W., West, L. T., and Miller, W. P. (1992). Soil surface modification by biomass inputs affecting rainfall infiltration. *Soil Sci. Soc. Am. J. 56*: 1614–1620.

Bruce, R. R., Wilkinson, S. R., and Langdale, G. W. (1987). Legume effects on soil erosion and productivity. In *The Role of Legumes in Conservation Tillage Systems* (J. F. Power, ed.), Soil Conserv. Soc. Am., Ankeny, IA, pp. 127–138.

Bubenzer, G. D. (1979). Rainfall characteristics important for simulation. In *Proc. Rainfall Simulator Workshop, USDA-SEA ARM-W-10*, pp. 22–34.

Bui, E. N., and Box, J. E., Jr. (1992). Stemflow, rain throughfall, and erosion under canopies of corn and sorghum. *Soil Sci. Soc. Am. J. 56*: 242–247.

Bui, E. N., and Box, J. E., Jr. (1993). Growing corn root effects on interrill soil erosion. *Soil Sci. Soc. Am. J. 57*: 1066–1070.

Chapman, G. (1948). Size of raindrops and their striking force at the soil surface in a red pine plantation. *Am. Geophys. Union Trans. 29*: 664–670.

Cooke, R. U. (1970). Stone pavements in deserts. *Ann. Assoc. Am. Geographers 60*: 560–577.

Dadkhah, M., and Gifford, G. F. (1980). Influence of vegetation, rock cover, and trampling on infiltration rates and sediment production. *Water Resources Bull. 16*: 979–986.

Duley, F. L. (1939). Surface factors affecting the rate of intake of water by soils. *Soil Sci. Soc. Am. Proc. 4*: 60–64.

Epstein, E., and Grant, W. J. (1966). Rock and crop-management effects on runoff and erosion in a potato-producing area. *Trans. ASAE 9*: 832–833.

Foster, G. R. (1982). Modeling the erosion process. In *Hydrologic Modeling of Small Watersheds* (C. T. Hann, H. P. Johnson, and D. L. Brakensiek, eds.), Am. Soc. Agric. Eng., St. Joseph, MI, pp. 296–380.

Foster, G. R., Johnson, C. B., and Nowlin, J. D. (1980). A model to estimate sediment yield from field sized areas: Application to planning and management for control of nonpoint source pollution. In *CREAMS: A Field Scale Model for Chemicals, Runoff, and Erosion from Agricultural Management Systems*, USDA Conservation Res. Rep. 26, pp. 193–281.

Foster, G. R., Young, R. A., Römkens, M. J. M., and Onstad, C. A. (1985). Processes of soil erosion by water. In *Soil Erosion and Crop Productivity* (R. F. Follett and B. A. Stewart, eds.). ASA, CSSA, and SSSA, Madison, WI, pp. 137–162.

Hargrove, W. L., Reid, J. T., Touchton, J. T., and Gallaher, R. N. (1982). Influence of tillage practices on the fertility status of an acid soil double-cropped to wheat and soybeans. *Agron. J. 74*: 684–687.

Hudson, N. (1981). *Soil Conservation*, 2nd ed. Cornell Univ. Press, Ithaca, New York.

Kladivko, E. D., Griffith, D. R., and Mannering, J. V. (1986). Conservation tillage effects on soil properties and yield of corn and soya beans in Indiana. *Soil Tillage Res. 8*: 277–287.

Koon, J. L. (1968). Some effects of soil surface cover on infiltration, Ph.D. dissertation, Auburn University, Auburn, AL.

Koon, J. L., Hendrick, J. G., and Hermanson, R. E. (1970). Some effects of surface cover geometry on infiltration rate. *Water Resources Res. 6*: 246–253.

Laflen, J. M., Baker, J. L., Hartwig, R. O., Buchele, W. F., and Johnson, H. P. (1978). Soil and water loss from conservation tillage systems. *Trans. ASAE 21*: 881–885.

Lamb, J., and Chapman, J. E. (1943). Effect of surface stones on erosion, evaporation, soil temperature, and soil moisture. *J. Am. Soc. Agron. 35*: 567–578.

Langdale, G. W., West, L. T., Bruce, R. R., Miller, W. P., and Thomas, A. W. (1990). Restoration of eroded soil with conservation tillage. In *Proc. Int. Conf. Soil Conserv. Environ.* and *Czech. Soil Sci. Conf. 7th*, Piestany Spa, Czechoslovakia, pp. 91–90.

McIntyre, D. S. (1958). Permeability measurements of soil crusts formed by raindrop impact. *Soil Sci. 85*: 185–189.

Meyer, L. D. (1981). How rain intensity affects interrill erosion. *Am. Soc. Agric. Eng. Trans. 24*: 1472–1475.

Meyer, L. D. (1985). Interrill erosion rates and sediment characteristics. In *Soil Erosion and Conservation* (S. A. El-Swaify, W. C. Moldenhauer, and Andrew Lo, eds.), Soil Conserv. Soc. Am., Ankeny, IA, pp. 167–177.

Meyer, L. D., Foster, G. F., and Römkins, M. J. (1975). Source of soil eroded by water from upland slopes. In *Present and Prospective Technology for Predicting Sediment Yields and Sources*, USDA-ARS, ARS-S-40, U.S. Gov. Printing Office, Washington, DC, pp. 177–189.

Meyer, L. D., Johnson, C. B., and Foster, G. R. (1972). Stone and woodchip mulches for erosion control on construction sites. *J. Soil Water Conserv. 27*: 264–269.

Mills, W. C., Thomas, A. W., and Langdale, G. W. (1986). Estimating soil loss probabilities for southern Piedmont cropping-tillage systems. *Trans. ASAE 29*: 948–955.

Mills, W. C., Thomas, A. W., and Langdale, G. W. (1988). Rainfall retention probabilities computed for different cropping-tillage systems. *Agric. Water Mgt. 15*: 61–71.

Morin, J., and Benyamini, Y. (1977). Rainfall infiltration into bare soils. *Water Resour. Res. 13*: 813–817.

Mutchler, C. K., and Carter, C. E. (1983). Soil erodibility variation during the year. *Trans. ASAE 26*: 1102–1104.

Mutchler, C. K., and Hansen, L. M. (1970). Splash of a waterdrop at terminal velocity. *Science 169*: 1311–1312.

Nearing, M. A., Foster, G. R., Lane, L. J., and Finker, S. C. (1989). A process-based soil erosion model for USDA-Water Erosion Prediction Project technology. *Trans. ASAE 32*: 1587–1593.

Poesen, J., Ingelmo-Sanchez, F., and Mucher, H. (1990). The hydrological response of soil surfaces to rainfall as affected by cover and position of rock fragments in the top layer. *Earth Surf. Process. Landforms 15*: 653–671.

Rawls, W. J., Onstad, C. A., and Brakensiek, D. L. (1993). Seasonal effects of agricultural practices on infiltration. *Proc. Industrial and Agricultural Impacts on the Hydrologic Environment*, Am. Inst. of Hydrology, Alexandria, VA, pp. 71–77.

Shaw, C. F. (1929). Erosion pavement. *Geograph. Rev. 19*: 638–641.

Simanton, J. R., Rawitz, E., and Shirley, E. D. (1984). Effects of rock fragments on erosion of semiarid rangeland soils. In *Erosion and Productivity of Soils Containing Rock Fragments* (J. D. Nichols, P. L. Brown, and W. J. Grant, eds.), *Soil Sci. Soc. Am. Spec. Pub. No. 13*: 65–72.

Van Doren, D. M., Jr., Moldenhauer, W. C., and Triplett, G. B., Jr. (1984). Influence of long-term tillage and crop rotation on water erosion. *Soil Sci. Soc. Am. J. 48*: 636–640.

West, L. T., Miller, W. P., Langdale, G. W., Bruce, R. R., Laflen, J. M., and Thomas, A. W. (1991). Cropping system effects on interrill soil loss in the Georgia Piedmont. *Soil Sci. Soc. Am. J. 55*: 460–466.

Wischmeier, W. H. (1975). Estimating the soil-loss equation's cover and management factor for undisturbed areas. In *Present and Prospective Technology for Predicting Sediment Yields and Sources*, USDA-ARS, ARS-S-40, U.S. Govt. Printing Office, Washington, DC, pp. 118–124.

Wischmeier, W. H., Johnson, C. B., and Cross, B. V. (1971). A soil erodibility nomograph for farmland and construction sites. *J. Soil Water Conserv. 26*: 189–193.

Wischmeier, W. H., and Smith, D. C. (1978). Predicting rainfall erosion—a guide to conservation planning. *USDA Agric. Handb. No. 537*, U.S. Govt. Printing Office, Washington, DC.

Yoo, K. H., and Touchton, J. T. (1988). Surface runoff and sediment yield from various tillage systems of cotton. *Trans. ASAE 31*: 1154–1158.

7

Interrill Erosion

Padam Prasad Sharma
Land Reclamation Research Center, North Dakota State University, Mandan, North Dakota

I. DEFINITION AND CONCEPT

The process of erosion of soil by water starts with the detachment and transport of soil particles by impact force of raindrops and drag force of overland flow (Ellison, 1947). The dominance of one force over the other or a vector combination of both the forces determines the controls on the processes of detachment and transport. Raindrop impact provides the primary force needed to initiate detachment of soil particles from the soil mass. From the point of origin, the raindrop splash and the overland flow water transport the sediments in a downslope direction. Sediments must be detached from the soil mass or be in a detached state before they can be transported.

The water erosion process can be detachment limited or transport limited (Foster, 1982). The process is transport efficient or detachment limited if all particles generated in a unit eroding area move across the width of the lower slope boundary. The process is transport limited if all particles detached in an upslope unit area are not carried across the downslope width. Generally, all particles that are detached are not transported out of the eroding area. The process of settling of detached and transported sediments, at a position different from their original location within the spatial unit, is called deposition. Detachment, transport and deposition of sediments are three integral processes of soil erosion (Rose, 1985).

To understand the definition of interrill erosion, it is imperative that we understand the domain of rill erosion. Rill erosion occurs when overland flow, though often considered to be of uniform depth across the slope, tends to concentrate in numerous small channels. Rills develop in localized depressions of sloping lands where overland flow accelerates as it traverses downslope. When the sediment load of the overland flow is not enough to satisfy its transport capacity, the excess force drags sediments on the bed, causing scouring of the bed for more sediments. Small channels begin to develop across the slope as a result of continuous longitudinal scours on the soil surface (Moss et al., 1982). As the process continues, the flow converges and travels downslope in these small well-defined channels. These channels, resulting from either erosion or tillage, become the main pathways of downslope runoff, and they are called rills regardless of whether erosion occurs in them or not. Any flow entrainment that occurs in these channels is called rill erosion (Foster et al., 1985).

The sediment producing areas on upslope nonrilled spaces and downslope in between rill spaces are called interrill areas. Formation of a rill network, with its intervening interrill catchment, can profoundly change the sediment transporting mechanism (Moss, 1988). Although the mean rill direction is downslope, unless overall slope is high, individual rill reaches can deviate considerably from this direction. As a result, most interrill catchments slope downward toward rill segments and supply their sediment load to the rill system (Foster, 1982). Efficient water drainage by the rill network ensures that interrill catchments are partially covered by very shallow water, facilitating the processes of rainfall detachment, air-splash, and rain-flow transportation (Moss, 1988).

The idea of spatial demarcation of rill and interrill erosion started with a necessity to model the process of erosion by mass balance equations (Meyer and Wischmeier, 1969). There was a need to estimate soil erosion by water at a level above the empiricism of the universal soil loss equation so that erosion rates from individual storms could be predicted based on the hydrology of raindrop impact and overland flow (Lane et al., 1992). Figure 1 shows the flowchart of the proposed conceptual model by Meyer and Wischmeier (1969) which required evaluations of soil detachability and sediment transportability parameters. This requirement resulted in the experimental design and collection of sediment yield data from spatially separate rainfall detachment dominated (interrill erosion) and flow entrainment dominated (rill erosion) plots. As we will discuss later, this led to the concept of sediment mass balance in a spatial context which often clouds the understanding of basic processes of water erosion.

II. PROCESSES OF INTERRILL EROSION

All four component processes of water erosion (detachment by raindrops and flow, and transport by raindrops and flow) shown in Fig. 1 are active in an

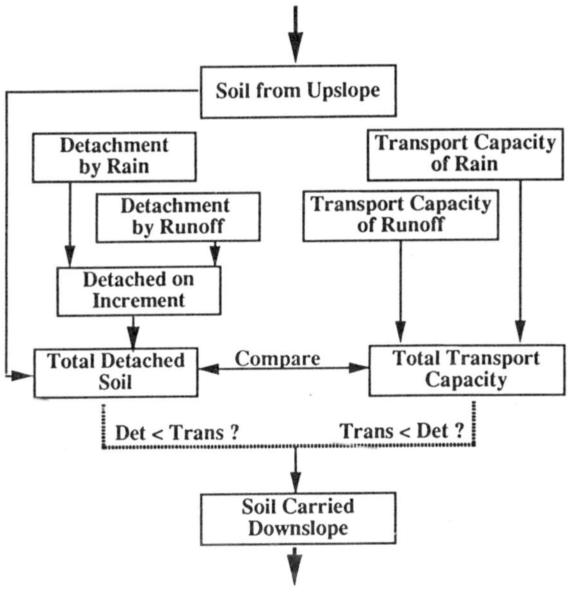

Fig. 1 The conceptual water erosion model of Meyer and Wischmeier (1969) which forms the basis for the rill-interrill erosion model. (From Foster and Meyer, 1975.)

interrill erosion area (Ellison, 1947; Rowlison and Martin, 1971). However, raindrop-induced detachment and flow transport are the two dominant processes. For each process, interaction with surface cover, slope and depth of flow are important. While Foster (1982) treated deposition as a default process due to lack of transport, Rose et al. (1983) emphasized that deposition is an integral process of interrill erosion.

A. Detachment

1. Raindrop Detachment

The process of soil detachment by raindrops is best understood by studying the mechanism of soil detachment due to impact of a single drop (Ghadiri and Payne, 1977). With high-speed photography, Mutchler (1967) and Al-Durrah and Bradford (1982) describe the mechanism of raindrop impact on the soil surface and the resulting soil detachment and sediment splash. When a raindrop impacts a saturated soil surface, a hemispheric cavity is formed on the surface due to the impulsive load of the spherical drop. The vertical compressive stress of the drop is then transformed into lateral shear stress of radial flow of water jetting away from the center of the cavity. At this stage, soil particle detachment is caused by the shear stresses of the radial flow acting on the bottom and sides of the cavity. The amount of soil detachment from the cavity sides will be determined by the magnitude of soil deformation that takes place in the earlier stages of cavity development and by the cohesive forces resisting the shear stresses (Al-Durrah and Bradford, 1982).

Park et al. (1982) describe the mechanism of raindrop detachment and splash as a case of fluid movement by process of impingement and Rayleigh jets in a drop-liquid-solid domain. A measure of the force or stress or energy of the impacting drop is called erosivity of the drop, while the ease with which the soil matrix yields to the raindrop impact is called detachability of the soil. As the strength of the soil to withstand the erosive force of the impacting drop increases, the detachability of soil decreases.

2. *Flow Detachment*

Detachment by surface flow or flow entrainment of sediments is primarily considered in the domain of the rill erosion process. For interrill erosion prediction purposes, the detachment of soil by shallow overland flow alone is assumed negligible (Foster, 1982). Detachment of soil particles by flow of water on the surface of the soil is induced when the shear stresses of flow exceed the tensile strength of the soil particles (Nearing, 1991). For soil surfaces, the shallow overland flow alone has shear stresses in the order of pascals while the soil tensile strength is on the order of kilopascals (Rowlison and Martin, 1971; Nearing, 1991). Yet, the overland flow detachment takes place due to driving force provided by a combination of turbulent events called ''bursts'' which are very localized in time and space (Nearing et al., 1991). Such turbulence is created by the impact of raindrops and by the presence of microdepressions on the soil surface. Compared to the bedload transport mechanisms (Bagnold, 1977) by flow of water over sand surfaces, the detachment rate of soil by surface flow cannot be expressed as a simple function of average shear stress or stream power (Nearing, 1991; Nearing, et al., 1991).

B. Sediment Transport

1. *Transport by Raindrops (Splash or Air-Splash)*

Commonly known as splash erosion, air-splash occurs immediately after raindrop impact initiates soil detachment. The droplets generated after impact radiate outward from the center of the impact while encapsulating solids and carrying them to the landing points. Air-splashing of solids declines rapidly as depth of water covering the soil surface increases, reaching near negligibility by a depth of about 2 mm (Moss, 1988). For normal impacts on horizontal surfaces, splash produces only random particle movement. Air-splash can cause net transportation in one direction, firstly, under the influence of slope or wind and, secondly, due to preferential movement of solids from areas of high activity to those of low activity. The apparent absence of thick, extensive deposits, attributable to air-splash, suggests that the mechanism is seldom, if acting alone, a major transporting agent (Moss, 1988). Wright (1986, 1987) has proposed a model to describe the dispersion of splash droplets from a single raindrop impact on a sloping soil surface. The component of the raindrop velocity parallel to the surface of the slope is identified as the main factor determining the degree and the direction of the asymmetry in the splash droplet dispersion.

2. *Transportation by Flow (Sheet Erosion, Wash Erosion)*

As soon as the process of runoff starts from the time of ponding, the overland flow which is inherently more unidirectional than air-splash, begins to transport solids downslope. The solid load of flow transportation is split into suspended and bed loads. Depending on the slope and surface roughness, flow alone can only transport small-size particles in suspended load. A small proportion of the solid load is provided by incoming air-splash debris, and the rest is due to combined action of raindrops and flow (Moss, 1988).

Bedload particles not transported by flow alone remain on the bed until lifted back in the flow by force of raindrops impacting the shallow overland flow (Kinnell, 1988). Raindrops impacting along the flow line provide a series of linked events which cause sediment to be transported downstream with the flow. Because the raindrop impact induces the flow to transport particles it would otherwise be incapable of transporting, the transport process associated with the raindrop impact is called rain-flow transportation (Moss, 1988) or rain-induced flow transport (RIFT) (Kinnell, 1990, 1991). Kinnell (1990) describes the RIFT mechanism of sediment transport and its consequential effect on sediment distribution as the net sediment flux moves downstream.

If local depth-slope products are low, rills may not develop and rain-flow transportation will dominate the whole area, bringing an almost imperceptible lowering of the surface with no obvious features manifesting its action (Moss, 1988). Because entrainment is caused by rain, this mechanism can operate at near-zero flow velocities. Rain-flow transportation thus bears a complementary relationship to overland-flow transportation. Where the RIFT mechanism is dominant, sheet flows tend to persist. The RIFT can operate at water depths of about 1 mm, but is most effective at depths of 2–3 drop diameters. It thus also bears a complementary relationship to raindrop detachment and air-splash, both of which decrease as depth of flow increases.

Kinnel (1990) has described the mechanism of sediment transport by RIFT. The RIFT mechanism explains how soil particles detached by drop impact move downstream within a flow under the stimulus of impact by subsequent drops.

Central to the RIFT theory is the recognition of an active zone where particles lifted to flow by drop impact pass across the downstream boundary of the zone without further turbulence by raindrop impact. Analyses of the concept on sand and soil surfaces indicate that drop size and flow depth variations have an interactive effect on the entrainment of particles by drop impact. The downstream movement of particles by RIFT is constrained when flows shallower than 3 drop diameters are impacted by medium-to-large-size raindrops.

C. Deposition

When the stream power of the overland flow is obstructed due to surface roughness, plant stalks, and stubble mulches, or when flow turbulence is lowered due to decrease in slope steepness or frequency of rainfall impact, the sediments in bedload and some suspended load transport settle on the surface. Depending on the depth of water on the surface and the extent of RIFT, the process of deposition is highly selective (Hairsine and Rose, 1991). The settling velocity of an aggregate or primary particle is a function of its size, shape, and density. The rate of deposition is indirectly proportional to the velocity of flow and directly related to the concentration and density of a given sediment size (Hairsine and Rose, 1991).

III. FACTORS AFFECTING INTERRILL EROSION

As defined earlier, interrill erosion comprises the subprocesses of detachment and transport of sediments by action and interaction of raindrops and overland flow and by deposition. Each subprocess is affected by (a) rainfall properties, (b) soil properties, and (c) surface properties (Park et al., 1982). Rainfall properties determine the erosivity of raindrops and flow. Soil properties affect the soil detachability, transportability (jointly called interrill erodibility), and infiltrability, while the surface properties affect all three subprocesses of interrill erosion.

A. Erosivity of Raindrops and Flow

1. Raindrop Size Distribution, Kinetic Energy, and Intensity

Since erosion starts with the process of soil detachment by raindrops, the basic unit of raindrop erosivity is represented by the force, stress, momentum, or kinetic energy of a single raindrop. These parameters are functions of drop size, drop shape, and impact velocity of drops (Gilley and Finkner, 1985; Riezebos and Epema, 1985; Sharma and Gupta, 1989). Of these, kinetic energy of a single drop is the most commonly used unit of raindrop erosivity.

The total energy of a rainfall event is calculated by summing individual kinetic energies of raindrops with the aid of drop-size distribution information for a rainstorm (Wischmeier and Smith, 1958; Sharma et al., 1993). Since measurement of drop-size distribution is cumbersome and subject to storm-type variability, rainfall energy and raindrop-size distribution are generally estimated as functions of rainfall intensity (Wischmeier and Smith, 1958; Park et al., 1983). These estimates are dependent upon regional climatic variability (McIssac, 1992). Successful attempts have been made to generalize these relationships for more "universal" application (Kinnell, 1981; Mualem and Assouline, 1986; Brown and Foster, 1987; Assouline and Mualem, 1989).

2. *Overland Flow Depth and Velocity*

The presence of a water layer on the soil surface influences the detachment (Palmer, 1965; Mutchler, 1967) process. Ponded water deeper than a critical depth cushions the impact of raindrops and diminishes the rate of detachment. Palmer (1965) observed that detachment and transport by raindrops increased to a critical depth (y_0) approximately equal to the drop diameter (d) but decreased sharply as water depth (y) increased beyond y_0. Mutchler and Young (1975) suggested that $y \geq 3d$ essentially eliminated detachment by raindrop impact. Profitt et al. (1991) have schematically shown that as depth of surface water rises, the detachability of the original soil and the redetachability of the deposited layer decreases exponentially.

Transport of particles by interrill flow is directly related to depth and velocity of flow, which, along with land slope and raindrop impact, determine the shear stress of flow (Gilley et al., 1985). Similar to detachment by flow, shallow flows alone are less effective in transporting solids than when they are impacted by raindrops (Podmore and Merva, 1971; Foster and Meyer, 1972; Walker et al., 1978; Moss, 1988; Kinnell, 1990).

The transportation of particles by the RIFT mechanism depends on particle size, drop size, and depth and velocity of flow (Moss and Green, 1983; Kinnell, 1988; Moss, 1988). Results from flume experiments indicate that transport rates peak when drops impact flows that are between 2 and 3 drop diameters deep (Moss and Green, 1983), and that transport rates tend to vary linearly with flow velocity (Moss, 1988). On the other hand, Kinnell (1991) showed that transport rate of particles declines linearly with flow depth when flows are deeper than about 2–3 mm and shallower than 3 drop diameters. Kinnell (1991) stated that the product of rainfall intensity, slope gradient and runoff rate may provide a basis for determining the effect of rainfall on erosion by rain-impacted flow when factors such as flow depth and velocity are unknown.

3. *Slope*

For interrill erosion, the local slope (for example, slope toward the rill or side slope of a crop row or a tillage furrow) and not the general land slope is the effective slope (Foster, 1982). Slope has the most direct effect on the erosivity of overland flow by determining its stream power, which is a product of hydraulic shear stress and average flow velocity (Nearing et al., 1991), both of which, in turn, are functions of slope. To a lesser extent, slope influences the erosivity of raindrops by determining the number of raindrop impacts per unit area and the angle of impact. A raindrop falling normal (impact angle $\theta = 90°$) to a horizontal surface with an impact force F will have all of its force dissipated as compressive force. Any increase in land slope or decrease in impact angle due to effect of wind will decrease the compressive force and increase the shearing force of raindrops (Ekern, 1950) by $F \cos(\theta)$. The steepness of the slope (ϕ) will increase the net transport of sediments by drop splash into downslope direction by a magnitude of $\sin(\phi)$. The increase in land slope, on the other hand, will also proportionately decrease the number of drops impacting the soil surface (Wang, 1988).

Rowlison and Martin (1971) have conceptually analyzed the effect of slope on detachment and transport by raindrops and overland flow. Hairsine and Rose (1991) proposed that three distinct regimes, each with a different dependency on slope steepness, exist for shallow-flow, low-slope erosion. When (i) the flow depth is less than or equal to a breakpoint depth and flow-driven processes are inactive, erosion is independent of slope; when (ii) flow depth is greater than the breakpoint depth but flow-driven processes are inactive, erosion is slightly dependent on slope; and when (iii) the combination of slope and discharge is such that flow-driven processes are active, then erosion is strongly dependent on slope.

4. *Effect of Surface Cover and Roughness*

Surface cover and canopy directly affect the soil detachment process by intercepting raindrops and dissipating their kinetic energies before impact on the soil surface. The effective energy of raindrops after the canopy interception depends on the height of plants and type of leaf structure. A shorter, bushy canopy (such as annuals and shrubs) generally dissipates the falling raindrops completely. Taller canopies produce large-size recoalesced drops which are generally prolate in shape. Such drops may be more erosive due to their higher detachment efficiency than similar size raindrops falling at terminal velocities (Chapman, 1948; Moss and Green, 1986; Sharma and Gupta, 1989).

Mulch reduces interrill erosion in two ways (Foster, 1982). First, mulch protects a portion of the interrill area from direct raindrop impact. Second, mulch reduces overland flow velocity, and the extra depth of water on the surface cushions raindrop impact, resulting in decreased raindrop-impact detachment and raindrop-induced flow transport. An added factor related to flow depth is the effect of mulch rate on infiltration. Higher mulch rates protect the soil surface from sealing, resulting in higher infiltration rates on mulched soil than on bare soil (Mannering and Meyer, 1963). This in turn results in lower runoff rates and lower energy available for sediment transport.

Surface microrelief or surface roughness is generally a function of tillage-induced clod size, clod distribution, and plant stubble distribution. Surface microrelief influences the extent of sediment generation and distribution, surface seal formation, infiltration, and soil loss (Moldenhauer and Koswara, 1968; Linden et al., 1988; Wang, 1988; Freebairn et al., 1991; Roth and Helming, 1992). Increase in surface roughness increases the surface area over which the raindrop impact occurs, resulting in a decrease in energy/m^2. An increase in the number of bigger clods on the surface increases the proportion of surface area where drops will impact at angles >90°, generating less splash. Abundance of large clods also increases production of depositional seals on microrelief depressions (Freebairn et al., 1991). Accumulation of water in these depressions lowers the detachment of sediments due to dissipation of drop impact energy and reduces the sediment transport due to obstruction of the flow by clods.

B. Detachability of Soil

The susceptibility of soil to shear forces of detachment primarily depends on soil strength (Cruse and Larson, 1977; Ghadiri and Payne, 1977; Al-Durrah and Bradford, 1982; Nearing and Bradford, 1985; Sharma et al., 1991). Various measures of soil strength, such as shear strength by unconfined compressive test (Cruse and Larson, 1977), shear strength by fall-cone device (Al-Durrah and Bradford, 1982; Sharma et al. 1991), friction angle from triaxial compression test (Nearing and Bradford, 1985), and aggregate stability by wet sieving (Young and Onstad, 1978; Young, 1984; Truman et al., 1990) have been related to raindrop detachment, splash erosion, and interrill erosion. Both soil strength and aggregate stability are in turn affected by aggregate size and density, amount and type of clay, organic carbon content, and inorganic constituents such as iron, sodium, calcium, and magnesium (Foster et al., 1985).

Water content of the soil at the onset of rainfall affects the rate of soil detachment and splash (Truman and Bradford, 1990), resulting in significant changes in interrill erosion rates (Luk, 1985). The effect of antecedent water content is directly manifested into status of soil strength (Kemper et al., 1987) and aggregate stability (Truman et al., 1990). Depending on the clay content and the nature of cations present in the soil, both aggregate stability and soil strength increase when water content increases from air-dry condition. On the other hand, aggregate stability and soil strength decrease as the matric potential approaches zero near saturation water contents (Francis and Cruse, 1983; Sharma and Gupta, 1989).

In addition to the impact energy of raindrops, water quality and exchangeable sodium percentage (ESP) of soil enhance aggregate breakdown by increasing dispersion and swelling of clay (Agassi et al., 1981; Shainberg and Latey, 1984). Interrill erodibility increases with increase in surface soil dispersibility, which in turn is enhanced by ESP (Singer et al., 1982), low electrolyte concentration in soil solution, and the presence of smectite clay (Stern, Ben-Hur, and Shainberg, 1991; Ben-Hur et al., 1992; Agassi et al., 1994a, 1994b).

C. Transportability of Sediments

The transportability of sediments and the critical shear stress that needs to be overcome for a particle to be lifted and transported or dragged downstream are functions of particle diameter and density (Foster and Meyer, 1975). Detached sediments comprise both primary particles and aggregates. Sand-size particles and aggregates are mainly transported as bedload sediments, which upon a relaxing of the transporting energy of flow and raindrop impact settle on the soil surface. Some aggregates may break down during transport and release primary particles. The silt and clay-size particles travel the farthest as suspended load.

IV. INTERRILL EROSION AND SURFACE SEAL FORMATION

Most soil erosion prediction models do not account for surface sealing and soil crusting effects on soil detachment (Bradford et al., 1986) and infiltration. Surface seal formation during rainfall decreases infiltration rates (Sharma et al., 1981), and the resulting crust increases soil strength (Bradford et al., 1986). Chen et al. (1980) described the erosion processes accompanying crust formation into three stages. During Stage I (prerunoff), drop impact effects on physical dispersion, compaction, and movement of particles (splash processes) represent the dominant crust-forming mechanisms. During Stage II (increasing runoff), the dispersion, entrainment, transport, and deposition of particles by flowing water (wash processes) play increasing roles; detachment and transport processes are moderated by a film of water covering a portion of the soil surface. Stage III (steady state) is characterized by an equilibrium crust; structural and depositional processes each contribute to the maintenance of the crust.

Figure 2 shows the observations by Moore and Singer (1990), who explained the interaction of erosion and surface seal formation in the context of the three stages described by Chen et al. (1980). Erosion rates increased in the beginning, reached a maximum during the period of increasing runoff, decreased, and then reached near equilibrium rates during the final, steady-state runoff period. These observations suggest that, under steady raindrop erosivity conditions, crust formation decreases soil detachment. Splash erosion rates peaked during the period of increasing runoff, while wash erosion rates displayed a gradual increase over

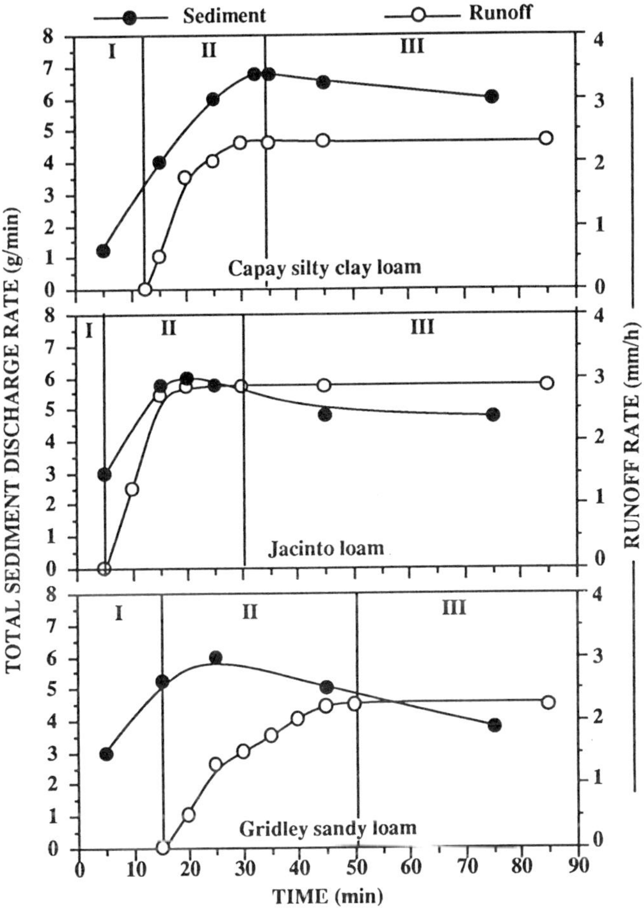

Fig. 2 Runoff and sediment loss rates vs. time and crusting Stages I, II, and III for Capay silty clay loam, Jacinto loam, and Gridly sandy loam. (From Moore and Singer, 1990.)

the periods of increasing and steady-state runoff. These trends are related to the decreasing size and degree of aggregation of surface materials available for detachment and to the buildup of a layer of shallow overland flow that enhances raindrop-induced flow transport while retarding splash transport.

Although surface sealing is detrimental in that it generates runoff, it is beneficial in that it stabilizes the soil surface. Surface seal formation seems to reduce soil loss from an interrill area, at least as long as runoff does not exert shear strengths greater than the cohesive forces acting on the seal (Roth and Helming,

1992) to initiate the rilling process. Crusted soils after rainfall impact have lower detachability than uncrusted soils due to increase in soil strength during crust formation (Bradford et al. 1986). The extent to which these single processes counteract or compensate each other depends on the type of force and flow conditions prevailing at the soil surface.

V. EVALUATION OF INTERRILL EROSION PARAMETERS

A. Experiments in the United States

The main objective of interrill erosion experiments in the United States has been to understand the detachment process and evaluate interrill erodibility by measuring sediment delivery rates from small interrill erosion plots. Figure 3 shows the schematics of typical ridged and flat plots used in interrill experiments for

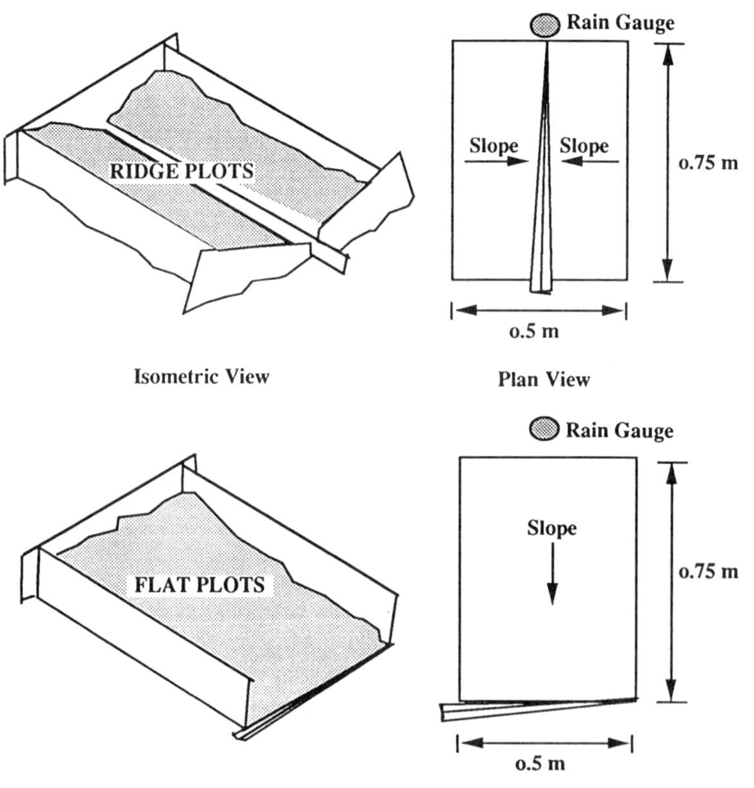

Fig. 3 Schematic diagrams of typical interrill erosion plots used in the United States. (From Elliot et al., 1989.)

the Water Erosion Prediction Project field investigations conducted in the United States (Liebenow et al., 1990). In the laboratory, various sizes of metal, wooden, or acrylic boxes have been used with varying depths of disturbed soil generally underlain by a layer of sand. Almost always, whether for surface seal formation studies or for interrill erosion studies, these boxes are packed with sieved and air-dry soil samples.

Wang (1988) investigated the effect of interrill pan size on splash erosion. The results showed that for first erosion run on air-dry samples, soil splash per unit area decreased sharply with increase in impact area. After the first run, due to surface seal formation and removal of loose sediments in the first run, the effect of pan size was minimal. Besides the pan size, initial water content and surface roughness influenced the measured erosion rates from such interrill plots (Wang, 1988). As a result of the experimental variability, success in relating interrill erodibility with routine soil physical and chemical properties of soils is still illusive.

B. Experiments in Australia

The objective of interrill erosion experiments in Australia has been to understand the transport mechanisms associated with rainfall and flow. Most of the published data come from a flume apparatus as shown in Fig. 4. The apparatus, described in detail by Moss and Green (1983), includes exposition of sand or soil bed under a flume-rainfall simulator (Walker et al., 1977). A water-saturated target sand bed is installed in the downstream part of the flume. Immediately downstream of the box is a stilling pool which terminates into a combined height-adjustable horizontal weir and flow collector. A tube drains the collector and allows total discharge to be monitored. A removable subaqueous-sediment

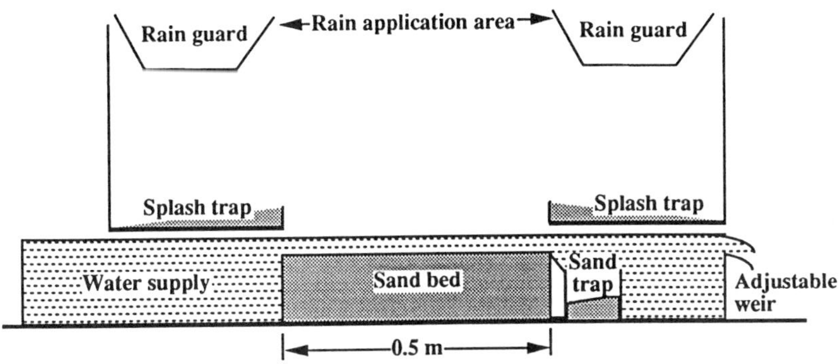

Fig. 4 Schematic longitudinal section of the flume apparatus used in laboratory soil erosion studies in Australia. (From Moss and Green, 1983.)

trap placed in the stilling pool against the bed wall catches particles moved by rain-flow transportation. The medium is dominantly sand, and observations on repacked or on natural columns of soil (Kinnell and McLachlan, 1989) are few.

VI. INTERRILL EROSION MODELS

Detachment, transport, and deposition occur simultaneously in interrill areas. For a given rainstorm event, a physically based interrill erosion model first considers accounting of total sediment pool generated by raindrop detachment and redetachment. The pool then needs to be instantaneously partitioned into those transported by transporting agents across the downslope flux boundary. The difference, between the amount of total detachment and the amount transported out of the reference area, is deposition which forms surface seal and becomes original soil for the next rainstorm event. Since the process is complex and unsteady, the accounting of sediments needs to be done in small temporal and spatial increments.

Park et al. (1982) and Gilley et al. (1985) modeled interrill erosion on the basis of single-raindrop kinetic energy dissipation, raindrop-size distribution, and soil detachment and transport parameters. Due to lack of single-drop detachment data, splash cup and interrill erosion data available in the literature were used to derive fitted parameters of the models. During the past decade, understanding the soil detachment mechanism by single drops has advanced to the extent that the erosivity of raindrops (Epema and Riezebos, 1983; Gilley and Finkner, 1985; Nearing et al., 1986), the mechanical resistance of soil to erosive forces (Nearing and Bradford 1985; Sharma, et al., 1991), and the interaction with depth and velocity of overland flow (Hairsine and Rose, 1991; Kinnell, 1990; Proffitt et al., 1991) on detachment and transport can be adequately described by various physioempirical equations.

Total interrill erosion for a storm can be calculated by integrating fundamental detachment and transport equations over time for a single-raindrop impact, over all the raindrops occurring in a storm, and over space for each hydrologic element in a land unit (Hagen and Foster, 1990). Unfortunately, neither science nor computational power was sufficient to allow practical application of this very basic approach during the 1970s and 1980s. Thus, erosion prediction technology such as the WEPP (Nearing et al., 1989) being developed today still involves empirical equations for the fundamental interrill erosion processes of detachment and transport.

From the principal of mass conservation of sediments in a planar land element, a comprehensive mathematical model for water erosion process is expressed as (Bennett, 1974; Rose et al., 1983)

$$\frac{\delta q_{si}}{\delta x} + \frac{\delta (c_i y)}{\delta t} = e_i + r_i - d_i \qquad (1)$$

where

q_{si} = sediment flux for a sediment size class i per unit width of plane
c_i = sediment concentration
y = depth of flow
e_i = rainfall detachment rate
r_i = sediment entrainment rate by overland flow
d_i = sediment deposition rate

Figure 5 shows the flowchart of the conceptual model presented by Eq. (1). When reduced to a steady-state condition, the sediment mass balance can be expressed by the following ordinary differential equation, which can be solved analytically (Rose et al., 1983) for the sediment flux (q_s):

$$\frac{dq_{si}}{dx} = e_i - d_i + r_i \tag{2}$$

Assuming negligible entrainment by flow alone, Hairsine and Rose (1991) added e_{di} (as the rate of redetachment of deposited layer) to the right-hand side of Eq. (2) to represent the mass balance for transport-limited, low-slope, depositional areas. Rose et al. (1983), Hairsine and Rose (1991), and Proffitt et al. (1991) show examples of analytical solutions of Eq. (2). The solution estimates

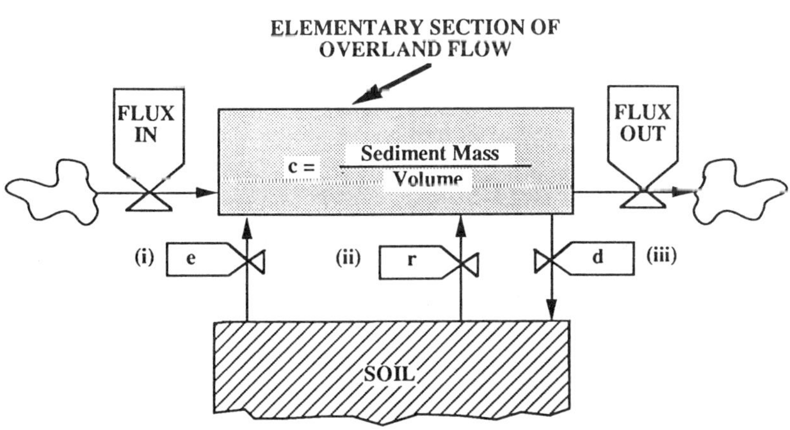

Fig. 5 Flowchart representing the processes of (i) rainfall detachment (e), (ii) flow entrainment (r), and (iii) deposition (d) in a water erosion model. (From Rose et al., 1983.)

sediment concentration at a given time as a function of distance down the plane in terms of soil factors, surface cover, rainfall intensity, and runoff rates.

The modeling approach most resourcefully pursued in the United States was to group the processes according to a sediment source area concept (Foster and Meyer, 1975; Foster et al. 1977; Foster, 1982). This approach allowed a way of spatially integrating point processes to represent large areas such as fields and farms, and it minimized distortion of parameter values determined from small idealized laboratory and field experiments (Hagen and Foster, 1990). As discussed earlier, interrill and rill areas represent the principal source areas for an upland erosion process (Foster and Meyer, 1975). The quasi-steady-state mass balance equation for the spatially delineated model is (Foster et al., 1977)

$$\frac{dq_s}{dx} = D_i + R_i \tag{3}$$

where D_i is the sediment delivery rate by interrill erosion, and R_i is the entrainment rate by rill erosion. Rose (1985) analyzed the similarities and differences between the above model and Eq. (2).

A. Water Erosion Prediction Project (WEPP) Model

Based on the interrill erosion experiments of Meyer (1981) and subsequent comprehensive review by Foster (1982), Liebenow et al. (1990) presented the following equation for interrill erosion component of the WEPP model:

$$D_i = K_i I^b S_f \tag{4}$$

where

K_i = interrill erodibility assumed entirely by raindrop impact
I = rainfall intensity
S_f = interrill slope steepness factor defined by Eq. (5).

Both contemporary models (Foster, 1982; Rose et al., 1983) use the exponent $b = 2$ to estimate rainfall detachment in the interrill area. This value was arrived at by deduction. Based on experimentally observed linear proportionality of soil splash loss versus EI, the product of rainfall kinetic energy and maximum 30-min intensity (Free, 1960), Meyer and Wischmeier (1969) deduced that soil detachment by rainfall was proportional to I^2. Subsequent work by Meyer (1981) related interrill erosion with the square of intensity, even though the exponent ranged from 1.65 to 2.16, depending on the texture of the soil. Watson and Laflen (1986) found that the exponents varied from 1.36 to 2.54.

Based on the threshold linear relationship between single-drop soil detachment and raindrop kinetic energy (Sharma and Gupta, 1989; Sharma et al., 1991) and a drop-size distribution function, Sharma et al. (1993) predicted rainfall detachment rates from an interrill area. When the predicted detachment rates were nonlinearly related to rainfall rates, the exponent b ranged from 1.08 for a low-strength soil to 1.44 for a high-strength clay soil. Proffitt et al. (1991) derived fitted b values of less than 1.0 from interrill erosion data with various depths of flow. They argue that it would greatly simplify comparison of the susceptibility of a soil to rainfall detachment or redetachment if b could be standardized and set equal to 1.0.

Numerous experiments have been conducted on interrill erosion plots to evaluate the effects of slope on interrill erosion (Latanzi et al., 1974; Singer and Blackard, 1982; Meyer and Harmon, 1985; Watson and Laflen, 1986) and define a slope factor. For use in the proposed WEPP model, Liebenow et al. (1990) proposed a slope factor (S_f) of the form

$$S_f = d - f \exp(- g \sin(\phi)) \quad (5)$$

where ϕ is the slope angle and d, f, and g are constants taken as 1.05, 0.85, and 4.0, respectively.

Equation (4) is a lumped parameter empirical model which does not explicitly delineate the soil detachment and sediment transport processes. The use of erosion rate as a power function of intensity fails to differentiate hydrologic and erosion processes prevalent in interrill areas (Nearing et al., 1990). The interrill erodibility parameter, K_i in Eq. (4), is not independent of infiltration and runoff because the sediment delivery from the interrill area is a function of detachment and transport. This is probably the main reason why predictive relationships between the measured K_i and basic soil properties cannot be successfully established.

USDA-WEPP research scientists conducted erosion experiments on cropland, rangeland, and forest soils around the United States during 1987–1988 (Foster and Lane, 1987). From the measured interrill erosion data, Elliot et al. (1989) and Laflen et al. (1991) have tabulated the derived K_i values for 33 cropland and 20 rangeland soils. In order to apply the WEPP technology to other soils, K_i needs to be experimentally evaluated or estimated using regression equations developed from properties of the benchmark soils.

The experimental evaluation of K_i depends upon the ability of a researcher to design interrill erosion plots and apply artificial rainfall of characteristics similar to those used in the WEPP experiments. Otherwise, K_i will be affected by differences in initial conditions of soil (Truman and Bradford, 1990), the geometry of the interrill erosion plots (Wang, 1988), and rainfall characteristics (Meyer and Harmon, 1992). Attempts to relate the WEPP-measured interrill erodibility coefficient with routine soil physical and chemical properties (Elliot et al., 1989; Elliot et al., 1990; Gilley et al., 1992) and to validate such relationships on erosion data obtained from independent field experiments have attained limited success (Bajracharya et al., 1992).

B. Soil Detachment and Transport-Based Models

Sharma et al. (1993) presented algorithms for simulating total soil detachment during a rainstorm by linking a threshold linear soil detachment model (Sharma et al., 1991) of single raindrop impact with the kinetic energy and drop-size distribution (Assouline and Mualem, 1989) of a rainstorm. For steady-state conditions, the rate of soil detachment is expressed as

$$D_r = k_d I(E - E_0) \tag{6}$$

where

D_r = rainfall detachment rate (kg m^2 h^{-1})
k_d = raindrop detachability (kg J^{-1})
I = rainfall rate (mm h^{-1})
E = unit kinetic energy of rainstorm (J m^2 mm^{-1})
E_0 = unit threshold kinetic energy needed to initiate soil detachment (J m^2 mm^{-1}).

The term $E - E_0$ is defined as the unit effective kinetic energy of a rainstorm. Equation (6) is similar in format to the basic rill entrainment equation (Elliot and Laflen, 1992), which separates the soil erodibility and flow erosivity components. Equation (6) clearly differentiates the rainfall erosivity (I and E) and the soil detachability components (k_d and E_0) of the raindrop-induced soil detachment model.

Sharma et al. (1993) presented cumulative and analytical expressions for the evaluation of E and E_0 from raindrop-size distribution and single-drop soil detachment data. The unit threshold kinetic energy, E_0, is derived by accounting single-drop threshold kinetic energies needed to initiate soil detachment. From single drop experiments on seven soils at various strength conditions, Sharma et al. (1991) showed that the threshold kinetic energy needed to initiate the soil detachment process is directly related to shear strength and clay content.

Equation (6) estimates the total pool of rain-detached sediments available for transport by splash, overland-flow, and rain-flow mechanisms prevalent in the interrill area (Moss, 1988). Under transport limiting conditions, a fraction of the rain-detached sediments will be deposited, redetached, and redistributed within the interrill area. For conditions where the interrill transport mechanism is sufficient to remove all sediments generated by raindrop impact, the process is detachment limited. Thus, Eq. (6), which estimates the steady-state rate of primary detachment, also appraises the sediment delivery rate across the lower slope boundary of the interrill area.

Assuming negligible detachment by flow alone, the sediment mass balance of an interrill area can be expressed as

$$D_r = T_r + R_r \tag{7}$$

where

D_r = primary detachment rate by raindrops (Eq. [6])
T_r = sediment transport rate by interrill transport mechanisms of air-splash, wash, and rain-induced flow transport
R_r = sediment redistribution (deposition, redetachment, and redeposition) rate within interrill area

For an experimental interrill erosion plot of a given geometry (slope and area) with negligible net transport by air-splash outside of the upslope and lateral boundaries of the plot, the measured interrill sediment delivery rate at the lower slope boundary, D_i, represents the interrill sediment transport rate, T_r for the particular plot geometry. Dividing Eq. (7) by the product of intensity (I) and the effective kinetic energy ($E - E_0$), the relationship between the coefficients of detachment, transport, and redistribution can be derived as

$$k_d = k_t + (k_r + \epsilon) \tag{8}$$

where

k_d = soil detachability (kg J^{-1})
k_t = interrill transportability (kg J^{-1})
k_r = sediment redistributability (kg J^{-1})
ϵ = error associated with estimation of D_r and measurement of D_i

Sharma et al. (1991) showed that the soil detachability (k_d) was inversely related to clay content and shear strength. To derive k_t, Sharma et al. (1995) represented the WEPP measured interrill erosion data (Elliot et al., 1989) in the form

$$D_i = k_t I(E - E_0)S_f \tag{9}$$

where D_i is interrill sediment delivery rate (kg m^2 h^{-1}) and S_f is the slope factor defined by Eq. (5). The difference between D_r and D_i is defined as the rate of sediment redistribution within the interrill area (Sharma et al., 1995). The main components of the redistribution are deposition and the redetachment of previously deposited layer.

By expressing the interrill erosion data in the above format, it was possible to isolate k_t as a simple linear function of clay content. The relationships between clay content versus soil detachability (k_d) and k_t due to raindrop impact on the bare surface of 33 cropland soils used in the WEPP experiments in the United States are shown in Fig. 6. Soils low in clay content show high detachability, but most of the detached sediments are likely to be redistributed within the interrill area due to their low transportability. On the other hand, soils high in clay have low detachability, but due to their high potential for transport most of the detached sediments are likely to be removed from the interrill area. This

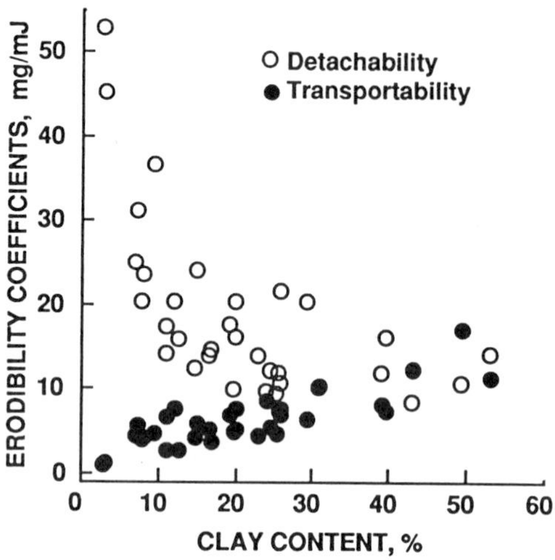

Fig. 6 Clay content vs. detachability and transportability coefficients of 33 soils used in the Water Erosion Prediction Project experiments in the United States. (From Sharma et al., 1995.)

analysis quantitatively confirms the conventional wisdom that ‘‘sands are easy to detach but difficult to transport; soils high in clay content are difficult to detach, but once detached, silt and clay size particles are more easily transported.’’

Equation (9) does not consider the interaction of the raindrop impact with depth and velocity of flow on detachment and transport of sediments from the interrill areas. Kinnell (1993a,b) analyzed the WEPP measured interrill erosion data from the perspective of transport of sediments by raindrop-induced flow turbulence. Assuming uniform impact energy and flow depth, he postulated that the transport of sediments from the interrill areas is determined by the interaction of intensity, flow rate, and slope. Kinnell (1993a) presented the WEPP data in the form

$$D_i = k_1 Q I S_f \tag{10}$$

where Q is runoff rate and k_1 is an estimate of interrill erodibility. The coefficient k_1 was shown to have lower within-soil variation than the WEPP tabulated K_i, suggesting that any attempt to isolate the soil parameter from rainfall and flow-dependent properties will improve the predictability of sediment delivery rates from interrill areas.

There is a difference between the two alternative sediment delivery models represented by Eqs. (9) and (10). The model used by Kinnell (1993a,b) to fit the WEPP data does not consider the detachment process and assumes that the sediments are at a predetached state all the time. Due to the expected shallow depth of overland flow specifically on the high-slope ridge plots used in the WEPP experiments, the effect of soil strength to control the rate of detachment throughout the period of rainfall application cannot be ignored. Though a slight improvement over WEPP tabulated K_i, the interrill sediment delivery coefficient, k_1 (Eq. (10)), was not shown to be a generic-soil parameter.

The threshold linear models of soil detachment (Eq. (6)) and sediment transport (Eq. (9)) proposed by Sharma et al. (1995) assume that interrill sediment delivery is primarily a raindrop-impact-dominated process controlled by clay content and variability of soil strength during the rainstorm event. Analyses of soil loss data from relatively steep and small interrill plots used in the WEPP experiments establish the predominant role of raindrop impact on both soil detachment and sediment transport processes. Literature on sediment transport from low slope and relatively large areas show that overland flow dominates the transport process. In order to improve the general applicability of the WEPP or any other physically based erosion model, we need to clarify the interactive roles of rainfall characteristics (intensity, energy, drop size), flow properties (depth, velocity), soil and sediment characteristics (soil strength, sediment size), and surface features (slope, cover, roughness, area) on soil detachment and sediment transport from the interrill areas.

Current consensus on the WEPP algorithm is towards using Eq. (10) to represent the fundamental interrill erosion model. While this approach is an improvement over the empiricism of Eq. (4), analysis of the WEPP data by Sharma et al. (1995) and evidence in the literature (Moss, 1988; Kinnell, 1993a) suggest that without the raindrop impact, overland flow alone is inefficient to transport the detached sediments. In other words, the premise that interrill sediment delivery is only a flow dominated transport process and intensity alone adequately represents the raindrop impact phenomena is tentative. Furthermore, exclusion of raindrop energy from the basic detachment and transport model propagates error while using soil erodibility data collected from artificial rainfall into natural rainstorms and among seasonally and physiographically variable natural rainstorms (McIssac, 1990; Meyer and Harmon, 1992).

VII. CONCLUSION

The most important soil parameter that influences interrill soil detachment and transport is soil strength, which is a manifestation of its texture (size distribution and clay content), initial matrix and pore characteristics (bulk density, water content, aggregate stability) and subsequent changes in soil strength brought about by addition of rainfall. There are enough evidences that soil chemistry and the chemistry of the applied water also have a considerable effect on interrill soil detachment. However, these factors are not yet included in the equations of interrill erosion. In addition to interrill erosion, the basic processes of soil detachment, transport, and deposition also determine the extent of surface seal formation and the amount and rate of runoff, both of which in turn affect the rate of sediment removal. Since soil strength changes with time during the rainfall event and between two subsequent rainfall events, the whole process is very dynamic.

Due to difficulty of experimentation, most of the evaluations of interrill detachment and transport parameters have been done under steady-state conditions with constant-intensity rainfall. Past experiments have indicated a strong dependence of the interrill erosion process on rainfall duration (Moldenhauer and Koswara, 1968; Schultz et al., 1985; Moore and Singer, 1990; Truman and Bradford, 1990) and rainfall intensity (Gimenez et al., 1992). Along with the steady-state spatial context, the need to study interrill erosion subprocesses in a temporal context with rainfall of varying intensities and drop-size distributions cannot be overemphasized (Meyer and Harmon, 1992).

In its present form, the WEPP algorithm requires a spatially definable separation of rill and interrill geometry. Except when formed by tillage mechanics or previous erosive events on freshly tilled soil surfaces, or dispersive sodic soils on slopes, the occurrence of a priori distribution of a distinguishable rill network is rare. At the initiation of water erosion process from a freshly tilled soil, the interrill processes of detachment and transport become active. The continuation of rainfall, sediment generation by raindrop impact, and the movement of sediment laden overland flow in downslope direction eventually start the rill formation and channel erosion processes.

This review suggests that interrill erosion is a complex process, and it cannot be fully understood if studied in isolation. Both rainfall detachment and rain-induced flow entrainment contribute to sediment concentration from ''interrill areas'' (Kinnell, 1990; Proffitt and Rose, 1991). The identity of interrill erosion only with rainfall detachment or interrill sediment delivery as only a flow transport problem may not be valid or may only apply in special circumstances. Since rain-induced soil detachment as well as sediment entrainment into the flow is occurring in interrill areas, rill erosion should be treated as a successive process contingent upon the continuation of interrill erosion from highly erodible soils. Thus, it seems undesirable to continue to use the terms interrill and rill processes as though they describe a single set and not a combination of erosion processes (Proffitt and Rose, 1991; Huang and Bradford, 1993).

REFERENCES

Agassi, M., Bloem, D., and Ben-Hur, M. (1994a). Effect of drop energy and soil water chemistry on infiltration and erosion. *Water Resources Res. 30*:1187–1193.

Agassi, M., Shainberg, I., and Morin, J. (1981). Effect of electrolyte concentration and soil sodicity on the infiltration rate and curst formation. *Soil Sci. Soc. Am. J. 45*: 848–851.

Agassi, M., Shainberg, I., and Van der Merwe, D. (1994b). Effect of water salinity on inter-rill erosion: laboratory study. *Aust. J. Soil Res. 32*: 595–601.

Al-Durrah, M. M., and Bradford, J. M. (1982). The mechanism of raindrop splash on soil surfaces. *Soil Sci. Soc. Am. J. 46*: 1086–1090.

Assouline, S., and Mualem, Y. (1989). The similarity of regional rainfall: A dimensionless model of drop size distribution. *Trans. ASAE 32*: 1216–1222.

Bagnold, R. A. (1977). Bedload transport by natural rivers. *Water Resources Res. 13*: 303–311.

Bajracharya, R. M., Elliot, W. J., and Lal, R. (1992). Interrill erodibility of some Ohio soils based on field rainfall simulation. *Soil Sci. Soc. Am. J. 56*: 267–272.

Ben-Hur, M., Stern, R., Van der Merwe, A. J., and Shainberg, I. (1992). Slope and gypsum effects on infiltration and erodibility of dispersive and nondispersive soils. *Soil Sci. Soc. Am. J. 56*: 1571–1576.

Bennett, J. P. (1974). Concepts of mathematical modeling of sediment yield. *Water Resour. Res. 10*: 485–492.

Bradford, J. M., Remley, P. A., Ferris, J. E., and Santini, J. B. (1986). Effect of soil surface sealing on splash from a single waterdrop. *Soil Sci. Soc. Am. J. 50*: 1547–1552.

Brown, L. C., and Foster, G. R. (1987). Storm erosivity using idealized intensity distributions. *Trans. ASAE 30*: 379–386.

Chapman, G. (1948). Size of raindrops and their striking force at the soil surface in a red pine plantation. *Trans. Am. Geophys. Union. 29*: 664–670.

Chen, Y., Tarchitzky, J., Brouwer, J., Morin, J., and Banin, A. (1980). Scanning electron microscope observations on soil crusts and their formation. *Soil Sci. 130*: 49–55.

Cruse, R. M., and Larson, W. E. (1977). Effect of soil shear strength on soil detachment due to raindrop impact. *Soil Sci. Soc. Am. Proc. 41*: 777–781.

Ekern, P. C. (1950). Raindrop impact as the force initiating soil erosion. *Soil Sci. Soc. Am. Proc. 15*: 7–13.

Elliot, W. J., and Laflen, J. M. (1992). A process based rill erosion model. *Trans. ASAE 36*: 65–72.

Elliot, W. J., Laflen, J. M., and Kohl, K. D. (1989). Effect of soil properties on soil erodibility. Paper 89-2150, ASAE, St. Joseph, MI.

Elliot, W. J., Liebenow, L. J., Laflen, J. M., and Kohl, K. D. (1989). A compendium of soil erodibility data from WEPP cropland soil field erodibility experiments 1987 and 1988. *NSERL Report* No. 3, National Soil Erosion Research Laboratory, West Lafayette, IN.

Elliot, W. J., Olivieri, L. J., Laflen, J. M., and Kohl, K. D. (1990). Predicting soil erodibility from soil strength measurements. Paper 90-2009, ASAE, St. Joseph, MI.

Ellison, W. D. (1947). Soil erosion studies. *Agric. Eng. 28*: 145, 197, 245, 297, 349, 402, 442.

Epema, G. R., and Riezebos, H. Th. (1983). Fall velocity of water drops at different heights as a factor influencing erosivity of simulated rain. In *Rainfall Simulation,*

Runoff and Soil Erosion (J. de Ploey, ed.), Catena Suppl. 4, Braunschweight, pp. 1–17.

Foster, G. R. (1982). Modeling the erosion process. In *Hydrologic Modeling of Small Watersheds* (C. T. Haan, ed.), ASAE Monograph No. 5, St. Joseph, MI, pp. 297–379.

Foster, G. R., and Lane, L. J. (1987). User requirements, USDA-water erosion prediction project (WEPP). *NSERL Report* No. 1, National Soil Erosion Research Laboratory, West Lafayette, IN.

Foster, G. R., and Meyer, L. D. (1972). Transport of soil particles by shallow flow. *Trans. ASAE 15*: 99–102.

Foster, G. R., and Meyer, L. D. (1975). Mathematical simulation of upland erosion using fundamental erosion mechanics. In *Proc. Sediment Yield Workshop*, Oxford, MS. USDA Rep. ARS-S-40, pp. 190–207.

Foster, G. R., Meyer, L. D., and Onstad, C. A. (1977). An erosion equation derived from basic erosion principles. *Trans. ASAE. 20*: 683–687.

Foster, G. R., Young, R. A., Romkens, M. J. M., and Onstad, C. A. (1985). Processes of soil erosion by water. In *Soil Erosion and Crop Productivity* (R. F. Follett and B. A. Stewart, eds.), ASA-CSSA-SSSA Publ., Madison, WI, pp. 137–162.

Francis, P. B., and Cruse, R. M. (1983). Soil water matric potential effects on aggregate stability. *Soil Sci. Soc. Am. Proc. 47*: 578–581.

Free, G. R. (1960). Erosion characteristics of rainfall. *Agric. Eng. 41*: 447–449, 455.

Freebairn, D. M., Gupta, S. C., and Rawls, W. J. (1991). Influence of aggregate size and microrelief on development of surface soil crusts. *Soil Sci. Soc. Am. J. 55*: 188–195.

Ghadiri, H., and Payne, D. (1977). Raindrop impact stress and breakdown of soil crumbs. *J. Soil Sci. 28*: 247–258.

Gilley, J. E., and Finkner, S. C. (1985). Estimating soil detachment caused by raindrop impact. *Trans. ASAE 28*: 140–146.

Gilley, J. E., Kincaid, D. C., Elliot, W. J., and Laflen, J. M. (1992). Sediment delivery on rill and interrill areas. *J. Hydrology 140*: 313–341.

Gilley, J. E., Woolhiser, D. A., and McWhorter, D. B. (1985). Interrill soil erosion. I: Development of model equations. *Trans. ASAE 28*: 147–153.

Gimenez, D., Dirksen, C., Miedema, R., Eppink, L. A. A. J., and Schoonderbeek, D. (1992). Surface sealing and hydraulic conductances under varying-intensity rains. *Soil Sci. Soc. Am. J. 56*: 234–242.

Hagen, L. J., and Foster, G. R. (1990). Soil erosion prediction technology. In *Proceedings of Soil Erosion and Productivity Workshop* (W. E. Larson, ed.), Bloomington, MN, pp. 117–135.

Hairsine, P. B., and Rose, C. W. (1991). Rainfall detachment and deposition: Sediment transport in the absence of flow driven processes. *Soil Sci. Soc. Am. J. 55*: 320–324.

Huang, C., and Bradford, J. M. (1993). Analyses of slope and runoff factors based on the WEPP erosion model. *Soil Sci. Soc. Am. J. 57*: 1176–1183.

Kemper, W. D., Rosenau, R. C., and Dexter, A. R. (1987). Cohesion development in disrupted soils as affected by clay and organic matter content and temperatures. *Soil Sci. Soc. Am. J. 51*: 860–867.

Kinnell, P. I. A. (1981). Rainfall intensity–kinetic energy relationship for soil loss prediction. *Soil Sci. Soc. Am. J. 45*: 153–155.

Kinnell, P. I. A. (1988). The influence of flow discharge on sediment concentrations in raindrop induced flow transport. *Aust. J. Soil Res. 26*: 575–582.
Kinnell, P. I. A. (1990). The mechanics of raindrop induced flow transport. *Aust. J. Soil Res. 28*: 497–516.
Kinnell, P. I. A. (1991). The effect of flow depth on sediment transport induced by raindrops impacting shallow flows. *Trans. ASAE 34*: 161–168.
Kinnell, P. I. A. (1993a). Interrill erodibilities based on the rainfall intensity–flow discharge erosivity factor. *Aust. J. Soil Res. 31*: 319–332.
Kinnell, P. I. A. (1993b). Runoff as a factor influencing experimentally determined interrill erodibilities. *Aust. J. Soil Res. 31*: 333–342.
Kinnell, P. I. A., and McLachlan, C. (1989). Shallow soil monoliths for laboratory studies on soil erosion from undisturbed areas. *Aust. J. Soil Res. 27*: 227–233.
Laflen, J. M., Elliot, W. J., Simanton, J. R., Holzhey, C. S., Kohl, K. D. (1991). WEPP soil erodibility experiments for rangeland and cropland soils. *J. Soil Water Conserv. 46*:39–44.
Lane, L. J., and Nearing, M. A. (Eds.) (1989). Water Erosion Prediction Project landscape profile model documentation. NSERL Report No. 2. Natl. Soil Erosion Res. Lab., USDA-ARS, Purdue Univ., West Lafayette, IN.
Lane, L. J., Renard, K. G., Foster, G. R., and Laflen, J. M. (1992). Development and application of modern soil erosion prediction technology the USDA experience. *Aust. J. Soil Res. 30*: 893–912.
Latanzi, A. R., Meyer, L. D., and Baumgardner, M. F. (1974). Influence of mulch rate and slope steepness on interrill erosion. *Soil Sci. Soc. Am. Proc. 38*: 946–950.
Liebenow, A. M., Elliot, W. J., Laflen, J. M., and Kohl, K. D. (1990). Interrill erodibility: Collection and analysis of data from cropland soils. *Trans. ASAE 33*: 1882–1888.
Linden, D. R., Van Doren, D. M., and Allmaras, R. R. (1988). A model of the effect of tillage induced soil surface roughness on erosion. Int. Soil Tillage Res. Org., Proc. 11th Int. Conf. Vol. 1, pp. 373–378.
Luk, S. H. (1985). Effect of antecedent soil moisture content on rainwash erosion. *Catena 12*: 129–139.
Mannering, J. V., and Meyer, L. D. (1963). The effects of various rates of surface mulch on infiltration and erosion. *Soil Sci. Soc. Am. Proc. 27*: 84–86.
McIssac, G. F. (1992). Apparent geographic and atmospheric influences on raindrop sizes and rainfall kinetic energy. *J. Soil Water Conserv. 45*: 663–666.
Meyer, L. D. (1981). How rain intensity affects interrill erosion. *Trans. ASAE 24*: 1472–1475.
Meyer, L. D., and Harmon, W. C. (1985). Sediment losses from cropland furrows of different gradients. *Trans. ASAE 18*: 905–911.
Meyer, L. D., and Harmon, W. C. (1992). Interrill runoff and erosion: Effects of row sideslope shape, rain energy, and rain intensity. *Trans. ASAE 35*: 1199–1203.
Meyer, L. D., and Wischmeier, W. H. (1969). Mathematical simulation of the process of soil erosion by water. *Trans. ASAE 12*: 754–758.
Moldenhauer, W. C., and Koswara, J. (1968). Effect of initial clod size on characteristics of splash and wash erosion. *Soil Sci. Soc. Am. Proc. 32*: 875–879.
Moore, D. C., and Singer, M. J. (1990). Crust formation effects on soil erosion processes. *Soil Sci. Soc. Am. J. 54*: 1117–1123.

Moss, A. J. (1988). Effects of flow velocity variation on rain-driven transportation and the role of rain impact in the movement of solids. *Aust. J. Soil Res. 26*: 443–450.

Moss, A. J., and Green, P. (1983). Movement of solids in air and water by raindrop impact: Effect of drop size and water depth variations. *Aust. J. Soil Res. 21*: 257–269.

Moss, A. J., Green, P., and Hutka, J. (1982). Small channels: their experimental formation, nature, and significance. *Earth Surf. Process Landforms 7*: 401–415.

Moss, A. J., and Green, T. W. (1986). Erosive effects of the large water drops (gravity drops) that fall from plants. *Aust. J. Soil Res. 26*: 9–20.

Mualem, Y., and Assouline, S. (1986). Mathematical model of rain drop distribution and rainfall kinetic energy. *Trans. ASAE 29*: 494–500.

Mutchler, C. K. (1967). Parameters for describing raindrop splash. *J. Soil Water Conserv. 22*: 91–94.

Mutchler, C. K., and Young, R. A. (1975). Soil detachment by raindrops. In *Proc. Sediment Yield Workshop*, Oxford, MS. USDA Rep. ARS-S-40, pp. 113–117.

Nearing, M. A. (1991). A probabilistic model of soil detachment by shallow turbulent flow. *Trans. ASAE 34*: 81–85.

Nearing, M. A., and Bradford, J. M. (1985). Single waterdrop splash detachment and mechanical properties of soils. *Soil Sci. Soc. Am. J. 49*: 547–552.

Nearing, M. A., Bradford, J. M., and Holtz, R. D. (1986). Measurement of force vs. time relations for waterdrop impact. *Soil Sci. Soc. Am. J. 50*: 1532–1536.

Nearing, M. A., Bradford, J. M., and Parker, S. C. (1991). Soil detachment by shallow flow at low slopes. *Soil Sci. Soc. Am. J. 55*: 339–344.

Nearing, M. A., Foster, G. R., Lane, L. J., and Finkner, S. C. (1989). A process-based soil erosion model for USDA Water Erosion Prediction Project technology. *Trans. ASAE 32*: 1587–1593.

Nearing, M. A., Lane, L. J., Alberts, E. E., and Laflen, J. M. (1990). Prediction technology for soil erosion by water: Status and research needs. *Soil Sci. Soc. Am. J. 54*: 1702–1711.

Palmer, R. S. (1965). Waterdrop impact forces. *Trans. ASAE 8*: 69–70, 72.

Park, S. W., Mitchell, J. K., and Bubenzer, G. D. (1982). Splash erosion modeling: Physical analyses. *Trans. ASAE 25*: 357–361.

Park, S. W., Mitchell, J. K., and Bubenzer, G. D. (1983). Rainfall characteristics and their relation to splash erosion. *Trans. ASAE 26*: 795–804.

Podmore, T. H., and Merva, G. E. (1971). Silt transport by thin film flow. *Trans. ASAE 14*: 1065–1067, 1072.

Proffitt, A. P. B., and Rose, C. W. (1991). Soil erosion processes. I: The relative importance of rainfall detachment and runoff entrainment. *Aust. J. Soil Res. 29*: 671–683.

Proffitt, A. P. B., Rose, C. W., and Hairsine, P. B. (1991). Rainfall detachment and deposition: Experiments with low slope and significant water depths. *Soil Sci. Soc. Am. J. 55*: 325–332.

Riezebos, H. Th., and Epema, G. S. (1985). Drop shape and erosivity. II: Splash detachment, transport and erosivity indices. *Earth Surf. Processes Landforms 10*: 69–74.

Rose, C. W. (1985). Developments in soil erosion and deposition models. *Adv. Soil Sci. 2*: 1–63.

Rose, C. W., Williams, J. R., Sander, G. C., and Barry, D. A. (1983). A mathematical model of soil erosion and deposition processes. I: Theory for a plane land element. *Soil Sci. Soc. Am. J. 47*: 991–995.

Roth, C. M., and Helming, K. (1992). Dynamics of surface sealing, runoff formation and interrill soil loss as related to rainfall intensity, micro-relief and slope. *Z. Pflanzenernahr. Bodenk. 155*: 209–216.

Rowlison, D. L., and Martin, G. L. (1971). Rational model describing slope erosion. *J. ASCE (IRI) 97*: 39–50.

Schultz, J. P., Jarrett, A. R., and Hoover, J. R. (1985). Detachment and splash of a cohesive soil by rainfall. *Trans. ASAE 28*: 1878–1884.

Shainburg, I., and Letey, J. (1984). Response of soils to sodic and saline conditions. *Hilgardia 52*: 1–57.

Sharma, P. P., Gantzer, C. J., and Blake, G. R. (1981). Hydraulic gradients across simulated rain-formed soil surface seals. *Soil Sci. Soc. Am. J. 45*: 1031–1034.

Sharma, P. P., and Gupta, S. C. (1989). Sand detachment by single raindrops of varying kinetic energy and momentum. *Soil Sci. Soc. Am. J. 53*: 1005–1010.

Sharma, P. P., Gupta, S. C., and Foster, G. R. (1993). Predicting soil detachment by raindrops. *Soil Sci. Soc. Am. J. 57*: 674–680.

Sharma, P. P., Gupta, S. C., and Foster, G. R. (1995). Raindrop-induced soil detachment and sediment transport from interrill areas. *Soil Sci. Soc. Am. J. 59*: 727–734.

Sharma, P. P., Gupta, S. C., and Rawls, W. J. (1991). Soil detachment by single raindrops of varying knit energy. *Soil Sci. Soc. Am. J. 55*: 301–307.

Singer, M. J., and Blackard, J. (1982). Slope angle-interrill soil loss relationships for slopes up to 50 percent. *Soil Sci. Soc. Am. J. 46*: 1270–1273.

Singer, M. J., Janitzky, P., and Blackard, J. (1982). The influence of exchangeable sodium percentage on soil erodibility. *Soil Sci. Soc. Am. J. 46*: 117–121.

Stern, R., Ben-Hur, M., and Shainberg, I. (1991). Clay mineralogy effect on rain infiltration, seal formation and soil losses. *Soil Sci. 152*: 455–461.

Truman, C. C., and Bradford, J. M. (1990). Effect of antecedent soil moisture on splash detachment under simulated rainfall. *Soil Science 150*: 787–798.

Truman, C. C., Bradford, J. M., and Ferris, J. E. (1990). Influence of antecedent water content and rainfall energy on soil aggregate breakdown. *Soil Sci. Soc. Am. J. 54*: 1385.

Walker, P. H., Hutka, J., Moss, A. J., and Kinnell, P. I. A. (1977). Use of a versatile experiment system for soil erosion studies. *Soil Sci. Soc. Am. J. 41*: 610–612.

Walker, P. H., Kinnell, P. I. A., and Green, P. (1978). Transport of a noncohesive sandy mixture in rainfall and runoff experiments. *Soil Sci. Soc. Am. J. 42*: 793–801.

Wang, Y. (1988). *Effect of slope steepness, bulk density and surface roughness on interrill soil erosion.* M.S. thesis. University of Minnesota.

Watson, D. A., and Laflen, J. M. (1986). Soil strength, slope, and rainfall intensity effects on interrill erosion. *Trans. ASAE 29*: 98–102.

Wischmeier, W. H., and Smith, D. D. (1958). Rainfall energy and its relationship to erosion losses. *Trans. Am. Geophys. Union 39*: 285–91.

Wright, A. C. (1986). A physically-based model of the dispersion of splash droplets ejected from a water drop impact. *Earth Surf. Process. Landforms 11*: 351–367.

Wright, A. C. (1987). A model of the redistribution of disaggregated soil particles by rain-splash. *Earth Surf. Process. Landforms 12*: 583–593.
Young, R. A. (1984). A method of measuring aggregate stability under waterdrop impact. *Trans. ASAE 27*: 1351–1354.
Young, R. A., and Onstad, C. A. (1978). Characterization of rill and interrill eroded soil. *Trans. ASAE 21*: 1126–1130.

8

Rill and Gullies Erosion

Earl H. Grissinger
National Sedimentation Laboratory, Agriculture Research Service, U.S. Department of Agriculture, Oxford, Mississippi (Retired)

I. INTRODUCTION

The coupling of rill erosion with gully erosion in this discussion may at first glance appear to lack coherence. Although rill and interrill erosion involve distinctly different controlling variables, they have often been considered together due to their complementary contribution to the soil erodibility factor as used in equations such as the USLE or RUSLE. Additionally, both rill and interrill factors contribute to erosing measurements from standard size erosion plots. Gully and rill erosion, however, are both driven by hydraulic processes related to shear of the water flow at the soil-water interface. Rill to concentrated flow (ephemeral gully) to classical gully erosion represents a continuum and any classification of hydraulically related erosion into separate factors is, to some degree, subjective. Certainly, individual variables have differing significance in this continuum and potential changes in the hierarchy of controlling variables will be addressed in this review. Well-documented processes, relations, etc. will be summarized but not rehashed. Rather, attention will be directed toward areas poorly understood and/or areas needing additional clarification.

II. RILL EROSION

Rills are usually described as small, intermittent water courses that present no obstacles or impediments to tillage operations using conventional equipment.

Rills also carry the connotation that once obliterated, they will not inherently reform at precisely the same location. Rills may form naturally in response to small-scale topographic or vegetative features, or they may result from cropping conditions and practices. In some instances, rill characteristics are relatively uniform whereas in many instances, the rill shape and slope are nonuniform. Small-scale headcuts and knickpoints are common.

The advent of process oriented approaches to evaluation of erosional characteristics of soils [1–4] has fostered numerous studies concerned with rill erosion, particularly in relation to cropping conditions and practices. Research on rill erosion has documented the effects of discharge, furrow slope, crop canopy, vegetative residue and sundry other controls [5–12]. The relatively rapid increase in erosion rate with increasing furrow slope and/or length [12–14] will be explored in a subsequent discussion. Topographic analyses have been used to develop new slope length–steepness influences on rill erosion [15] and alternate topographic options have been included in the overland flow profile version of WEPP [16]. Topographic influences on rill formation have been described [17, 18] with respect to landscape characteristics. Prior to Geographic Information System (GIS) type topographic analyses, these descriptions have by necessity been primarily qualitative.

Hydraulic aspects of overland flow have also been the subject of detailed analyses. Obviously, this type of flow is highly complex. As stated by W. W. Emmett [19]:

> Overland flow is both unsteady and spatially varied since it is supplied by rain and depleted by infiltration, neither of which is necessarily constant with respect to time and location. The flow may be either laminar or turbulent or a combination of these two conditions. Flow depths may be either below or above critical, or the depths may change from subcritical to supercritical. Under certain conditions the flow may become unstable and may give rise to the formation of roll waves, or rain waves as they are often called.

Raindrop-induced perturbations further complicate the situation. Nevertheless, progress has been achieved in development of simulation models for rill erosion. Lane, Shirley, and Sing [20] summarized model evolution from empirical types through conceptual to physically based models, and presented an approximating kinematic wave equation for shallow overland flow on a plane. Details of furrow-flow hydraulics have also been documented by Foster [21] and Lu, Foster, and Smith [22]. Sediment transport relations have been presented by Young, Onstad, and Bosch [23] and Alonso, Neibling, and Foster [24]. In most models, particle detachment is expressed as a function of the excess of applied shear over the material critical shear stress [20, 25, 26], where the applied shear is the total shear adjusted by the ratio of the soil friction factor divided by the total rill friction factor. The total shear is calculated as the product of the flow weight density, the hydraulic radius, and the bed slope. Unit stream power and potential energy dissipation rate per unit rill area have also been applied to particle detachment and sediment discharge [27].

The model of Li, Simons, and Carder [25] also has a subroutine for calculating rill density as an area percentage. Detailed descriptions of process simulations are presented in the series of publications from the National Soil Erosion Research Laboratory [16, 28–31] documenting the Water Erosion Prediction Project (WEPP). The process overview by Foster et al. [26] is also recommended to readers interested in additional detail.

III. GULLY EROSION

Gullies have historically been described as oversized or entrenched stream channels. They may be present as extensions of the drainage net and in this situation described as continuous. Alternately, they may be discontinuous. Obviously, the continuity of the gully systems with the drainage net significantly influences sediment yield and hence is important in conservation planning to reduce offsite effects. Gullies are also usually classified as being either valley-floor or upland. The distinction between valley-floor gullies and stream channels is again subjective and for the purpose of this discussion, valley-floor gullies will include channels whose behavior is strongly modified by mass failure processes.

Within the last decade, the term *ephemeral gullies* has been coined to describe channels intermediate in size between rills and classical gullies. As with classical gullies, the ephemeral gullies may be continuous or discontinuous. They are larger than rills but are small enough to be obliterated by usual farming practices. Ephemeral gully locations are topographically defined, that is they will reform at the same locations year after year. For ease of discussion, ephemeral gullies, classical upland gullies and valley-floor gullies will be discussed separately.

A. Ephemeral Gullying

In the early 1980s USDA Soil Conservation Service (SCS) personnel became concerned that soil erosion assessments were underestimating total erosion by neglecting concentrated flow erosion larger than rill erosion but less than classical gully erosion. Measurements by the SCS in Alabama indicated that ephemeral gully erosion (per gully area) was 10 to 15 times the USLE estimated annual erosion from the contributing watershed area [32]. SCS studies in Aroostook County, Maine established that ephemeral gullying imparted 42% of a 1930 hectare study area but that sediment production was highly variable between fields [33]. About 55% of the gullies in this study area originated for implement tracks or from row breakover where cultivation was across slope. Relatively large variations in the ratio of ephemeral gully to sheet-rill erosion have been reported from several study locations. Laflen, Watson, and Franti [34] reported ratios ranging from 0.24 to 1.47 for study sites in Alabama and Iowa, respectively. The ratios ranged from 0.13 to 0.70 for three nearby sites in Mississippi for the 1986 crop year [35], and about 0.70 for a site in Georgia [36]. Stereo aerial photogrammetric methods were employed in this study. In one gaged Mississippi watershed, ephemeral gully development equaled about 60% of the annual soil loss [37].

Various topographic indices have been proposed as indicators for ephemeral gully development [38–40]. Indices related to drainage area, slope, platform and profile curvature, and measures of soil saturation have been applied primarily as surrogates for the magnitude of concentrated flow. A more process-oriented approach to ephemeral gullying has been presented by Foster [41]. This approach is generally comparable to that for rill erosion. The detachment capacity (calculated as a function of the excess of applied shear over the critical shear of the material) is multiplied by the ratio of the sediment deficit (the transport capacity minus the calculated sediment load) divided by the transport capacity. As stated by Foster [41] the "erosion and sedimentation of an ephemeral gully cannot be evaluated without considering the fields hydrology and its rill and interrill erosion." Additional factors discussed by Foster [41] include the partitioning of the shear stress between the soil boundary and surface roughness elements, nonerodible layers, nonflow detachment, headcut advance, and changes in the soil material critical shear strengths.

B. Upland Gullying

Upland gullies are perhaps the most visible eyesores related to erosion and this visibility inspired early studies in the United States to document the magnitude of the problem. In 1928 Bennett and Chapline [42] gave several examples of counties in which tens of thousands of hectares had been "permanently ruined by erosion." They stated that such areas had been "dissected by gullies, and bedrock is exposed in thousands of places." Bennett [43] in 1939 estimated that 20 million ha (50 million A) had been rendered "virtually useless" because of being "stripped of topsoil or riddled with gullies. . ." and that the agricultural usefulness of another 60 million ha (150 million A) had been impaired "far enough to make farming difficult or unprofitable." A more recent estimate (1976) of the upland gully problem in the United States stated that about 200,000 ha (500,000 A) were affected to such a degree as to require conservation treatment [44]. The severity of the current problem in Mexico has been addressed by Bocco and Garcia-Oliva [45].

Articles describing gully evolution and controlling factors are largely restricted to morphometric studies of gully development and evolution. One of the earliest of these reports is the classical study by Ireland, Sharpe, and Eargle [46] of gully development in the Piedmont of South Carolina. They identified four evolutionary stages, including (a) initial channel cutting, (b) incision into weaker subsurface materials, (c) grade adjustment, and (d) stabilization or natural healing. Of these four stages, stage (b) was the time of rapid and violent gully growth. Sidewall slumping was rapid as was headward migration of the overfall. Sediment production was also a maximum. Comparable results were reported by Bariss [47, 48] for gully development in the loess area of central Nebraska. For both studies, the stage of development was reflected by characteristic cross-sectional profiles.

Literature relevant to factors controlling or related to the rate of gully development is less abundant. Woodburn [49] reported a rate of gully growth of 5 cm (2 in) per year for areas of moderate relief. Relations between gully erosion and a rainfall intensity function [50] and drainage area [51] have been documented. Both of these factors have been incorporated into the equation used by the United States Department of Agriculture, Soil Conservation Service [52] to predict gully head advance. More complex relations have been presented by Thompson [53] and by Beer and Johnson [54]. Thompson [53] developed a complex power function involving drainage area, a rainfall index, slope of the approach channel and clay content of the eroding soil profile. This relation was developed from a study of 210 gullies in six locations of the eastern United States. Beer and Johnson [54] developed a somewhat more complex relation for gullies in western Iowa. Their relation included an exponential term for the deviation from normal of the annual precipitation coupled with complex power functions involving an index of surface runoff, the terraced area of the watershed, the gully length at the beginning of the study, and the distance from gully head to the watershed divide. These relations are empirical and inherently, have limited application [55–57]. A more detailed discussion of processes and controls will be discussed in the following section.

C. Valley-Floor Gullying

Literature pertinent to valley-floor gullies is relatively abundant. The evolutionary sequence for discontinuous valley-floor gullies has been reported by Blong [58, 59], Brice [60] and Tuckfield [61]. The location of initial development of discontinuous valley-floor gullies has been related to areas of reduced vegetative cover by Leopold and Miller [62] and Graf [63] and to areas of above average valley-floor slope by Schumm and Hadley [64] and Patton and Schumm [65]. Such areas have been termed "critical locations" by Heede [66, 67]. Critical locations may also result from "cutoff of the down valley toe of alluvial fill" or from piping conditions according to Hamilton [68] and Bocco and Garcia-

Oliva [45]. Brice [60] and Leopold and Miller [62] both described the coalescence of discontinuous gullies into a single continuous system with time. In general, this merging into a single system resulted from the upstream migration of the knickpoint from the more downstream set of gullies into the headward set. Parker [69] has reported that the rate of migration for such knickpoints is a function of stream order, the rate of migration decreases as stream order decreases. Additional details and examples may be found in the publication *Incised Channels* by Schumm, Harvey, and Watson [17].

Literature pertinent to gullying processes falls generally into two groups, that concerned with gully head or knickpoint migration, and that concerned with bank stability. Whereas knickpoints are simply reaches with excessive slopes that can be simulated using standard procedures, gully headcut migration is extremely complex, with hydraulic considerations generally comparable to those for drop structures. Stability of the drop varies with the ratio of approach flow depth to drop height and with the depth of submergence (that is, backwater). Certainly, stability is also dependent upon the properties of subsurface layers [70] and a cropout of excessively erodible material of the headcut toe will lead to catastrophic failures such as described by Ulmer [71]. A deterministic approach has been used in the development of procedures to calculate headcut migration in earthen emergency spillways [72–74]. In other studies, Begin, Meyer, and Schumm [75, 76] and Watson, Harvey, and Schumm [77] used a version of the heat flux equation wherein the diffusion coefficient was the migration rate, Kemp née Marchington [78] used as Musgrave-type approach, and de Ploey [79] proposed a linear relation between the headcut rate of recession and the total discharge. In a series of papers, Piest, Bradford, Grissinger and coworkers established that gully-wall (that is channel bank) instability is a two-stage process [80–87]. In the first stage, gravity induced mass failures deliver bank-derived slough materials to the active transport areas of the channel. In the second stage, the slough materials are subsequently entrained into the flow and incorporated into the sediment load. In this scenario, hydraulic features determine the rate of slough removal from the failure site and the rate of toe removal and, hence, bank oversteepening. Material controls include the ratio of cohesion to mass loading and the degree of tension crack development [88]. Obviously, these material strength controls are not consistent but vary with long-term climatic factors, suggesting that a conditional probability element should be considered in stability–instability relations for gullies. This element then is the probability that a storm event of a certain size will occur at the precise time that a bank material is at or near the worst-case condition. In the worst-case bank condition, materials are effectively saturated, with minimum cohesion, maximum loading, and full development of a tension crack. At this time, however, the climatic and site controls of tension crack development are not well understood.

IV. DISCUSSION

The relatively recent advancements in terrain analysis have opened new horizons in approaching erosion problems. Martz and Garbrecht [89] and Garbrecht, Martz, and Starks [90] have applied terrain analysis to drainage basin and watershed scale features and Zevenbergen [91] has used comparable analyses for field-sized areas. The potential coupling of detailed terrain analyses of select relatively small areas with the more general basin and watershed scale analyses offers great utility in theoretical studies of landscape features such as scale effects and erosion thresholds (see [92] and [93]). This coupling also facilitates the application of field-sized watershed gaged data to studies of off-site problems. The coupling of terrain analyses with erosion models further expands the capabilities for both erosion assessment and for site specific design of erosion control measures. ANSWERS [94, 95], AGNPS [96], and SWAT [97] have all been coupled with terrain analyses, usually in a GIS environment, and efforts are under way to produce a gridded version of WEPP. GRASS is the terrain analyses most commonly used. Certainly, other stimulation models can and will be coupled with terrain-type GIS inputs. For models such as described by de Ploey [98] "which stress the reality of variability and discontinuities, taking into account many site specific conditions, which sometimes completely override the theoretical relationships . . . ," distributive, terrain based models are mandatory.

As mentioned previously, erosion rates have been found to increase rapidly with increased slope and or length [12–14]. Meyer [13] noted the nonlinear relation between slope length and erosion rate for several mulch types and briefly discussed the nature of the error introduced by forging a linear or average relationship. His discussion emphasized the importance of high discharge events. Comparable results presented by Edwards and Owens [99] documented the effects of large storms on soil loss by watersheds. Obviously, simulation of large storm effects require realistic flow routing within the study area, proper distribution of controlling properties and conditions, and accurate functional expressions for the controlling properties. In the author's opinion, several aspects concerned with calculating the applied shear, the material critical shear stress, and the material mass stability need some degree of clarification.

Partitioning of the flow roughness is critical for accurate quantification of the applied shear, that is the shear acting on the soil surface [26, 29]. For rill erosion the applied shear as used in WEPP is calculated as the product of the total shear adjusted by the ration of the soil Darcy-Weisbach friction factor divided by the total Darcy-Weisbach rill friction factor [28]. For grass-lined open channels, Manning's roughness coefficient is used and the total shear is adjusted by the square of the roughness ratio to calculate the applied shear [100]. As observed by Abdel-Rahman [101], materials eroding at steady-state conditions exhibit a roughness characteristic of the material. For rill erosion, the characteristic roughness has been expressed as a function of clay contact [28], reflecting the influence of clay content on aggregate structure. This influence of clay content on aggregation, however, is restricted to surface soils where the soil materials have been thoroughly mixed by cultivation [102]. For ephemeral and larger gullies, consolidated materials typically comprise the channel boundary materials and these materials typically have surface roughness characteristics reflecting structural aspects of the material. These structural properties may or may not be related to clay content.

For shallow flows, such roughness frequently imposes turbulent conditions on the flow [103]. In the author's opinion, it would be extremely interesting to use a flow visualization technique similar to that developed by Allen [104] to qualitatively assess the influence of material roughness on bed-interface flow conditions. This type of qualitative assessment would have great utility as an aid to the efficient design of experiments to quantify both the influence of consolidated material structure on surface roughness and the effects of the eroded surface roughness on the near-surface flow properties. Obviously, this type of data is prerequisite to accurate hydrologic simulation in gridded field-scale models involving concentrated flow erosion. Both aspects, that is the roughness elements and their functional relation to the applied shear, would have to be distributed to the concentrated flow channels bounded by consolidated materials.

In addition to influencing surface roughness and hence the relation of applied shear to total shear, the structure of consolidated materials also affects material erodibility. In the author's opinion, this is one of the critical issues concerning consolidated material erodibility. Grissinger [105] previously reviewed consolidated cohesive material stability and discussed this subject in greater detail. Salient points of this review included the importancc of spatial and temporal variations in near-bed flow properties associated with material surface conditions [106], and the significance of size of the material structural elements in relation to the size of the test materials [107]. Another critical issue is the time of consolidation. Although it is convenient to classify fine-grained materials as either consolidated or as unconsolidated (loose, friable, etc.), consolidation of a given material can vary with time from a slight to a relatively dense condition. Reported temporal variations in erosion [26, 37, 41, 108, 109] reflect, at least partly, this tendency of fine-grained materials to consolidate with time. Certainly, consolidation under no-till management systems is axiomatic. The sensitivity of erosion resistance for loamy materials to bulk density changes has been well documented in laboratory studies [110] and field studies have shown significantly reduced erosion due to wetting-drying cycles between cultivations and erosive events [37, 109]. Consolidation associated with wetting and drying was interpreted as the controlling process.

This suggests that conditional probabilities should be developed to quantify the chance that an erosive event of a given magnitude might occur as the first storm since cultivation or seedbed preparation. Surface seals and crusts represent a special case in the consolidation continuum where the surface of the material is made sufficiently dense to impart properties that are distinctly different from those of the underlying materials. Again surface roughness and erodibilities are affected (personal communication M. J. M. Römkens, National Sedimentation Laboratory, Oxford, MS).

Mass stability of channel and gully banks is the last area identified for further study. In most analyses, stability varies directly with the ratio of cohesion to mass loading and average or worst-case ratios are used for stability predictions. Neither of the terms in the ratio, however, is constant under actual field conditions. Both vary through time for a given bank material. Relatively short-term variations result from water content changes, whereas longer-term variations result from vegetative development. Increases in cohesion due to root development have been documented primarily under forest conditions [111] but data on grass- and shrub-induced cohesion is sparse. Since failure frequence is dependent upon the recurrence interval of worst-case conditions, quantification of probability density functions for the cohesion to mass loading ratio is needed for design decisions. This type of information would also have utility in clarifying soil strength changes under no-till management. Additional information is also needed for quantifying the development of tension cracks.

V. SUMMARY

The coupling of Geographic Information System technology with runoff models has created a powerful tool for management of soil and water resources. With proper scaling, off-site effects can be addressed within the same spatial domains as employed for on-site studies. Moreover, the gridded structure of the Geographic Information System facilitates the spatial representation of controlling parameters and processes realistic to actual field conditions. In turn, this capability to distribute variables and processes rather than simply use average conditions, has focused attention on several areas that, in the author's opinion, need further study. The three areas most in need of additional research are

1. The significance of surface roughness of consolidated materials on near-interface flow conditions and the dependency of such roughness on material structure
2. The influence of structure of consolidated materials on erodibility
3. The functional definition of probability density functions for short-term variations in material strength due to water content changes and of long-term variations in cohesion associated with plant root development

REFERENCES

1. W. G. Knisel (ed.), *CREAMS: A Field Scale Mode for Chemicals, Runoff, and Erosion from Agricultural Management Systems.* Conservation Research Report No. 26, Superintendent of Documents, Washington, DC, p. 643 (1980).
2. D. B. Beasley, L. F. Huggins, and E. J. Monke, *Trans. ASAE*, 23(4):938–944 (1980).
3. G. R. Foster, L. J. Lane, J. D. Nowlin, J. M. Laflen, and R. A. Young, *Trans ASAE*, 24(5):1253–1262 (1981).
4. G. R. Foster, Modeling the erosion process. In *Hydrologic Modeling of Small Watersheds* (C. T. Haan et al., eds.), ASAE Monograph No. 5, St. Joseph, Michigan, pp. 297–380 (1982).
5. G. R. Foster, W. R. Osterkamp, L. J. Lane, and D. H. Hunt, ASAE Paper #82-2572, p. 23 (1982).
6. L. C. Brown, G. R. Foster, and D. B. Beasley, *Trans. ASAE*, 32(6):1967–1978 (1989).
7. M. A. Nearing, L. T. West, and L. C. Brown, *Trans. ASAE*, 31(3):696–700 (1988).
8. J. E. Gilley, S. C. Finkner, R. G. Spomer, and L. N. Mielke, *Trans. ASAE*, 29(1): 161–164 (1986).
9. M. H. Hussein and J. M. Laflen, *Trans. ASAE*, 25(5):1310–1315 (1982).
10. M. W. Van Liew and K. E. Saxton, *Trans. ASAE*, 26(6):1738–1743 (1983).
11. R. B. Bryan (ed.), *Catens Supplement*, No. 8, p. 160 (1987).
12. L. D. Meyer and W. C. Harmon, *Trans. ASAE*, 28(2):448–453, 461 (1985).
13. L. D. Meyer, Erosion processes and sediment properties for agricultural cropland. In *Hillslope Processes* (A. D. Abrahams, ed.), Allen and Unwin, Boston, pp. 55–76 (1986).
14. L. D. Meyer, G. R. Foster, and M. J. M. Romkens, Source of soil eroded by water from upland slopes, Proceedings of Sediment-Yield Workshop, National Sedimentation Lab., Oxford, Mississippi, pp. 177–189 (1975).
15. D. K. McCool, G. O. George, M. Freckleton, C. L. Douglas, Jr., and R. I. Papendick, *Trans. ASAE*, 36(4):1067–1071 (1993).
16. *User Requirements, USDA-Water Erosion Prediction Project (WEPP)*, NSERL Report No. 1, USDA-ARS National Soil Erosion Research Laboratory, West Lafayette, Indiana, p. 43 (1987).
17. S. A. Schumm, M. D. Harvey, and C. C. Watson, *Incised Channels*, Water Resources Publications, Littleton, Colorado, p. 200 (1984).
18. B. E. Frazier, D. K. McCool, and C. F. Engle, *J. Soil and Water Conservation*, 38(2):70–74 (1983).
19. W. W. Emmett, Overland flow. In *Hillslope Hydrology* (M. J. Kirkby, ed.), John Wiley and Sons, New York, pp. 145–176 (1978).
20. L. J. Lane, E. D. Shirley, and V. P. Singh, Modeling erosion on hillslopes. In *Modeling Geomorphological Systems* (M. G. Anderson, ed.), John Wiley and Sons, New York, pp. 287–307 (1988).
21. G. R. Foster, *Hydraulics of Flow in a Rill*, Ph.D. thesis, Purdue University, Lafayette, Indiana, p. 222 (1975).
22. J. Y. Lu, G. R. Foster, and R. E. Smith, *Trans. ASAE*, 30(4):969–976 (1987).

23. R. A. Young, C. A. Onstad, and D. D. Bosch, Sediment transport capacity in rills and small channels, Proceedings 4th Federal Interagency Sedimentation Conference, Las Vegas, Nevada, pp. 6-25–6-33 (1986).
24. C. V. Alonso, W. H. Neibling, and G. R. Foster, *Trans. ASAE*, 24(5):1211–1220 and 1226 (1981).
25. R-M. Li, D. B. Simons, and D. R. Carder, Mathematical modeling of soil erosion by overland flow, Proceedings of National Conference on Soil Erosion, Purdue University, Lafayette, Indiana, pp. 210–216 (1977).
26. G. R. Foster, R. A. Young, M. J. M. Römkens, and C. A. Onstad, Processes of soil erosion by water. In *Soil Erosion and Crop Productivity* (R. F. Follett and B. A. Stewart, eds.), American Society of Agronomy, Inc., Madison, Wisconsin, pp. 137–162 (1985).
27. G. F. McIsaac, J. K. Mitchell, J. W. Hummel, and W. J. Elliot, *Trans. ASAE*, 35(2): 535–544 (1992).
28. *USDA-Water Erosion Prediction Project: Hillslope Profile Version*, NSERL Report No. 2, USDA-ARS National Soil Erosion Research Laboratory, West Lafayette, Indiana, p. 263 (1989).
29. *WEPP Second Edition*, NSERL Report No. 4, USDA-ARS National Soil Erosion Research Laboratory, West Lafayette, Indiana, p. 34 (1990).
30. *WEPP Version 91.5*, NSERL Report No. 6, USDA-ARS National Soil Erosion Research Laboratory, West Lafayette, Indiana, p. 40 (1991).
31. *WEPP Hillslope Profile Erosion Model, Version 91.5 User Summary*, NSERL Report No. 7, USDA-ARS National Soil Erosion Research Laboratory, West Lafayette, Indiana, p. 28 (1991).
32. H. A. Miller, Estimating erosion on gullied land, Presented at the USDA, SCS State Conservation Engineers Meeting, Greenville, South Carolina, p. 8 (1982).
33. USDA, SCS, Study of erosion from ephemeral gullies on cropland—1983, Report to St. John-Aroostook RC&D Steering Committee, Central Aroostock Soil and Water Conservation District by USDA Soil Conservation Service, Orono, Maine, p. 22 (1984).
34. J. M. Laflen, D. A. Watson, and T. G. Franti, Ephemeral gully erosion, Proceedings of the 4th Federal Interagency Sedimentation Conference, Las Vegas, Nevada, pp. 3-29–3-37 (1986).
35. P. Forsythe, J. S. Parkman, and R. Ulmer, Evaluation of concentrated flow erosion estimation procedures, Mississippi, 1986, Report to the SNTC, SCS, Fort Worth, Texas, p. 29 (1986).
36. A. W. Thomas and R. Welch, *Trans. ASAE*, 31(6):1723–1728 (1988).
37. E. H. Grissinger and J. B. Murphey, Ephemeral gully erosion in the loess uplands, Goodwin Creek watershed, northern Mississippi, USA, Proceedings of the 4th International Symposium on River Sedimentation, Beijing, pp. 51–58 (1989).
38. C. R. Thorne, Prediction of soil loss due to ephemeral gullies in arable fields, Report to the USDA, ARS, National Sedimentation Lab., Oxford, Mississippi, p. 79 (1984).
39. R. D. Lentz, R. H. Dowdy, and R. H. Rust, *J. Soil Water Conservation*, 48(4): 354–361 (1993).

40. I. D. Moore, G. J. Burch, and D. H. Mackenzie, *Trans. ASAE*, 31(4):1098–1107 (1988).
41. G. R. Foster, Understanding ephemeral gully erosion. In *Soil Conservation*, Volume 2, National Academy of Science Press, Washington, DC, pp. 90–125 (1986).
42. H. H. Bennett and W. R. Chapline, *Soil Erosion a National Menace*, USDA Circular 33, U. S. Government Printing Office, Washington, DC, p. 35 (1928).
43. H. H. Bennett, *Soil Conservation*, McGraw-Hill, New York, p. 993 (1939).
44. U. S. Department of Agriculture, Agricultural Research Service, *ARS National Research Program*, U. S. Government Printing Office, Washington, DC, p. 58 (1976).
45. G. Bocco and F. Garcia-Oliva, *J. Soil Water Conservation*, 47(5):365–367 (1992).
46. H. A. Ireland, C. F. Sharpe, and D. H. Eargle, *Principles of Gully Erosion in the Piedmont of South Carolina*, U. S. Department of Agriculture Technical Bulletin No. 633, U. S. Government Printing Office, Washington, DC, p. 142 (1939).
47. N. Bariss, *Rocky Mountain Society Science J.*, 8:47–57 (1971).
48. N. Bariss, *Great Plains-Rocky Mountain Geography J.*, 6:125–132 (1977).
49. R. Woodburn, *Soil Conservation* XV(1):11–22 (1949).
50. L. L. McDowell, G. C. Bolton, and M. F. Ryan, Sediment production from a Lafayette County, Mississippi gully, Proceedings of the Second Mississippi Water Resources Conference, Water Resources Research Institute, Mississippi State University, Mississippi State, pp. 87–102 (1967).
51. I. Seginer, *J. Hydrology*, 4:236–253 (1966).
52. U. S. Department of Agriculture, Soil Conservation Service, *Procedures for Determining Rates of Land Damage, Land Depreciation and Volume of Sediment Produced by Gully Erosion*, SCS Technical Release No. 32, Soil Conservation Service, Washington, DC, p. 18 (1966).
53. J. R. Thompson, *ASAE Paper No. 62-1713*, p. 7 (1962).
54. C. E. Beer, and H. P. Johnson, Factors related to gully growth in the deep loess area of western Iowa, Proceedings of the 1963 Interagency Sedimentation Conference, USDA Miscellaneous Publication 970, U. S. Government Printing Office, Washington, DC, pp. 37–43 (1963).
55. R. F. Piest and A. J. Bowie, Gully and streamback erosion, Proceedings of the 29th Annual Meeting of the Soil Conservation Society of America, Ankeny, Iowa, pp. 188–196 (1974).
56. R. F. Piest and E. H. Grissinger, Gully erosion. In *Volume III. Supporting Documents. CREAMS—A Field Scale Model for Chemical Runoff and Erosion from Agricultural Management Systems*, USDA Conservation Research Report No. 26, Chapter 9, pp. 455–462 (1980).
57. E. H. Grissinger, J. B. Murphey, and N. L. Coleman, Planned gully research at the USDA Sedimentation Laboratory, Proceedings of the Natural Resources Modeling Symposium, Pingree Park, Colorado, pp. 475–478 (1985).
58. R. J. Blong, *New Zealand J. Hydrology*, 5:87–99 (1966).
59. R. J. Blong, *American J. Science*, 268:369–384 (1970).
60. J. C. Brice, *Erosion and Deposition in the Loess Mantled Great Plains, Medicine Creek Drainage Basin, Nebraska*, U. S. Geological Survey Professional Paper 352-H, U. S. Government Printing Office, Washington, DC, pp. 255–339 (1966).

61. C. G. Tuckfield, *American J. Science*, 262:795–807 (1964).
62. L. B. Leopold and J. P. Miller, *Ephemeral Streams, Hydraulic Factors, and Their Relation to the Drainage Net*, U. S. Geological Survey Professional Paper 282-A, U. S. Government Printing Office, Washington, DC, p. 37 (1956).
63. W. L. Graf, *Earth Surface Processes*, 4:1–14 (1979).
64. S. A. Schumm and R. F. Hadley, *American J. Science* 225:161–174 (1957).
65. P. C. Patton and S. A. Schumm, *Geology*, 3:88–90 (1975).
66. B. H. Heede, *International Association of Scientific Hydrology Bulletin*, 12:42–50 (1967).
67. B. H. Heede, *Zeitschr. Geomorph.*, 18:260–271 (1974).
68. T. M. Hamilton, *Channel-Scarp Formation in Western North Dakota*, U. S. Geological Survey Professional Paper 700-C, U. S. Government Printing Office, Washington, DC, pp. C229–C232 (1970).
69. R. S. Parker, *Experimental Study of Basin Evolution and its Hydrologic Implications*, Ph.D. thesis, Colorado State University, Fort Collins, Colorado, p. 331 (1977).
70. J. Poesen and G. Govers, Gully erosion in the loam belt of Belgium: Topology and control measures. In *Soil Erosion on Agricultural Land* (J. Boardman, I. D. L. Foster, and J. A. Dearing, eds.), John Wiley and Sons, New York, pp. 513–530 (1990).
71. R. L. Ulmer, ASAE Paper No. 85-2624, p. 17 (1985).
72. K. M. Robinson, ASAE Paper No. 91-2066, p. 20 (1991).
73. K. M. Robinson and G. J. Hanson, ASAE Paper No. 92-2638, p. 16 (1992).
74. D. M. Temple, J. A. Brevard, J. S. Moore, G. J. Hanson, E. H. Grissinger, and J. M. Bradford, Analysis of vegetated earth spillways, Proceedings of the 10th Annual Conference of the Association of State Dam Safety Officials, Kansas City, Missouri, pp. 225–230 (1993).
75. Z. B. Begin, D. F. Meyer, and S. A. Schumm, *Trans. ASAE*, 23(5):1183–1188 (1980).
76. Z. B. Begin, D. F. Meyer, and S. A. Schumm, *J. ASCE*, 106(WW3):369–388 (1980).
77. C. C. Watson, M. D. Harvey, S. A. Schumm, and D. I. Gregory, *Geomorphic Study of Oaklimeter Creek, Burney Branch, and Muddy Creek in Benton, Lafayette, and Tippah Counties, Mississippi*, Report to the USDA Soil Conservation Service, Jackson Mississippi, p. 372 (1986).
78. A. C. Kemp ńee Marchington, Towards a dynamic model of gully growth, Proceedings of the Jerusalem Workshop, Jerusalem, pp. 121–134 (1990).
79. J. de Ploey, *Catena Supplement*, 14:81–86 (1989).
80. J. M. Bradford, D. A. Farrell, and W. E. Larson, *Soil Science Society of America Proceedings*, 37(1):103–107 (1973).
81. J. M. Bradford, R. F. Piest, and R. G. Spomer, *Soil Science Society of America Proceedings*, 42(2):323–328 (1978).
82. J. M. Bradford and R. F. Piest, *Soil Science Society of America Proceedings*, 41(1): 115–122 (1977).
83. R. F. Piest, C. E. Beer, and R. G. Spomer, Entrenchment of drainage systems in western Iowa and northwestern Missouri, Proceedings of the 3rd Federal Interagency Sedimentation Conference, Denver, Colorado, pp. 48–60 (1976).

84. R. F. Piest, J. M. Bradford, and G. M. Wyatt, *J. ASCE*, 101(HY1):65–80 (1975).
85. R. F. Piest and R. G. Spomer, *Trans. ASAE*, 11(6):850–853 (1968).
86. E. H. Grissinger and J. B. Murphey, Bank stability of Goodwin Creek Channel, northern Mississippi, USA, Proceedings of the 4th International Symposium on River Sedimentation, Beijing, pp. 59–66 (1989).
87. E. H. Grissinger, A. J. Bowie, and J. B. Murphey, Goodwin Creek bank instability and sediment yield, Proceedings of the 5th Federal Interagency Sedimentation Conference, Las Vegas, Nevada, pp. PS-32–PS-39 (1991).
88. E. H. Grissinger and W. C. Little, Similarity of bank problems on dissimilar streams, Proceedings of the 4th Federal Interagency Sedimentation Conference, Las Vegas, Nevada, pp. 5-51–5-60 (1986).
89. L. W. Martz and J. Garbrecht, *Computers and Geoscience*, 18(6):747–761 (1992).
90. J. Garbrecht, L. W. Martz, and P. Starks, Automated watershed parameterization from digital landscapes: capabilities and limitations, to be included in the Proceedings of Hydrology Days, Fort Collins, Colorado, p. 13 (1994).
91. L. W. Zevenbergen, *Modelling Erosion Using Terrain Analysis*, Ph.D. thesis, Department of Geography and Earth Science, Queen Mary College, University of London, London, p. 345 (1989).
92. D. R. Montgomery and W. E. Dietrich, *Science*, 255:826–830 (1992).
93. W. E. Dietrich, C. J. Wilson, D. R. Montgomery, J. McKean, and R. Bauer, *Geology*, 20:675–679 (1992).
94. A. P. J. de Roo, *Modelling Surface Runoff and Soil Erosion in Catchments Using Geographic Information Systems*, Ph.D. thesis, NGS 157, Nederlands Geografische Studies, University of Utrecht, Utrecht, The Netherlands, p. 295 (1993).
95. C. C. Rewerts and B. A. Engel, ASAE Paper No. 91-2621, p. 8 (1991).
96. R. Srinivasan and B. A. Engel, ASAE Paper No. 91-7574, p. 8 (1991).
97. R. Srinivasan and J. G. Arnold, Basin scale water quality modeling using GIS. In *Application of Advanced Information Technologies for Management of Natural Resources*, American Society of Agricultural Engineers, St. Joseph, Michigan, p. 12 (1993).
98. J. de Ploey, *Catena*, 17(2):175–183 (1990).
99. W. M. Edwards and L. B. Owens, *J. Soil Water Conservation*, 46(1):75–78 (1991).
100. *Stability Design of Grass-Lined Open Channels*, Agricultural Handbook No. 667, U. S. Department of Agriculture, Agricultural Research Service, Superintendent of Documents, Washington, DC, p. 167 (1987).
101. N. M. Abdel-Rahman, *The Effect of Flowing Water on Cohesive Beds*, Ph.D. thesis, Technical University of Zurich, Zurich, p. 114 (undated).
102. W. D. Kemper, *Aggregate Stability of Soils from Western United States and Canada*, Technical Bulletin 1355, U. S. Department of Agriculture, Agricultural Research Service, Superintendent of Documents, Washington, DC, p. 52 (1966).
103. C.-C. Wu, *Transport of Noncohesive Sediment by Shallow Flow*, Ph.D. thesis, School of Engineering, University of Mississippi, University, Mississippi, p. 225 (1991).
104. J. R. L. Allen, *Current Ripples*, North-Holland, Amsterdam, p. 433 (1968).
105. E. H. Grissinger, Bank erosion of cohesive materials. In *Gravel-Bed Rivers* (R. D. Hey, J. C. Bathurst, and C. R. Throne, eds.), John Wiley and Sons, New York, pp. 273–287 (1982).

106. I. F. Karasev, *Soviet Hydrology*, 6:551–579 (1964).
107. E. H. Grissinger, W. C. Little, and J. B. Murphey, *Trans. ASAE*, 24(3):624–630 (1981).
108. L. D. Meyer and W. C. Harmon, *Trans. ASAE*, 35(2):459–464 (1992).
109. S. M. Dabney, C. E. Murphree, and L. D. Meyer, *Trans. ASAE*, 36(1):87–94 (1993).
110. G. Govers, W. Everaert, J. Poesen, G. Rauws, J. de Ploey, and J. P. Lautridou, *Earth Surfaces Processes and Landforms*, 15:313–328 (1990).
111. N. J. Coppin and I. G. Richards, *Use of Vegetation in Civil Engineering*, Butterworths, London, p. 292 (1990).

9

Soil Loss Estimation

Kenneth G. Renard and Leonard J. Lane
Southwest Watershed Research Center, Agricultural Research Service, U.S. Department of Agriculture, Tucson, Arizona

George R. Foster
National Sedimentation Laboratory, Agricultural Research Service, U.S. Department of Agriculture, Oxford, Mississippi

John M. Laflen
National Soil Erosion Research Laboratory, Agricultural Research Service, U.S. Department of Agriculture, West Lafayette, Indiana

I. HISTORY OF EROSION PREDICTION

Zingg (1940) is often credited with the development of the first erosion-prediction equation used to evaluate erosion problems and select conservation practices to reduce excessive erosion. Zingg's equation was a simple expression that related soil erosion to slope steepness and slope length. Smith and Whitt (1948) added terms to Zingg's equation to reflect the influence of cover and management on soil erosion.

Relative differences among conservation practices do not consider important differences among locations caused by differences in rainfall erosivity or soil. Thus, rainfall-erosivity and soil-erodibility terms were added to the Zingg and the Smith and Whitt equations (Musgrave, 1947; Wischmeier and Smith, 1958; Meyer, 1984). Concurrent with the development of these erosion-prediction equations was the development of a soil loss tolerance concept (Stamey and Smith, 1964; ASA, 1982). These terms along with the soil loss tolerance concept allowed users to consider differences among site characteristics and to consider the severity of erosion relative to a measure of how much erosion a soil could "tolerate" before experiencing excessive damage. By the early 1950s a set of regional equations had been developed that used soil-erodibility terms reflective of major soils in each region. Even though these equations proved to be quite useful, the United States Department of Agriculture (USDA), Soil Conservation Service (SCS), needed a more "universal" soil loss equation than these regionally based equations. Beginning in the mid-1950s W. H. Wischmeier, D. D. Smith, and their associates began to assemble and analyze an extensive quantity of available plot data. The result was the universal soil loss equation (USLE) (Wischmeier and Smith, 1965), which became by far the most widely used equation for estimating interrill and rill erosion.

Development of the USLE continued after 1965, resulting in a major revision of the equation in 1978 (Wischmeier and Smith, 1978). Many of the modifications between 1965 and 1978 used data collected from rainfall simulators (Meyer, 1960). In the 1960s much of the field-erosion research shifted from natural runoff plots to rainfall simulator plots.

The basis of a soil-erodibility nomograph and the cover-management factor values in the 1978 USLE version were derived from rainfall simulators. Another important USLE concept introduced in 1970 was the subfactor method for estimating cover-management factor values (Wischmeier, 1975). This method was originally introduced for computing factor values for range, woodland, and similar land uses where plot data were not available, but where agencies needed to apply the USLE. This method has since been extended to all land uses (Laflen, Foster, and Onstad, 1985) and is central to the revised universal soil loss equation (RUSLE) (Renard et al., 1991).

The USLE and RUSLE are empirically based technologies that compute soil erosion by assigning values to indices that represent the major factors of climate, soil, topography, and land use. An alternative approach based on fundamental hydrologic and erosion processes is emerging in a form that can be easily used to estimate soil loss by sheet and rill erosion and erosion by concentrated flow in field-sized areas. This technology, known as the USDA Water Erosion Prediction Project (WEPP), is intended as 20th-century erosion-prediction technology (Lane and Nearing, 1989). It is based on concepts and relations developed by Ellison (1947). In the late 1960s, Meyer and Wischmeier (1969) utilized Ellison's concepts using computer programming and showed the potential of this approach, especially for dealing with the spatial variation of erosion and deposition along a complex landscape profile. In the early 1970s, the concept of rill-interrill erosion was developed to provide a powerful model structure for representing and connecting the major erosion processes of detachment by raindrop impact, detachment by surface flow, sediment transport by flow, and deposition by flow (Foster and Meyer, 1975).

Concern for the impact of agricultural practices on surface-water quality in the 1970s led to the development of several models that included process-based erosion components. For example, CREAMS (chemicals, runoff, and erosion from agricultural management systems) (Knisel, 1980), which was a combination of process-based and empirically based components, became widely used for field-sized areas. Several new concepts were introduced in CREAMS, including the use of hydrologic elements to represent flow patterns on the landscape, sediment as a mixture of primary particles and aggregates, and the effect of a nonerodible layer on erosion by concentrated flow. This model is implemented in a computer program that can be run on desktop computers.

II. THE REVISED UNIVERSAL SOIL LOSS EQUATION

RUSLE is a major revision of the USLE. While retaining the equation structure of the USLE, several concepts from process-based erosion modeling have been incorporated in RUSLE to improve erosion predictions. These concepts provide a basis for estimating factor values for slope length, slope steepness, and supporting practice effects. RUSLE has been developed and distributed in the form of a computer program that readily runs on desktop computers.

The effort to upgrade the USLE was precipitated by recognition that the knowledge acquired after the 1978 USLE release needed to be incorporated to computerize erosion prediction. Thus, the RUSLE effort was initiated in 1985, and the effort to develop the RUSLE model in a computer program was initiated in 1987. Although RUSLE retains the basic six-factor product form of the USLE, the equations used to arrive at the factor values are significantly modified. Furthermore, the decision to computerize the technology permits calculations which address prototype conditions not possible with the USLE. For example, crops for which soil loss ratios were not available in the USLE can now be simulated based on fundamental crop measurements.

Like the USLE, RUSLE retains a regression relation to estimate soil loss. The conceptual equation is

$$A = R \cdot K \cdot L \cdot S \cdot C \cdot P \tag{1}$$

where

- A = computed average spatial and temporal soil loss per unit of area, expressed in units selected for K and for period selected for R (in practice, A is usually expressed in t ac^{-1} yr^{-1}, but other units can be selected (i.e., mt ha^{-1} yr^{-1})
- R = rainfall and runoff erosivity factor—the number of rainfall erosion index units plus a factor for runoff from snowmelt where such runoff is significant
- K = soil-erodibility factor—the soil loss rate per erosion index unit for a specified soil as measured on a unit plot, defined as a 72.6-ft (22.1-m) length of uniform 9% slope in continuous clean-tilled fallow
- L = slope-length factor—the ratio of soil loss from the field slope length to that for a 72.6-ft length (22.1-m) under the unit plot conditions as above

S = slope-steepness factor—the ratio of soil less from the field slope gradient to that from a 9% slope under unit-plot conditions

C = cover and management factor—the ratio of soil loss from an area with specified cover and management to that from a unit plot in tilled continuous fallow

P = supporting practice factor—the ratio of soil loss with a support practice like contouring, stripcropping, or terracing to that with straight-row farming up and down the slope

The schematic diagram of the RUSLE computer model is shown in Fig. 1. The RUSLE computer program is designed for inputs and outputs in English units. Foster et al. (1981) list English and SI units and conversions for USLE/RUSLE.

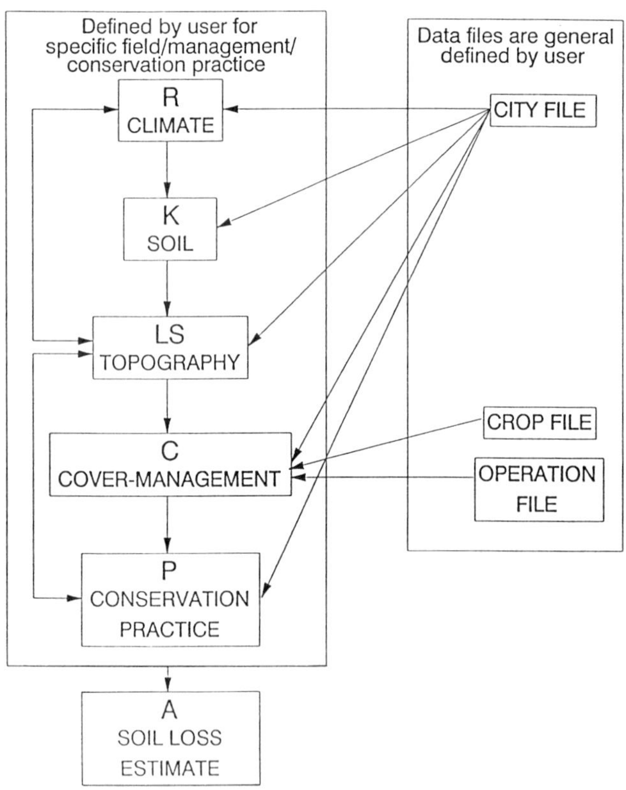

Fig. 1 Schematic or RUSLE soil loss calculations.

RUSLE (Fig. 1) uses three databases to simplify soil loss calculations: CITY, CROP, and OPERATION. The CITY DATABASE contains monthly average temperature and precipitation data used for predicting residue decomposition, R-factor, 10-yr-frequency maximum daily EI used to calculate the contouring subfactor in the support-practice factor, frost-free period used in the time-varying K-factor, and the twice-monthly distribution of the rainfall-runoff erosivity used to weight K- and C-values.

The CROP DATABASE contains information used in calculating the twice-monthly soil loss ratios for the cover-management (C) factor. Essential data include the root mass in the upper 4-in of the soil, fraction of land surface covered by canopy, and the canopy raindrop fall height at 15-day intervals following planting. The file also contains the mass of residue required to cover 30%, 60%, and 90% of the surface and a yield unit, residue/yield ratio, and the rate at which the residue decomposes.

The OPERATION DATABASE describes the impact of soil-disturbing practices and requires information on the percentage of the area disturbed, the amount of residue left following an operation, and the depth of soil disturbance. The database also initiates crop growth or death, specifies residue removal or additions to the field, and crop harvest.

A. Climate: Erosivity Factor (*R*)

The earlier procedure used to extrapolate limited calculations of R in the western United States using the two-yr-frequency, 6-h-duration precipitation values produced by the National Weather Service (NWS) was useful but not entirely satisfactory. In RUSLE, over 1000 NWS rain gauges with hourly precipitation amounts were used to calculate point values of R using the equation

$$R = \sum_{i=1}^{m} EI_{30} \tag{2}$$

where m is the number of storm events in a given year and EI_{30} is the product of kinetic energy times the maximum 30-min intensity of individual storm periods (Wischmeier, 1959).

Records for the western U.S. data varied from 5 years to over 20 years. A linear correction was used to adjust the hourly recorded amounts to those which might be obtained if a more conventional short-interval hyetograph were used. The new western U.S. map produced point estimates that vary over a much greater range than those in *Agricultural Handbook 537* (Wischmeier and Smith, 1978).

In the Pacific Northwest (PNW), where much of the erosion occurs from rainfall and melting snow on partially frozen soil, an equivalent R (R_{eq}) was obtained from

$$R_{eq} = \frac{A}{KLSCP} \tag{3}$$

where A, K, L, S, C, P were based on field measurements.

These R_{eq} values in turn were regressed against annual precipitation to produce isoerodent maps for use in the small-grain-growing areas of the PNW.

Minor changes were also made in the isoerodent map of the eastern United States, but, more significantly, a correction factor was developed to reduce R-values where low slopes occur in regions of long, intense thunderstorms. This correction factor is designed to account for the reduction in raindrop-impact erosivity due to ponded water on the soil. Moss and Green (1983) found that ponded water depths of 2 to 3 drop diameters considerably decreased detachment and transport of soil particles by raindrop impact.

To facilitate the soil loss calculations, a CITY DATABASE file is developed for each of 119 "climatic homogeneous" EI distribution areas of the United States. These data files include station identification codes plus monthly and annual precipitation, monthly average temperature, number of frost-free days, annual R- or R_{eq}-values, distribution of 15-day-period EI_{30}, and 10-yr-frequency annual maximum storm EI_{30}.

One of the problems of developing R-factor values for the CITY DATABASES is the paucity of data used to calculate time-intensity relationships and, in turn, kinetic energy and maximum 30-min intensity. Other authors have attempted to calculate R by correlating annual precipitation and monthly precipitation with known R-factors. Renard and Freimund* recently used U.S. CITY DATABASE information to present the following two regression relations (in SI units):

$$R = 0.0048P^{1.61} \tag{4}$$

and

$$R = 0.074F^{1.85} \tag{5}$$

where

R = rainfall and runoff factor (10^{-2} N h^1 yr^{-1}) [N is newton force]
P = annual precipitation (mm)
F = Fournier (1960) index (mm) = $\sum_{i=1}^{12} P_i^2 \Big/ P$
P_i = monthly precipitation (mm)

---

*Renard, K.G., and J.R. Freimund 1993 (pending). Estimating RUSLE R-factor and 10-yr-storm EI value for locations with monthly precipitation data.

Although both relations had high coefficients of determination ($r^2 = 0.81$), the standard error of estimate was 107 and 108, respectively. At this time, no recommendations can be made regarding the geographic areas for which these relations might best be applied.

The RUSLE CITY DATABASE also requires an estimate of the 10-yr-frequency maximum storm kinetic energy times maximum 30-min intensity. Renard and Freimund likewise developed a regression equation as follows:

$$(EI_{30})_{10} = 2.98R^{0.70} \tag{6}$$

where

$(EI_{30})_{10}$ = 10-yr-frequency maximum annual storm kinetic energy (E) times maximum storm intensity for 30-min (EI_{30}) (10^{-2} N h^{-1})
R = average annual rainfall-runoff value (10^{-2} N h^{-1} yr^{-1})

This equation has a coefficient of determination (r^2) of 0.90 and a standard error of estimate of 30 (10^{-2} N h^{-1}).

B. Soil-Erodibility Factor (*K*)

In addition to the soil-erodibility nomograph, RUSLE includes equations for estimating *K*-values where the nomograph does not apply (e.g., volcanic soils and soils with high organic matter). Erodibility data from around the world have been reviewed and an equation developed that gives a useful estimate of *K* as a function of an "average" soil particle diameter. This function is only recommended where the nomograph does not apply. An equation is also provided for use with volcanic soils such as occur in Hawaii.

Another change incorporated in RUSLE accounts for rock fragments on and in the soil profile. Rock fragments on the soil surface (i.e., erosion pavement) are treated like mulch in the *C*-factor, while the *K*-value (in the nomograph) is adjusted to reflect the effects of rock on permeability in the soil profile and, in turn, runoff. The rock fragments in the soil profile are assumed to reduce permeability and thereby increase runoff and soil erodibility.

Experimental data have shown that *K* varies with season, being highest in spring immediately following freeze-thaw actions. The lowest values occur in mid-fall and in winter. The seasonal variability is addressed by RUSLE with instantaneous estimates of *K* weighted in proportion to the twice-monthly *EI* estimates from the CITY DATABASE files. Instantaneous *K* estimates are obtained with equations relating *K* to the frost-free period and to the annual *R*-factor.

C. Topographic Factor (*LS*)

Users ask more questions and express more concern about selecting a slope length than nearly any other term in RUSLE and USLE. This involves judgment, and different users choose different slope lengths for similar field conditions. Although the RUSLE handbook provides guidelines which should give consistency among users, the concern is not warranted because soil loss predicted by RUSLE is less sensitive to slope length than slope steepness or other RUSLE factors. For example, a 10% error in slope-length determination results in about a 5% error in computed soil loss, whereas a 10% error in slope steepness results in a predicted soil loss difference of more than 10%.

RUSLE uses three slope-length relations that are functions of the soil's susceptibility to rill erosion relative to interrill erosion, and a separate slope-length relation is used for the small-grain-farming areas in the Pacific Northwest. Guides in the computer program and the RUSLE handbook help the user select the appropriate relationship for the particular field condition encountered.

RUSLE has a more nearly linear slope-steepness relationship than the USLE. Computed soil loss for sloper less than about 20% are similar in RUSLE and the USLE. However, on steep slopes, computed soil loss is significantly less with RUSLE. Experimental data and field observations do not support the USLE quadratic relationship for steep slopes. RUSLE provides a slope-steepness relationship for short slopes subject primarily to interrill erosion (such as might be experienced on bedded fields). RUSLE also incorporates a slope-steepness relationship developed for the small-grain-farming areas in the Pacific Northwest, where partially frozen soil and rain on snow lead to excessive rill erosion.

Of great significance in RUSLE is the ease with which a slope segment previously estimated as a single plane or uniform slope can represent the actual topography. A simple representation can often lead to gross errors in the topographic factor (*LS*). For example, a three-segment slope consisting of a 100-ft length at 6%, a 150-ft length at 10%, and a 50-ft length at 6% would be represented as a single 300-ft 8% slope segment in the USLE rather than as the three segments with RUSLE. Predicted average soil loss by RUSLE for these slope segments is 13% greater than the USLE for an Indiana cornfield and 12% less than by USLE for a southeastern Arizona rangeland. Thus, the differences cannot be readily generalized, but they can be quite large.

D. Cover Management Factor (C)

The soil loss ratios (SLR) used to calculate the *C*-factor are perhaps the most important terms in RUSLE because they represent conditions that can be managed most easily to reduce erosion. Furthermore, values of *C* can vary from near zero for a very well protected soil to about 1.5 for a finely tilled, ridge surface that results in much runoff and leaves the soil susceptible to rill erosion. The changes in RUSLE *C*-factor calculations are very significant over those of the USLE.

Values for *C* are average SLRs that represent the predicted soil loss for a given surface condition at a given time to that for a unit plot. SLRs vary during the year as soil, plant conditions, cover, and random roughness change. RUSLE computes C-values by weighing the 15-day SLRs according to the distribution of EI.

In RUSLE, a subfactor method is used to compute SLRs as a function of five subfactors given by the equation

$$SLR = PLU \cdot CC \cdot SC \cdot SR \cdot SM \tag{7}$$

where

PLU = prior-land-use subfactor
CC = canopy-cover subfactor
SC = surface-cover subfactor
SR = surface-roughness subfactor
SM = soil-moisture subfactor used in small-grain-farming areas in Pacific Northwest.

1. *Prior Land Use*

The prior-land-use subfactor (PLU) expresses (1) the influence on soil erosion of subsurface residual effects from previous crops and (2) the effect of previous tillage practices on soil consolidation. The relationship is of the form:

$$PLU = C_f \exp(-cB_u) \tag{8}$$

where

C_f = surface-soil consolidation factor
B_u = mass of live and dead roots and buried residue found in the upper 4 in of soil (lb ac^{-1})
c = coefficient representing effectiveness of buried roots and residue in controlling erosion (ac lb^{-1})

The variable C_f expresses the effect of tillage-induced surface-density changes on soil erosion. Tillage operations tend to break soil aggregate bonds, increasing the potential for erosion. This is reflected in the lower erosion rates associated with the undisturbed soils of rangeland or no-till systems. Based on the work of Dissmeyer and Foster (1981), the value of C_f for freshly tilled conditions is 1.0. If the soil is left undisturbed, this value decays exponentially to 0.45 after 7 yr. The impact of a field operation on this factor is determined by the portion of the surface disturbed. For example, if a planting operation disturbs only 30% of the surface which had already consolidated to the point where C_f = 0.6, then 70% of the field would have a value of C_f = 0.6, and the disturbed 30% would have a value of C_f = 1.0; the overall value would be [(70%)(0.6) + (30%)(1.0)]/100% = 0.72.

Incorporated residue and roots in the upper 4 in (100 mm) of the soil profile reduce soil erosion significantly, as given by the B_u term in Eq. (8). This residue not only directly reduces erosion, but it also indirectly lowers soil loss by providing energy for microorganisms that produce organic compounds that bond soil particles. Estimates of root mass at various times during the growing season for different agronomic crops are given in the new handbook and obtained from the CROP DATABASE in the computer program.

For many rangeland conditions, values of root mass are not available. Weltz et al. (1987) developed Eq. (9) for estimating root biomass (B_u) on rangelands:

$$B_u = B_a n_i \, u_i \tag{9}$$

where

B_a = annual aboveground biomass (lb ac^{-1})
n_i = ratio of root mass in the upper 4 in of soil to the total below-ground root biomass
u_i = ratio of root mass to the aboveground biomass

Suggested values of n_i and u_i for many plant communities in the western United States rangelands are found in the RUSLE manual and the computer program. Estimates of B_a can be made using standard biomass estimating techniques such as clipping, drying, and weighing, or using guides as the USDA's SCS range-site descriptions. On areas that have been grazed within the past six months, B_a should be estimated from published site-potential estimates as given by the SCS or the U.S. Department of the Interior, Bureau of Land Management (BLM). Practices like burning and/or mechanical treatments may remove B_a but leave B_b, below-ground biomass. The user must consider these effects when estimating B_b.

On croplands the amount of both above- and below-ground biomass present at a given time depends on the initial mass of the residue, the root mass, the fraction of crop residue buried by field operations, and the decomposition rate of residue and roots, values of which are presented in the pending RUSLE *Agriculture Handbook* or in the computer program or both.

For RUSLE, residue decomposition is estimated using a relation based on temperature, soil moisture, and plant characteristics (Stott et al., 1990; Stott, 1991).

Continuous pasture, meadow, and rangeland are assumed to be at a stable mature state. Because these lands are not usually disturbed by tillage tools, the blow-ground biomass, B_u, becomes the live and dead roots. Even though crop and residue values for these practices change slowly with time, they are considered constant in RUSLE. Therefore, residue decomposition is not used for these conditions.

2. *Canopy Cover*

The canopy-cover subfactor (CC) expresses the effect of vegetative canopy in reducing the rainfall energy impacting the soil surface. Although most rainfall intercepted by crop canopy eventually reaches the soil surface, it usually does so with much less energy than nonintercepted rainfall. Intercepted rain reforms in drops with less energy or travel down crop stems to the ground. The CC-factor is expressed as

$$CC = 1 - FC \exp(-0.1H) \tag{10}$$

where

CC = canopy-cover subfactor
FC = fraction of land surface covered by canopy
H = distance raindrops fall after striking canopy (ft)

Suggested values of FC and H are given for numerous crops in the handbook and in the CROP DATABASE of the computer program.

3. *Surface Cover*

The effect of surface ground cover (SC) on erosion has been observed to vary greatly in research studies. In some studies a 50% cover reduced soil loss by about 65%. In other studies a 50% cover reduced soil loss by 95%. To accommodate this varied effectiveness in RUSLE, the following equation for SC is used:

$$SC = \exp\left[-bS_p\left(\frac{0.24}{R_c}\right)^{0.08}\right] \tag{11}$$

where

b = coefficient
S_p = percent of land with surface cover
R_c = current surface roughness

The b coefficient is assigned a value of 0.025, the value in the present USLE; 0.035, the new "typical" value in the RUSLE, or 0.050 for small-grain conditions in the Pacific Northwest. The value of b is increased as the tendency for rill erosion to dominate interrill erosion for the soil increases. SC is the most sensitive of the subfactors and must be carefully treated to obtain reasonable SLRs.

The amount of residue cover can be estimated from residue weight by Gregory's relation (1982):

$$S_p = [1 - \exp(-\alpha B_s)] \cdot 100 \tag{12}$$

where

S_p = percent residue cover
α = ratio of area covered by a piece of residue to the mass of that residue (ac lb^{-1})
B_s = weight of crop residue on the surface (lb ac^{-1})

Typical values for α are given in the RUSLE handbook. If more than one type of residue is present, the resulting total surface cover is calculated by modifying Eq. (12) as

$$S_p = \left[1 - \exp\left(-\sum_{i=1}^{N} \alpha_i B_{si}\right)\right] \cdot 100 \tag{13}$$

where N is the number of residue types and α_i is the ratio of the area covered to the mass of that residue for each type encountered.

Within RUSLE, rather than entering a value for α, the computer program requires residue weights associated with specific values of residue cover and calculates the corresponding α-value. The program asks for residue weights at 30%, 60%, and 90% surface cover. Only one of these needs to be entered to calculated an α-value. If more than one weight is entered, the program will calculate an α-value for each and then average them.

4. *Surface Roughness*

Surface roughness (SR) has been shown to affect soil erosion directly (Cogo, Moldenhauer, and Foster, 1984) and indirectly through the impact on residue effectiveness implied in Eq. (11). In either case this is a function of the soil surface's random roughness, which is defined as the standard deviation of the surface elevations when changes due to land slope or nonrandom tillage marks (dead furrows, traffic marks, disk marks, etc.) are removed from consideration (Allmaras et al., 1966). A rough surface has many microdepressions and flow barriers. During a rainfall event, these trap water and sediment, causing rough surfaces to erode at lower rates than smooth surfaces under similar conditions. Increasing the surface roughness also decreases the transport capacity and runoff detachment by reducing the flow velocity.

Roughness and clodiness of soils also affect the degree and rate of soil sealing by raindrop impact. Rough, cloddy soils typically have high infiltration rates. Finely pulverized soils are usually smooth, seal rapidly, and have low infiltration rates.

Random-roughness (RB) values vary, depending on the type and degree of surface disturbance. Roughness conditions for a field may vary, depending on previous tillage, implement speed, and other field conditions. Additional information is provided in the *Agriculture Handbook* for RUSLE.

The impact of surface roughness on erosion is defined by a baseline condition, which sets SR equal to 1 for unit-plot conditions of clean cultivation smoothed by extended exposure to rainfall of moderate intensity. These conditions yield a random roughness of about 0.24 in. This makes it possible to obtain SR values greater than 1 for practices in which the soil is very finely pulverized and smoothed to a smaller random roughness, as might be the case for some rototilling operations or for repeated cultivations of silt loam soils under dry, fallow conditions.

If a field operation normally leaves a random roughness greater than 0.24 in, the amount of biomass within the top 4 in of the soil has a significant impact on the actual roughness. This effect is defined by the relation

$$R_a = 0.24 + (R_t - 0.24)\{0.8[1 - \exp(-0.0003B_u)] + 0.2\} \tag{14}$$

where

R_a = roughness after biomass adjustment (in)
R_t = original tillage roughness based on assumption of ample subsurface biomass (in)
B_u = total subsurface biomass in top 4 in of soil (lb ac^{-1})

For field operations that do not disturb the entire soil surface, the roughness following the operation should reflect both the roughness caused by the operation and that already existing in the rest of the field. This combination is handled through a simple weighting procedure, where

$$R_n = R_aF_d + R_uF_u \tag{15}$$

where

R_n = net roughness following the field operation (in)
R_u = roughness of the surface before disturbance and, therefore, also the roughness of the undisturbed portion of the surface (in)
F_d,F_u = fractions of surface disturbed and undisturbed, respectively, so their sum equals 1

Surface roughness has been shown to decay exponentially with the amount of rainfall since the last tillage (Onstad et al., 1984). The change is computed with the roughness decay coefficient (D_r), which decreases exponentially from a value of 1.0 with zero rainfall to asymptotically approach a value of 0.0 for high rainfall amounts. The decay follows the equation

$$D_r = \exp(-0.14P_t) \tag{16}$$

where P_t is the total rainfall since the last operation that disturbed the entire surface (inches).

We now use this roughness decay relation to determine the roughness of the undisturbed portion of the field (R_u), based on the accumulated rainfall since the previous field operation and the net roughness following that operation (R_{np}), as

$$R_u = 0.24 + \frac{R_{np} - 0.24}{D_r} \tag{17}$$

Putting this value of R_u into Eq. (15) gives a value for the net roughness following the field operation, and this value holds until the next operation.

If a tillage does not disturb the entire field, the precipitation since the last complete operation must be adjusted accordingly. This is done by first adjusting the roughness decay coefficient to reflect an overall average, using

$$D_r = D_u F_u + 1.0 F_d \tag{18}$$

where D_u is the decay coefficient for the field before the operation and thus also the decay coefficient for the undisturbed portion. Once this is calculated, we can determine the corresponding value of P_t as

$$P_t = \frac{-\ln(D_r)}{0.14} \tag{19}$$

With the passage of each time segment in the calculations, the total rainfall since tillage is incremented by the amount of rainfall in that segment, so the roughness decay coefficient (D_r) is recalculated and the current roughness (R_c) is recalculated as

$$R_c = 0.24 + D_r(R_n - 0.24) \tag{20}$$

This current roughness value is then used in calculating the surface roughness subfactor for each time segment from the equation

$$SR = \exp[-0.026(R_c - 0.24)] \tag{21}$$

If a field operation results in a random roughness of less than 0.24 in, the impacts of both subsurface biomass and rainfall smoothing are assumed to be negligible, and the surface roughness subfactor SR is defined as

$$SR = 1.17 \exp(-0.026 R_t) \tag{22}$$

where R_t is the random roughness produced by the tillage operation. In this case the value of R_c used in Eq. (11) is 0.24. Consolidation because of rainfall decreases surface roughness over time, which is reflected in Eq. (11) through the R_c term.

5. *Soil Moisture*

Antecedent soil moisture has a substantial influence on infiltration and runoff and, hence, on soil erosion. In general, antecedent-moisture effects are inherent components of continuous tilled fallow plots, and these effects are reflected in soil-erodibility variation throughout the year. In most of the continental United States, soil moisture is usually high during vulnerable crop stages in the spring and early summer, which is when much of the erosion occurs. Hence, the antecedent soil moisture on cropped plots parallels that on the continuous tilled fallow plots from which soil-erodibility factors are derived. The soil-moisture subfactor is not used for rangelands.

In the nonirrigated portions of the Pacific Northwest, such as eastern Oregon, eastern Washington, and Idaho, soil moisture during critical crop periods is dependent on crop rotation and management. Winter wheat may be seeded after a previous crop of winter wheat, a more-shallow-rooted crop, or summer fallow. When a full year of fallow is used in the rotation, part of the moisture stored during the previous winter is retained in the profile. This is particularly true when an effective mulch system is used, such as either a loose soil and residue mulch in conjunction with a rodweeder, or direct stubble seeding into an untilled residue mulch. This is in contrast to continuous cropping, where soil moisture is at or below the wilting point in the fall prior to the fall and winter precipitation. Addition of a soil-moisture factor (*SM*) is suggested for this region of the Pacific Northwest. The factor reflects these dry fall conditions and the soil-moisture accumulation during the winter. Its subsequent decrease through the summer depends on the crop rooting depth and soil depth, and its replenishment depends on the precipitation amount and soil depth. The *SM* subfactor is then accessed in the computer calculations.

6. *Subfactor Summary*

One reason for the subfactor approach in the RUSLE is to accommodate applications where SLR values are not available. For example, no experimental erosion data exist for many vegetable and fruit crops, such as asparagus and blueberries. Developing SLR values using the subfactor method in the RUSLE is easier and more accurate than making comparisons with values in *Agricultural Handbook 537*.

RUSLE has computer subroutines for many tillage operations and crops. In other instances, the user must input new data reflecting residue incorporated by a tillage operation and the surface-roughness residual following tillage. For crops not represented in the computer program, data are needed to reflect canopy characteristics and root mass in the upper 4 in of the soil profile. Thus, the user must specify the crops in a rotation; crop yield; and the date of operations, such as tillage and harvest. The computer calculates SLRs and the average annual C-factor.

Grazing effects on rangeland, pasture, and meadow are reflected by canopy height, ground cover, and root biomass. Finally, ground cover as used in the

USLE expressed vegetation and litter; in RUSLE, ground cover is given as 1.0 minus the amount of bare soil that reflects the addition of litter in the form of rock and stone besides the conventional vegetative litter.

E. Support Practice Factor (*P*)

Of the RUSLE/USLE factors, values for the support practice (*P*) factor are most uncertain. The *P*-factor mainly represents the effect of surface conditions on flow paths and flow hydraulics. For example, with contouring, tillage marks direct runoff around the slope at much reduced grades. However, slight changes in grade can change runoff erosivity greatly. In experimental field studies, small changes in such features as row grade and their effect on erosion are difficult to document, leading to appreciable scatter in measured data. For example, the contouring effectiveness in field studies conducted on a given slope have ranged from no reduction in soil loss to a 90% reduction. Likewise, identifying these subtle characteristics in the field is difficult when applying RUSLE. Thus, *P*-factor values represent broad, general effects of such practices as contouring and stripcropping.

In RUSLE, extensive data have been analyzed to reevaluate the effect of contouring. Furthermore, simulation studies have been conducted using the CREAMS model (Knisel, 1980). The results have been interpreted to give factor values for contouring as a function of ridge height, furrow grade, and climatic erosivity. New *P*-factor values for the effect of terracing account for grade along the terrace, and a larger array of stripcropping conditions are considered in RUSLE. Finally, *P*-factors have been developed to reflect conservation practices on rangeland. The practices require estimates of surface roughness and runoff reduction as with stripcropping.

F. Applications

The development of the RUSLE computer program permits application to situations not possible with the USLE technology. At the same time, the uncertainty of the simulated result can be increased because of the empirical basis of the equations used and inadequate data with which to verify the results.

Closure of data gaps for estimating *R*-factors, the time-varying *K*-factor, the new algorithms for the topographic factor, and the new technology developed for estimating support practices greatly enhance RUSLE and permit its application to modern farming practices used throughout the United States. The technology also shows promise for use in developing countries.

Of great significance is that *C*-factor values can be estimated with RUSLE for crops where SLRs are not available. Data are not available in tables of *Agriculture Handbook 537* to address many specialty crops and operations. Given that a user can obtain data for developing a CROP DATABASE to cover the specific conditions for a defined climate, SLRs with which to calculate a C-factor can be made for any crop. Furthermore, new tillage implements can be added to the OPERATIONS DATABASE to cover an extensive range of activities with which to simulate their effect on soil loss.

The RUSLE computer program is used to select factor values based on site-specific conditions and computes soil loss. Most of the input values for RUSLE are readily available in database files supplied with the program. Values in these files can be modified and values can be added as necessary.

Rainfall erosivity values for locations where a particular user might apply RUSLE are in a database that can be customized to that particular use. The CITY DATABASE on the computer disk provides values for many cities. Data for additional cities can be added, and unnecessary data can be deleted to accelerate operation of the RUSLE computer program.

Soil-erodibility values are selected from soil survey information available from local offices of the USDA SCS. The particular site is located on a soil survey map, and the erodibility of the soil mapping unit at the site is identified and entered into the RUSLE program.

Slope-length and -steepness values are determined during an on-site visit or from other available topographic information. These values are entered directly in the RUSLE program, and the program computes values for L and S.

The factors C and P are most important in RUSLE for conservation planning because they represent the land-use changes available to the land user to reduce erosion. To compute values of C, the user selects from the crops in the CROP DATABASE when operating the RUSLE program. The RUSLE program requires values that describe plant characteristics such as canopy. These values are stored in the CROP DATABASE.

In addition to information concerning crops, the RUSLE program uses information on tillage, harvest, and other operations that affect soil and cover conditions, such as roughness and residue cover. The OPERATIONS DATABASE contains information on these operations. Some of the information in this file includes depth of tillage, amount of residue incorporated, and roughness left by each operation. Operations can be added, and existing values can be modified to customize the file. The user selects appropriate operations from this file and enters the dates when the operations occur to represent a cover-management system.

The RUSLE predicts interrill and rill erosion from rainfall and the associated runoff. RUSLE is a tool useful in conservation planning, inventory, and assessment. Soil loss values computed by RUSLE should be used as a guide rather than being considered absolutely accurate erosion rates.

RUSLE is intended for use in field offices of land management agencies such as the Soil Conservation Service and Forest Service of the USDA and the Bureau of Land Management of the U.S. Department of Interior. As such, the technology must be user friendly and contain databases for the wide variety of conditions that occur in their applications.

RUSLE computes average annual interrill and rill erosion for a landscape profile. The soil loss value computed for that profile is representative of an area to the degree that the profile represents the area. It does not compute average interrill and rill erosion for a field unless soil loss is computed for several profiles and the results weighted according to the fraction of the field that each profile represents. RUSLE does not compute sediment yield.

III. WATER EROSION PREDICTION PROJECT

The technology in USLE and RUSLE greatly limits their utility for evaluating many natural resource problems associated with man's activities on the land. Empirical limitations and an inability to deal with deposition preclude using RUSLE/USLE technology on larger areas where sediment delivery estimates are needed. Soil erosion from practices such as contouring and ridge tillage is difficult to estimate (Foster, 1991). Practices that drastically change the hydrology are also difficult to address with the RUSLE/USLE technology. The WEPP is an effort to develop a technology for erosion prediction that extends into the next century (Foster and Lane, 1987).

A. Overview of WEPP

WEPP is a daily simulation model that predicts erosion and sediment delivery at different scales. Three versions applicable for different scales are being developed:

1. *Profile*: Computes sediment detachment and transport on the land and sediment delivery to a channel and is common to all versions. The model computes on a daily basis the surface, soil, and crop conditions important to the hydrology and erosion processes. For the soil, these conditions include bulk density, moisture status, and buried residue. For the crop, these conditions include canopy cover, canopy height, and above- and below-ground biomass accumulation. For the surface, these conditions include surface roughness and crop residue mass and cover (Fig. 2).
2. *Watershed*: Takes the sediment delivery computed by the profile version and routes it through the channel system to the exit from the watershed. The model output includes erosion and deposition in the channel system. A watershed will include one or more areas where the profile version is operated (Fig. 3).
3. *Grid*: Computes the sediment delivery from an area that has been divided into small or regular grid elements. Within each of these elements, the profile

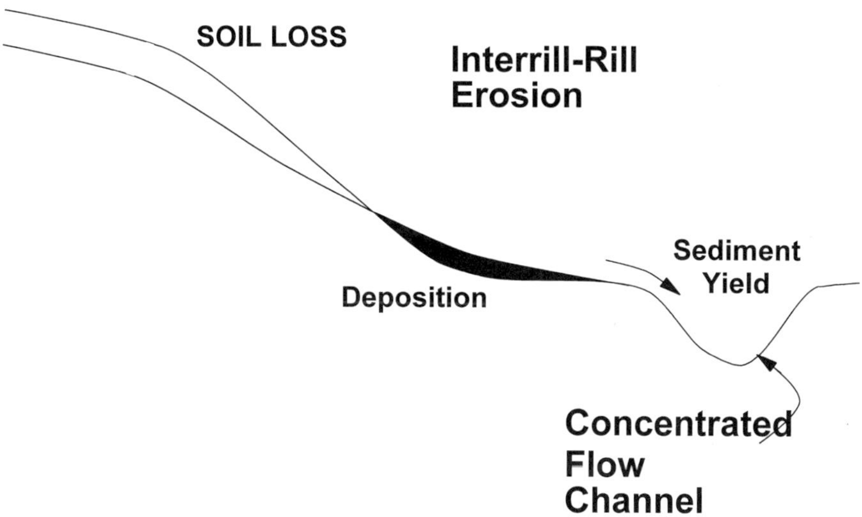

Fig. 2 Separation of a hillslope into overland flow elements (OFEs).

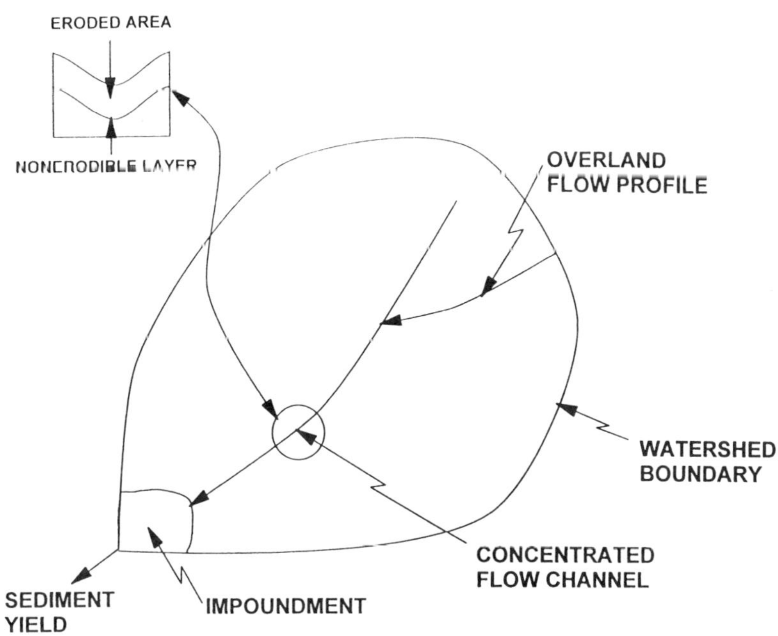

Fig. 3 Representation of watershed-version area of application.

version operates and, with the grid version, represents the transport, erosion, and deposition in the channel system within the area of interest (Fig. 4).

WEPP requires data input files that include soil topography, cropping, and management data. It also requires an input daily climate file. The Climate Generation program (CLIGEN) is part of the WEPP package. CLIGEN (Nicks and Lane, 1989) simulates daily precipitation, temperature, wind, and radiation, and has provision to stochastically disaggregate precipitation into an intensity distribution within a day.

WEPP computes new values for the soil characteristics, soil surface, and crop conditions on a daily basis. If no rainfall occurs, it then proceeds to the next day's conditions. If rainfall does occur, WEPP determines whether runoff (overland flow) occurs based on infiltration rates and rainfall distribution. If it does occur, it computes volumes and rates of runoff and the time over which it occurs and uses them to estimate soil detachment and sediment delivery to the channel system.

One of the advantages of the profile version is to represent a hillslope as a combination of homogeneous portions (Fig. 4) termed overland flow elements (OFE). Each OFE is assumed homogeneous with regard to soils and/or land use and is treated separately with regard to the status of the soil surface, soil characteristics, and crop. Within each OFE, sediment detachment is estimated for 100 points along the length of the OFE. Flow of soil and water from one OFE to another is also estimated.

B. Erosion Processes in WEPP

In the WEPP profile version, erosion consists of rill and interrill processes. Interrill erosion is the detachment and transport of soil particles by raindrops and very shallow flows, while rill erosion is the detachment and transport of sediment by flowing water. Calculations in WEPP are in SI units.

The sediment delivery to rills from interrill areas is estimated from the equation

$$D_i = K_i I_e^2 G_e C_e \, S_f \tag{23}$$

where

D_i = delivery of sediment from interrill areas to a nearby rill (kg m^{-2} s^{-1})
K_i = interrill erodibility (kg m^4 s^{-1})
I_e = effective rainfall intensity (m s^{-1})
G_e = ground-cover adjustment factor
C_e = canopy-cover adjustment factor
S_f = slope adjustment factor given by

$$S_f = 1.05 - .85 \exp(-4 \sin a) \tag{24}$$

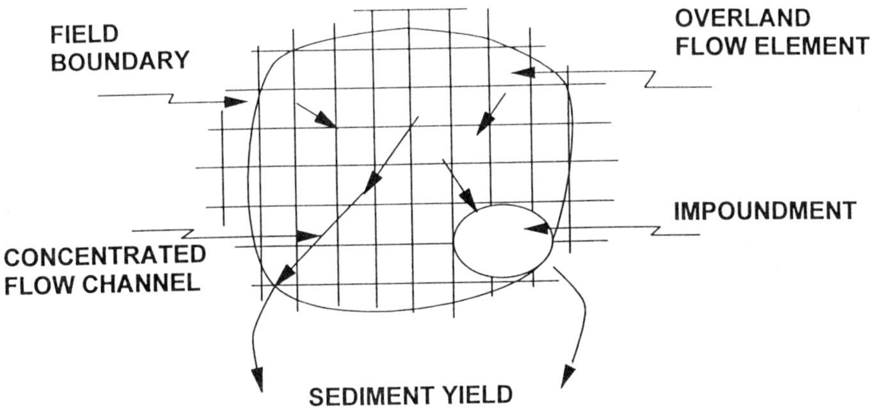

Fig. 4 Representation of grid-version area of application.

where a is the slope of the surface toward a nearby rill. The relationships expressed in Eqs. (23) and (24) are reasonable fits to data reported by Meyer (1981), Meyer and Harmon (1984, 1989), and Watson and Laflen (1986). Equation (23) lumps the processes of sediment detachment, transport, and deposition on interrill areas.

Rill erosion is the detachment and transport of soil particles by concentrated flowing water. Factors affecting rill erosion rates include hydraulic shear, soil resistance to hydraulic shear, sediment load in runoff water, and the sediment transport capacity of runoff. In WEPP, the detachment capacity (D_c) (kg s^{-1} m^{-2}) by flowing water is expressed as

$$D_c = K_r(\tau - \tau_c) \tag{25}$$

where

K_r = rill erodibility (s^{-1} m^{-1})
τ = hydraulic shear of flowing water (Pa)
τ_c = critical hydraulic shear that must be exceeded before rill detachment can occur (Pa)

Hydraulic shear is the force exerted on the channel bed and bank material by flowing water. The detachment capacity is the maximum rill detachment rate, which occurs when there is no sediment in the water. The rill detachment rate is less than detachment capacity when there is sediment in the runoff water. The detachment rate (D_r) of flowing water is

$$D_r = D_c\left(1 - \frac{G}{T_c}\right) \tag{26}$$

where G is the sediment load per unit width (kg s^{-1} m^{-1}) and T_c is the sediment transport capacity per unit width (kg s^{-1} m^{-1}). Sediment transport capacity is computed as

$$T_c = k_t\tau^{1.5} \tag{27}$$

where k_t is a transport coefficient (m$^{0.5}$ s^{-2} kg$^{-0.5}$). The Yalin (1963) equation is used to obtain the transport coefficient (Finker et al., 1989).

When the sediment transport capacity is exceeded by the sediment load, deposition is predicted as

$$D_r = \beta \frac{V_{\text{eff}}}{q}(T_c - G) \tag{28}$$

where

D_r = deposition rate
β = turbulence factor
V_{eff} = effective particle fall velocity (m s^{-1})
q = discharge per unit width (m^2 s^{-1})

C. Hydrologic Components in WEPP

WEPP must adequately forecast the hydrologic cycle if erosion and sediment delivery are to be accurately predicted. WEPP uses several climate variables, including storm rainfall volume and duration, ratio of peak rainfall intensity to average rainfall intensity, time that peak intensity occurs, daily maximum and minimum temperatures, wind velocity and direction, and solar radiation. These variables are required in components related to plant growth and residue decomposition, water balance, and in estimating volume, duration, and peak rate of runoff.

The Green and Ampt equation is used in WEPP to compute infiltration. Rainfall excess is computed as the difference between rainfall intensity and infiltration rate during the rainfall event. Overland flow routing procedures include analytic solutions to the kinematic wave equations. Equivalent planes are used to represent areas where flow depth is finite at the upper boundary of an area (stripcropping, for example). These techniques yield runoff rates necessary to compute hydraulic shears for sediment detachment and transport.

The winter component of the model computes soil frost, snowmelt, and snow accumulation. When frost is present, frost and thaw depth, infiltration capacity, and water balance are estimated. If snow is present, snowmelt, infiltration, and surface runoff are computed. These variables are used in the water-balance and deep-percolation components.

Input for the daily water balance is essential to estimate infiltration and runoff. The water-balance and percolation component uses the climate, plant growth, and infiltration components to compute the status of the soil water at each soil layer of interest, and percolation from the root zone. Daily potential evapotranspiration and soil evaporative and plant transpiration are computed in this component. This component also receives the estimates of infiltration of melted snow from the winter component.

D. Plant Growth and Residue Processes

Quantity and quality of plants and crop residue are vital to the accurate estimation of soil detachment and transport. The status of below- and aboveground biomass must be accurately estimated to evaluate the effect of different managements on soil erosion.

The EPIC plant-growth subroutines (Williams et al., 1984), simplified to include only moisture stress, are used to compute daily plant status. The decomposition and accumulation of residue and litter for both cropland and rangeland are computed on a daily basis. Residue decomposition routines in WEPP are based on the decomposition model of Stroo et al. (1989). Effects of grazing and tillage on residue are incorporated into WEPP.

Important plant-growth characteristics include canopy cover and height, mass of live and dead biomass below and above ground, leaf-area index and basal area, and residue cover. Information about dates and operations are input to the model. Parameters have been determined for many annual and perennial crops, management systems and operations that may occur on cropland, rangeland, forestlands, pastures, vineyards, and gardens.

E. Hydraulic Processes

The hydraulic component of WEPP estimates hydraulic shear for estimation of rill erosion. The hydraulic component uses information about surface runoff volumes, hydraulic roughness, and runoff duration and peak rate. Major differences in hydrology among OFEs create problems in dealing with the hydraulic variables when a hillslope contains several OFEs. It is possible that during a runoff event, runoff may not occur on all OFEs, and, in fact, runoff can occur on an upper-slope OFE, disappear through infiltration into a lower OFE, and then reappear in a lower OFE. This possibility is most likely for smaller storms.

To estimate rill erosion, runoff in individual rills, rill shape, rill width, and flow depth are estimated. Rill spacing is required because it determines flow rate in individual rills. A rectangular channel is assumed, and rill width is expressed as a function of flow discharge. Rill spacing is computed as the average plant spacing for rangelands, but it is never less than .5 m or greater than 5 m. For croplands, the rill-spacing parameter is set to a default value of 1 m if not specified as input.

F. Soil Processes

The soil component of the model addresses the temporal changes in soil properties affecting erosion and runoff. These properties include random surface roughness, ridge height, saturated hydraulic conductivity, soil erodibilities, and bulk density. The effects of tillage, weathering, consolidation, and rainfall are considered in estimating soil erodibility. Baseline rill erodibility and critical hydraulic shear for a freshly tilled condition are adjusted for changing conditions. Adjustments to interrill erodibility are based on live and dead roots in the upper 150 mm of the soil, and may be adjusted through time, rainfall, or consolidation since last tillage. Rill erodibility is adjusted based on incorporated residue in the upper 150 mm of the soil and time since last tillage. Critical hydraulic shear is adjusted based on time since last tillage. Rainfall effects on bulk density of freshly tilled soils are also estimated in the soil component.

Past efforts to model the erosion processes have often used USLE relation for estimating soil erodibility. Extensive field studies (Simanton et al., 1987; Elliot et al., 1989) were completed to develop new technology to predict erodibility values for WEPP based on soil properties. These efforts have not yet yielded a satisfactory prediction technology; interim equations for predicting soil erodibility have indicated that any prediction technology will likely include parameters related to mineralogy, texture, organic matter, and soil chemistry.

G. Power of WEPP

WEPP incorporates knowledge of soils, crops, tillage, residue, climate, and sediment transport into a soil-erosion prediction tool that is to be used at the farm and field levels. WEPP considers many of the temporal changes that occur on a land area. Because it is a simulation tool, it provides the potential to study many interactive effects of various conditions as they affect soil erosion. Because of its construction, WEPP can be used to estimate erosion caused by snowmelt and irrigation, as well as that caused by rainfall. WEPP does not consider stream or channel erosion processes such as gullying.

Some of the power of WEPP is demonstrated in Fig. 5 by comparing the average soil erosion rate versus slope length for two different soils. For comparison purposes, the RUSLE slope-length effect recommended for most cropland soils is shown for comparison. The simulation is for a 10-year continuous corn crop at Forest City, Iowa, on a 9% gradient. The soils were assume to differ in baseline rill and interrill erodibility and critical shear values. One of the soils had a very low rill erodibility rate, a high critical shear, and a high

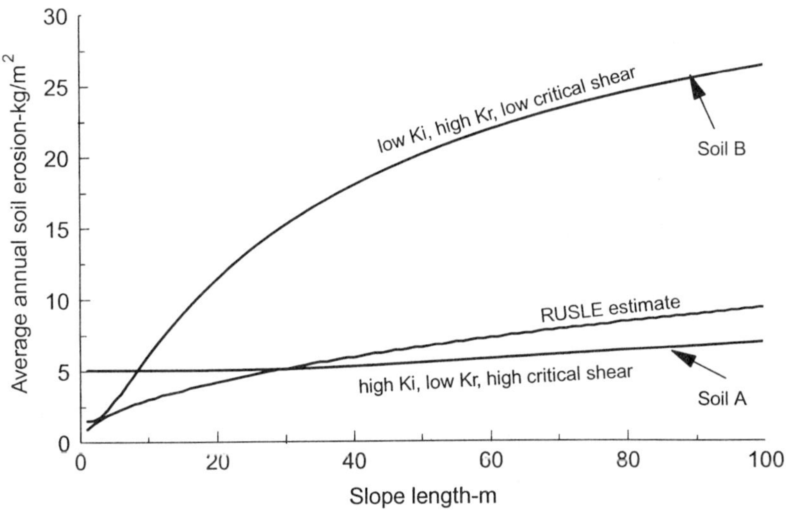

Fig. 5 Average annual soil erosion vs. slope length for a continuous corn crop at Forest City, Iowa (Ki = interrill erodibility and Kr = rill erodibility).

interrill erosion rate (SOIL A). The contrasting soil (SOIL B) had a very low interrill erodibility, a low critical shear, and a high rill erodibility (both of these soils were toward the extremes of the soils reported by Elliot et al., 1989). Both soils had average annual water yields over this 10-year simulation period of about 170 mm.

As shown in Fig. 5, the soil with high interrill erodibility, low rill erodibility, and high critical shear was (B) much less response to slope length than was either the RUSLE estimate or the other soil (A), although rill processes began to detach soil at about 30 m downslope. Interrill processes generally move soil to nearby rills (Young and Wiersma, 1973), and hence interrill contributions are little affected by slope length, as the WEPP application demonstrates. However, when the soil was susceptible to rill erosion, as indicated by a high rill erodibility and a low critical shear (B), soil erosion increased rapidly when hydraulic shear exceeded the critical shear, and sediment transport began to limit erosion rates at a fairly short distance downslope.

Another demonstration of the power of WEPP is indicated in Fig. 6. WEPP was used to simulate a slope with five OFEs, the OFEs are of uniform length (40 m), and the gradient was 9%. Each OFE is in the same rotation, corn-soybeans-wheat-alfalfa-alfalfa, but there is never a row crop grown in an OFE

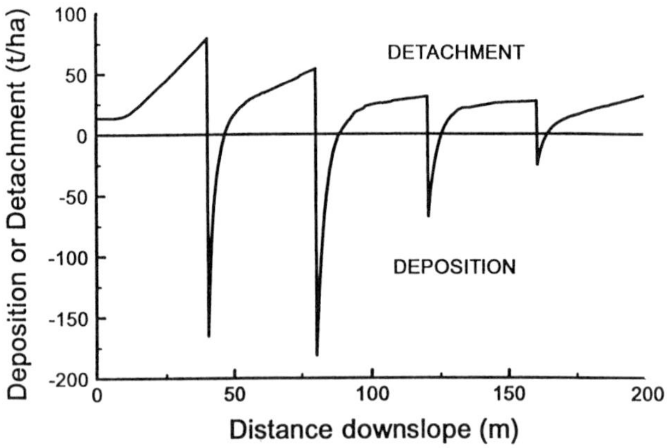

Fig. 6 Annual sediment detachment vs. distance downslope for five overland flow elements in a sod-based rotation. (Negative values indicate deposition.)

directly below an OFE with a row crop in the same year. Negative values indicate deposition. Figure 6 shows the average annual detachment rates (or deposition rates) for each of 100 points within an OFE for all storms that predicted runoff and erosion during the five-year rotation.

WEPP predicted an annual average of almost 20 t ha^{-1} of sediment delivered from the slope, but the soil detached was predicted to be more than 28 t ha^{-1} on areas where detachment occurred; deposition was predicted to average about 43 t ha^{-1} on areas where it occurred. Maximum detachment and deposition rates were predicted to be 79 and 181 t ha^{-1}, respectively. Both detachment and deposition areas may change from storm to storm, and a deposition area may become a detachment area, or vice versa, depending on circumstances.

Surprisingly, the uppermost OFE had the highest predicted erosion rate. The uppermost OFE was in corn the first year and soybeans the second year; predicted erosion was high for these years. The corn followed a meadow that was moldboard-plowed, so there was litter residue for erosion control, and tillage was quite intensive after corn harvest. The other OFEs were either in a close-grown crop or had considerable residue from a previous crop for the first year. Deposition is predicted when runoff velocity is reduced as runoff flows onto a lower OFE. The combination of no residue or cover and severe weather demonstrates the power of the model and the problem in selecting a representative period of record within a feasible computer run time for the simulation to give good estimates for long-term effects. Additionally, it demonstrates its power for determining probabilities for individual events and annual values. Other examples could be used that demonstrate the interactive effects on erosion and sediment delivery of topography, soils, crops, climate, and tillage.

H. Applications

WEPP applications include those of RUSLE as well as many others. Some of the applications might include

Anticipating where sediment detachment occurs on a slope, either for individual storms or for long time averages.

Evaluation of land treatment effects on sediment delivery from a field or farm.

Evaluation of range management and treatment alternatives, including grazing alternatives, on sediment delivery. Grazing alternatives would include timing and duration of grazing as well as stocking rates. Management alternatives might include various improvement practices including tillage, brush control through herbicides, and burning.

Recognizing the effect of forest road design, placement, and construction on sediment delivery from forest lands. Additional forest applications would include evaluation of the effect on sediment delivery of clear-cutting certain portions of small watersheds.

Recognizing the effect of ridge height and row direction on soil detachment and sediment delivery from a field. Evaluation of grassed waterways on sediment delivery is also possible.

Use of Natural Resource Inventory (NRI) sites for estimates of sediment delivery from fields and farms. NRI sites and real-time weather systems (currently available in some areas of the United States) could be used with WEPP to make same-day estimates of soil loss on fields and sediment delivery from fields, perhaps at county or state levels.

Recognizing the effect of stubble management and slope aspect on the capture of snow and its consequent effects on soil erosion.

WEPP is designed for use by local government organizations and natural-resource action agencies concerned with sediment transport and deposition. The model's ability to simulate erosion and sedimentation processes that occur and interactively to bring these into play in a modeling and predictive sense are important attributes. WEPP is also expected to become a major component of surface-water-quality models. It is likely that the databases used in WEPP and many of its components will become part of other natural resource models.

IV. INTRODUCTION TO DECISION SUPPORT SYSTEMS

Previously presented material describes past and current soil erosion, conservation, and rehabilitation from a USDA perspective with special emphasis on development and application of an erosion-prediction technology used by the USDA and other agencies. The following sections describe selected opportunities and implications of some possible future developments in the USDA natural-resource programs.

A. Computer-Based Prediction Technology

As technology evolves from handbook-type soil-erosion models, such as the USLE, to the modern water-erosion-prediction technology requiring the use of computers for their implementation (i.e., RUSLE, and WEPP), erosion modeling and soil conservation design will be computer based. Opportunities presented by these advancements include (1) the potential to utilize processes-based models, (2) the potential to consider rapidly a larger range of soil/climate/management alternatives and interactions, (3) user adoption of improved technology, and, as a result, (4) improved soil conservation. Implications include new responsibilities and methods of operation for research scientists, technology developers, and technology users.

Implications of computer-based erosion prediction and conservation design technology will affect all aspects of soil conservation. Research, technology development, technology transfer, and application will need to be planned, designed, developed, documented, verified, validated, maintained, and monitored as a "life-cycle" project. Research scientists, technology developers, and users should be directly involved in all phases from conception to implementation to eventual replacement of each new technology.

B. National Databases

Opportunities presented by national databases include those that (1) ensure standardized procedures on a national basis, (2) allow rapid extension of analyses and interpretations from local to county to state or national levels, and (3) enhance repeatability of analyses and results. Regional and national databases including climate, soils, topography, land use, management practices, economics, and regulations and policies governing soil conservation will enable technology users at the local, county, state, and national levels access to common data and information. This common access together with a nationally uniform methodology will allow users at all administrative levels to repeat the calculations, duplicate the erosion predictions, and thus meet the test of scientific defensibility.

Standardized handbook procedures allowed repeatability and defensibility in the past. National databases and nationally uniform erosion prediction and soil conservation technology will bring a similar repeatability and defensibility to the computer-based technology. Implications of this national standardization include the need to sustain the national databases and the technological infrastructure to sustain them.

C. Predicting Sediment Yield

Previously, the USLE was used to predict on-site erosion for parts of the landscape. Portions of the landscape where sediment deposition occurred were not addressed with USLE-type technology, nor could the USLE technology provide information about sediment properties and sediment yield downstream of the eroding portions of the upland areas. Although RUSLE is improved in many ways over the USLE, it still shares these weaknesses.

Hydrologically driven models such as CREAMS and WEPP are designed to deal with sediment detachment, transportation, and deposition. Users now have the opportunity to make on-site erosion predictions (as in the USLE) and to make sediment yield estimates "off-site," including sediment deposited and sediment transported downstream.

Technology that predicts on-site erosion and off-site sediment yield presents the opportunity to consider broader objectives for soil erosion and off-site sediment yields presents the opportunity to consider broader objectives for soil erosion, soil conservation, farm planning, pollution control, resource inventory, and environmental protection.

Implications of the broader objectives involving soil erosion will require a broad-based concept of acceptable on-site erosion and off-site sediment yield. With the new technology, the soil loss tolerance concept directly tied on on-site erosion and USLE is no longer appropriate in defining acceptable rates of on-site erosion. The soil loss tolerance concept needs to be replaced by concepts that are multiobjective and explicitly conclude both on-site erosion and off-site sediment yield.

D. Erosion Prediction, Soil Conservation, and Multiobjective Decision Making

Farm planning, land use, and management decisions should be made with multiobjective decision methodologies. Opportunities for research and application of the new technologies in a multiobjective decision-making context are significant. Soil erosion and soil protection will need to be considered along with decisions affecting water supply, surface water and groundwater quality, local and regional economic factors, biodiversity, wildlife habitat, recreation, aesthetics, and other factors of site-specific concern.

Implications of soil erosion and conservation in a multiobjective decision making context include the need to maintain the soil resource by having it explicitly included in the analyses. For long-range problems of soil protection and sustainable agriculture to receive equitable consideration in these multiobjective analyses, the concept of soil loss tolerance must be replaced by concepts that include on-site erosion and off-site sediment yield as important components.

V. OTHER APPROACHES

One of the more widely used approaches for estimating sediment yield not mentioned heretofore is MUSLE (modified universal soil loss equation) (Williams and Berndt, 1977; Williams, 1982). In this model, the rainfall runoff factor (R) of the USLE is replaced by a runoff term in the form $95(Qq_p)^{0.56}$, where Q is the runoff volume (acre-ft), and q_p is the peak flow rate ($ft^3\ s^{-1}$). This relation has been used successfully to simulate watershed sediment yield in models such as SWRRB (simulation for water resources in rural basins) (Williams, Nicks, and Arnold, 1985) and SPUR (simulation of production and utilization of rangeland) (Wight and Skiles, 1987).

In the early 1980s, two significant modeling efforts resulted in advancing the technology of erosion prediction and conservation planning and the impact of erosion on water quality. The ANSWERS model (Beasley et al., 1980) is a hydrologically driven model that used cell concepts to describe the heterogeneity of small watersheds on a storm basis to estimate sediment yield. The CREAMS model (Knisel, 1980; Foster et al., 1981) was also hydrologically driven and intended to show management impacts on water quality through continuous simulation using fundamental equations for rill and interrill erosion.

The kinematic runoff and erosion model KINEROS (Woolhiser et al., 1990) is an event-oriented, physically based model describing the processes of interception, infiltration, surface runoff, and erosion from small agricultural and urban watersheds. The watershed is represented by a cascade of planes and channels, and the partial differential equations describing overland flow, channel flow and erosion, and sediment transport are solved by finite difference techniques. Spatial variability of rainfall and infiltration, runoff, and erosion parameters can be accommodated. KINEROS may be used to determine the effects of various artificial features such as urban developments, small detention reservoirs, or lined channels on flood hydrographs and sediment yield.

The AGNPS (agricultural non-point source) (Young et al., 1981) model is intended for simulating hydrology, erosion, sediment, and chemical transport for larger river basins. The watershed is conceptualized as a series of cells (grid basis) with hydrologic, sediment, and chemical movement in each cell. The model predicts runoff volume and peak discharge, eroded and delivered sediment, and nitrogen, phosphorus, and chemical oxygen demand concentrations in the runoff and the sediment loss for single storm events at all points in the watershed.

Other erosion simulation efforts are ongoing in the United States and other places in the world. Many of these are more local in nature (site specific) and have not been validated widely or applied to a variety of climatic conditions, varied land uses, and physiographic areas. Their exclusion here is not intended to slight them but rather portrays ignorance on the part of the authors.

VI. SUMMARY AND CONCLUSIONS

Soil erosion (by water) knowledge and technology have undergone major changes in the past two decades. The RUSLE and WEPP developments are two examples of this technology. Essential to such advancement is the need for advances in field experimental and monitoring equipment to parallel explosive improvements in computer science. Soil-erosion model postulations and the formulation of hypotheses have progressed beyond our ability to perform in situ data-gathering experiments with which to test new hypotheses. For example, sampling for sediment concentration in a water system currently involves destructive sampling that, for example, precludes consecutive collection of samples without upsetting an energy grade line. We need to be able to sample repeatedly in the downslope direction to understand spatial variability problems that are important to resource management. Although remote-sensing techniques have been applied to many environmental scenarios, they have not been applied to water erosion problems.

The material of this chapter is presented in less detail than desired for full comprehension of the technology. We hope that the users of this technology will pursue the necessary details in the references and will continue to follow developments in soil erosion in the technical literature.

REFERENCES

Allmaras, R. R., Burwell, R. R., Larson, W. E., Hob, R. F., and Nelson, W. W. (1966). Total porosity and random roughness of the interrow zones influenced by tillage. *U.S. Dept. of Agric. Cons. Res. Rep. No. 7.*

ASA (Agronomy Society of America) (1982). Determination of soil loss tolerance. *ASA Special Publ. No. 45*, Madison, WI.

Beasley, D. B., Huggins, L. F., and Monke, E. J. (1980). ANSWERS: A model for watershed planning. *Trans. ASAE 23*: 938–944.

Cogo, N. D., Moldenhauer, W. C., and Foster, G. R. (1984). Soil loss reduction from conservation tillage practices. *Soil Sci. Soc. Am. J. 48*: 368–373.

Dissmeyer, G. E., and Foster, G. R. (1981). Estimating the cover-management factor (c) in the universal soil loss equation for forest conditions. *J. Soil Water Conserv. 36*: 235–240.

Elliot, W. J., Liebenow, A. M., Laflen, J. M., and Kohl, K. D. (1989). A compendium of soil erodibility data from WEPP cropland soil field erodibility experiments 1987 and 1988. *NSERL Report No. 3*. Ohio State University and USDA Agricultural Research Service. Available from USDA-ARS, National Soil Erosion Research Laboratory, West Lafayette, IN.

Ellison, W. D. (1947). Soil erosion studies. *Agric. Eng. 23*: 156–146, 197–201, 245–248, 297–300, 349–351, 402–405, 442–444.

Finker, S. C., Nearing, M. A., Foster, G. R., and Gilley, J. E. (1989). A simplified equation for modeling sediment transport capacity. *Trans. ASAE 32*: 1545–1550.

Foster, G. R. (1991). Advances in wind and water erosion prediction. *J. Soil Water Conserv. 46*: 27–29.

Foster, G. R., and Lane, L. J. (1987). User requirements: USDA-Water Erosion Prediction Project (WEPP). *NSERL Report No. 1*, National Soil Erosion Research Laboratory, USDA-ARS, W. Lafayette, IN.

Foster, G. R., Lane, L. J., Nowlin, J. D., Laflen, J. M., and Young, R. A. (1981). Estimating erosion and sediment yield on field-sized areas. *Trans. ASAE. 24*: 1253–1262.

Foster, G. R., McCool, D. K., Renard, K. G., and Moldernhauer, W. G. (1981). Conversion of the universal soil less equation to SI metric units. *J. Soil. Water Conserv. 36*: 355–359.

Foster, G. R., and Meyer, L. D. (1975). Mathematical simulation of upland erosion using fundamental erosion mechanics. In *Present and Predictive Technology for Predicting Sediment Yields and Sources*, Report No. ARS-S-40, U.S. Dept. of Agric., Agric. Res. Serv., Washington, DC, pp. 177–189.

Fournier, F. (1960). *Climat et Erosion*. Presses Universitaries de France, Paris.

Gregory, S. M. (1982). Soil cover prediction with various amounts and types of crop residue. *Trans. ASAE. 25*: 1333–1337.

Knisel, W. G. (1980). CREAMS: A field scale model for chemicals, runoff, and erosion from agricultural management systems. *U.S. Dept. of Agr. Cons. Res. Rep. No. 25*, Tucson, AZ.

Laflen, J. M., Foster, G. R., and Onstad, C. A. (1985). Simulation of individual storm soil loss for modeling impact of soil erosion on crop productivity. In *Soil Erosion and Conservation* (S.A. El-Swaify, W. C. Moldenhauer, and A. Lo, eds.), Soil Cons. Soc. Am., Ankeny, IA, pp. 285–295.

Laflen, J. M., Lane, L. J., and Foster, G. R. (1991). WEPP—A new generation of erosion prediction technology. *J. Soil Water Conserv. 46*: 34–38.

Lane, L. J., and Nearing, M. A. (Eds.) (1989). USDA-Water Erosion Prediction Project: Hillslope profile model documentation. NSERL Rep. No. 2. U.S. Dept. of Agr., Agr. Res. Serv., Natl. Soil Erosion Res. Lab., West Lafayette, IN.

Meyer, L. D. (1960). Use of the rainulator for runoff plot research. *Soil Sci. Soc. Am. Proc. 24*: 319–327.

Meyer, L. D. (1981). How rain intensity affects interrill erosion. *Trans. ASAE 24*: 1472–1475.

Meyer, L. D. (1984). Evolution of the universal soil loss equation. *J. Soil Water Conserv. 39*: 99–104.

Meyer, L. D., and Harmon, W. C. (1984). Susceptibility of agricultural soils to interrill erosion. *Soil Sci. Soc. Am. J. 48*: 1152–1157.

Meyer, L. D., and Harmon, W. C. (1989). How row-sideslope length and steepness affect sideslope erosion. *Trans. ASAE 32*: 639–644.

Meyer, L. D., and Wischmeier, W. H. (1969). Mathematical simulation of the process of soil erosion by water. *Trans. ASAE 12*: 754–755, 762.

Moss, A. J., and Green, P. (1983). Movement of solids in air and water by raindrop impact. Effect of drop size and water-depth variations. *Aust. J. Soil Res. 21*: 257–269.

Musgrave, G. W. (1947). The quantitative evaluation of factors in water erosion, a first approximation. *J. Soil Water Conserv. 2*: 133–138.

Nicks, A. D., and Lane, L. J. (1989). Weather generator. In *USDA-Water Erosion Prediction Project* (L. J. Lane, and M. A. Nearing, eds.), NSERL Rept. No. 2, National Soil Eros. Res. Lab., West Lafayette, IN, pp. 2.1–2.19.

Onstad, C. A., Wolfe, M. C., Larson, C. L., and Slack, D. C. (1984). Tilled soil subsidence during repeated wetting. *Trans. ASAE 27*: 733–736.

Renard, K. G., Foster, G. R., Weesies, G. A., and Porter, J. P. (1991). RUSLE: Revised universal soil loss equation. *J. Soil Water Conserv. 46*: 30–33.

Simanton, J. R., Weltz, M. A., West, L. T., and Wingate, G. D. (1987). Rangeland experiments for water erosion prediction project. Paper No. 87-2545, ASAE, St. Joseph, MI.

Smith, D. D., and Whitt, D. M. (1948). Evaluating soil losses from field areas. *Agric. Eng. 29*: 394–396.

Stamey, W. L., and Smith, R. M. (1964). A conservation definition of erosion tolerance. *Soil Sci. 97*: 183–186.

Stott, D. E. (1991). RESMAN: A tool for soil conservation education. *J. Soil Water Conserv. 46*: 332–333.

Stott, D. E., Stroo, H. F., Elliott, L. F., Papendick, R. I., and Unger, P. W. (1990). Wheat residue loss in fields under no-till management. *Soil Sci. Soc. Am. J. 54*: 92–98.

Stroo, H. F., Bristow, K. L., Elliot, L. F., Papendick, R. I., and Campbell, G. S. (1989). Predicting rates of wheat residue decomposition. *Soil Sci. Soc. Am. J. 53*: 91–99.

Watson, D. A., and Laflen, J. M. (1986). Soil strength, slope, and rainfall intensity effects on interrill erosion. *Trans. ASAE 29*: 98–102.

Weltz, M. A., Renard, K. G., and Simanton, J. R. (1987). Revised universal soil loss equation for western rangelands. In *US/Mexico Symp. on Strategies for Classification and Management of Native Vegetation for Food Production in Arid Zones*, U.S. Forest Serv. General Tech. Rep. RM-150, pp. 104–111.

Wight, J. R., and Skiles, J. W. (Eds.) (1987). *SPUR: Simulation of production and utilization of rangelands. Document and user guide.* ARS 63, USDA-ARS.

Williams, J. R. (1982). Sediment-yield prediction with universal equation using runoff energy factor. In *Proc. Workshop on Estimating Erosion and Sediment Yield on Rangelands*, ARM-W-26, USDA-ARS, pp. 244–252.

Williams, J. R., and Berndt, H. D. (1977). Sediment yield prediction based on watershed hydrology. *Trans. ASAE 20*: 1100–1104.

Williams, J. R., Jones, C. A., and Dyke, P. T. (1984). A modeling approach to determining the relationship between erosion and soil productivity. *Trans. ASAE 27*: 129–144.

Williams, J. R., Nicks, A. D., and Arnold, J. G. (1985). Simulator for water resources in rural basins. *J. Hydraul. Eng., ASCE 111*: 970–986.

Wischmeier, W. H. (1959). A rainfall erosion index for a universal soil loss equation. *Soil Sci. Soc. Am. Proc. 23*: 246–249.

Wischmeier, W. H. (1975). Estimating the soil loss equations cover and a management factor for undisturbed lands. In *Present and Prospective Technology for Predicting Sediment Yields and Sources*, ARS-S-40, Agr. Res. Serv., U.S. Dept. of Agr., Washington, DC, pp. 118–125.

Wischmeier, W. H., and Smith, D. D. (1958). Rainfall energy and its relationship to oil loss. *Trans. AGU 39*: 285–291.

Wischmeier, W. H., and Smith, D. D. (1965). Predicting rainfall-erosion losses from cropland east of the Rocky Mountains. *Agr. Handbook No. 282*, U.S. Dept. Agr., Washington, DC.

Wischmeier, W. H., and Smith, D. D. (1978). Predicting rainfall erosion losses. *Agr. handbook No. 537*, U.S. Dept. of Agr., Science and Education Administration.

Woolhiser, D. A., Smith, R. E., and Goodrich, D. C. (1990). *KINEROS, A Kinematic Runoff and Erosion Model: Documentation and User Manual.* ARS-77. USDA-ARS.

Yalin, Y. S. (1963). An expression for bed-load transportation. *J. Hydraul. Div., Am. Soc. Civil. Eng. 89*: 221–250.

Young, R. A., Onstad, C. A., Bosch, D. D., and Anderson, W. P. (1987). AGNPS, Agricultural non-point-source pollution model: A watershed analysis tool. *Conservation Research Report No. 35*, USDA-ARS, Washington, DC.

Young, R. A., and Wiersma, J. L. (1973). The role of rainfall impact in soil detachment and transport. *Water Resour. Res. 9*: 1629–1636.

Zingg, A. W. (1940). Degree and length of land slope as it affects soil loss in runoff. *Agric. Eng. 21*: 59–64.

10

Runoff Estimation on Agricultural Fields

Jeffry Stone, Kenneth G. Renard, and Leonard J. Lane
Southwest Watershed Research Center, Agricultural Research Center, U.S. Department of Agriculture, Tucson, Arizona

I. INTRODUCTION

Runoff is the primary driving variable in the water-induced erosion process. Although the impact of rain drops detaches soil particles, it is flowing water which transports the detached particles to areas of concentrated flow and depending on the sediment load, and the transport capacity of the flow, detaches or re-entrains previously detached particles or deposits particles being transported. Many of the conservation practices employed to reduce erosion are in fact practices to reduce the volumes and rates of runoff. Practices which increase the amount of residue can significantly reduce the volume of runoff by increasing the infiltration potential of the soil. Other practices such as tilling on the contour or installing terraces decrease the flow length, thus reducing the flow's velocity and discharge rate. Accurate estimation of the effects of an agricultural management system or individual practice on the volume and rate of runoff are essential in the evaluation and planning associated with soil and water conservation.

A. The Field Scale Runoff Process

Theories and experiments of the rainfall runoff process abound at scales ranging from those at a point which considers the vertical flow of water within the soil profile to those at a watershed which considers the integrated response of the entire watershed occurring at the outlet. Although there have been many studies of the rainfall-runoff process at the field scale, accepted methodologies specific to that scale are rare. In general, methodologies derived at smaller or larger scales are used to estimate runoff from fields. Application of infiltration equations, such as the Horton (1940), Green-Ampt Mein-Larsen (Mein and Larsen, 1973), or Philip's (1957), are examples of point models which have been applied to the field scale while the curve number approach and unit hydrographs are examples of watershed models which have been applied to the field scale. The reasons for this lack of methodology at the field scale can be classified into two broad categories: the characteristics of the process and the economics of information.

Representations of point processes such as infiltration are highly nonlinear, while watershed runoff response to rainfall can sometimes be represented as a linear system. The field scale process seems to represent a transit between the two. Runoff from rainfall occurs as a poorly defined shallow sheet flow grading into concentrated flow which can become a shallow sheet flow again further downslope. Spatial and temporal variability of characteristics which control rainfall-runoff response at the field scale can be significant. The measurement of runoff, because of its amorphous nature, is difficult at the field scale. At the point scale, one can accurately measure the amount of water applied and rates of water entering the soil. At the watershed scale, given a control section, one can measure the amount and rate of water leaving the watershed.

The economics of information also play a role in the dearth of field scale specific methodologies. Data collection at a point is inexpensive and easy to implement. Data collection at the watershed scale involves monitoring major stream channels, which, although expensive, can be justified by the potential benefits of the data or methodologies produced. However, the traditional purposes for predicting agricultural field scale runoff (terrace or culvert design, for example) need only a reasonable estimate of runoff. For example, the cost of adding a foot to the height of a terrace is insignificant as opposed to adding a foot to the height of a dam. In addition, the economic consequences of the failure of a terrace are insignificant when compared with the failure of a dam.

The field scale runoff process which is of interest in most erosion studies is known as Hortonian flow. It is characterized by rainfall excess dominated runoff occurring as shallow sheet flow or flow in small concentrated flow areas. The runoff response to rainfall is controlled by two basic factors: the rainfall intensity (and to a lesser extent, the rainfall total) and the soil characteristics. When the soil moisture is high and/or the rainfall intensity is high, the runoff response is controlled by the rainfall intensity distribution. When soil moisture is low and/or the rainfall intensity is low, runoff response is controlled by the spatial distribution of the infiltration capacity of the soil.

The actual time rate distribution of runoff is complicated by characteristics of the rainfall and soil surface. Rainfall, particularly that which causes runoff in semiarid regions, is highly variable both in space and time. In addition, the high intensities associated with runoff-producing rainfall can cause the formation of surface crusts which decrease the soil's ability to infiltrate water. The soil infiltration capacities are also highly variable in space and time due to both man's activities such as cultivation, climatic influences such as freeze-thaw of the soil, and changes of the soil surface due to residue decomposition and biological activity.

B. Runoff and Erosion

Many approaches have been taken into account for the runoff effects on the erosion process, ranging from ignoring the direct effects of runoff, to computing the water flow profile and velocity as it changes in the downslope direction. The *universal soil loss equation* (USLE) (Wischmier and Smith, 1978) is an example of ignoring the runoff process directly and concentrating on the effects of rainfall energy as the driving variable. Runoff is considered indirectly through the 30 minute intensity used in the rainfall factor and the slope-length factor of the USLE. The modified USLE (MUSLE) (Williams et al., 1983) substitutes the runoff amount and peak discharge for the rainfall energy factor of the USLE to compute sediment yield. The recently developed Water Erosion Prediction Project (WEPP) (Nearing et al., 1989) models take a different approach by using a rainfall intensity-based infiltration equation to compute rainfall excess and the peak discharge rate. Erosion is conceptualized as consisting of two steady state processes: interrill and rill erosion. Interrill erosion is driven by rainfall intensity during the period of rainfall excess (that is, the period when the rainfall rate is greater than the infiltration capacity of the soil) and the average runoff rate. Rill erosion or deposition is a function of the peak discharge rate, which is used to compute the transport capacity, the runoff volume, and the duration of runoff.

II. APPROACHES TO ESTIMATE RUNOFF

As mentioned in the introduction, there are few methodologies which have been specifically derived for the field scale. The two approaches described within this chapter are representative of methods used in erosion modeling for management decisions. The first method, the U.S. Soil Conservation Service Curve Number method, uses total storm rainfall and an index of initial abstractions to compute total storm runoff volume. In modified forms which adjust the curve number for soil moisture, it is the basis for runoff calculation for many erosion simulation models (see CREAMS, Kinsel, 1980; SWRRB, Williams et al., 1985, SPUR, Wight et al., 1987) which attempt to account for the effects of management systems on the runoff-erosion-sediment yield processes. The second method, the kinematic wave model coupled with an infiltration equation, has become more popular in recent times partially due to the increased power of computers and partially due to the potential of a closer approximation of the runoff process.

The type of approach taken will depend on the data available and the objectives of the analysis. For example, a study to screen management systems to replace a current system may only need an estimate of annual average runoff differences among the systems, while a study to design the capacity of a grassed waterway may need detailed information of the flow profile down the channel length.

A. Soil Conservation Service Total Storm Approach

The Soil Conservation Service (SCS) total storm approach involves computing direct storm runoff volume from total storm rainfall and watershed characteristics to estimate the runoff peak discharge. The SCS approach is the primary methodology used by the SCS and others to estimate runoff amounts and peak rates in the United States. It is popular because the method is simple to use, yields reasonable estimates of runoff, and is easy to parameterize.

1. Runoff

One of the most widely used methods to compute direct storm runoff is the SCS Curve Number method (SCS, 1972). The method is based on the concept that rainfall can be divided into runoff and losses, or initial abstractions which occur before runoff begins (interception, infiltration, and surface storage) and losses which occur after the start of runoff (infiltration). The relationship among storm rainfall, storm runoff, and initial abstractions can be written as

$$Q = \frac{(R - I_a)^2}{R - I_a + S} \qquad (1)$$

where Q = storm runoff (in.), R = storm rainfall (in.), I_a = initial abstractions (in.), and S = potential maximum losses after runoff begins plus initial abstraction (in.). An analysis of the rainfall runoff relationships of a number of small agricultural watersheds in the United States yielded the following relationship between I_a and S as

$$I_a = 0.2S \qquad (2)$$

which when substituted into Eq. (1) gives the curve number equation

$$\begin{aligned} Q &= \frac{(R = 0.2S)^2}{R + 0.8S} && \text{for } R > 0.2S \\ Q &= 0 && \text{for } R \leq 0.2S \end{aligned} \qquad (3)$$

where

$$S = \frac{1000}{CN} - 10 \tag{4}$$

The relationship of Eq. (3) is plotted in Fig. 1.

The relationship of Eq. (2) was derived from a plot of estimates of both I_a and S (Fig. 2) for a data set of rainfall and runoff pairs from watersheds less than 10 acres (SCS, 1972, p. 10.6). The wide amount of scatter in Fig. 2 is attributed to errors in estimating I_a due to difficulty in estimating the time of the start of rainfall and runoff and in estimating the water which infiltrated prior to the start of runoff and eventually became return flow.

The main criticism of the curve number method is that the amount of runoff computed is not sensitive to rainfall intensity. Thus, the method will compute the same amount of runoff given the same amount of total rainfall independent of the duration of the event or the distribution of rainfall intensity during the event.

Curve Number Parameter Estimation. The curve number is estimated from land use, treatment, hydrologic condition, and hydrologic soil group from the information listed in Table 1. Definitions of hydrologic condition are given in Table 2 and definitions of the hydrologic soil group are given in Table 3. The curve number estimated from Table 1 is termed the antecedent moisture condition (AMC) II curve number (CN_{II}). The AMC is based on the amount of

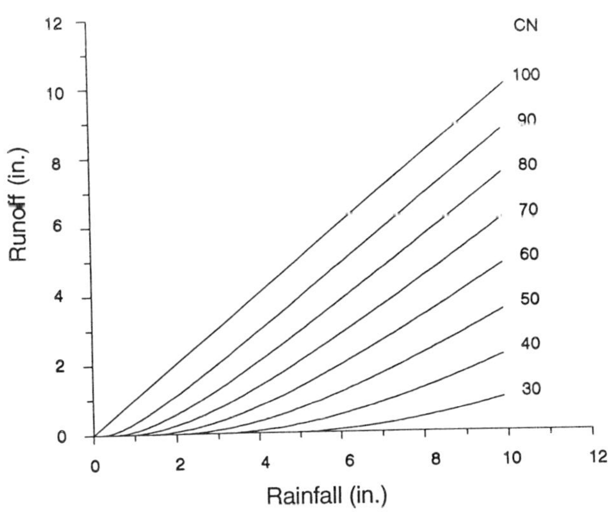

Fig. 1 Curve number rainfall–runoff relationship.

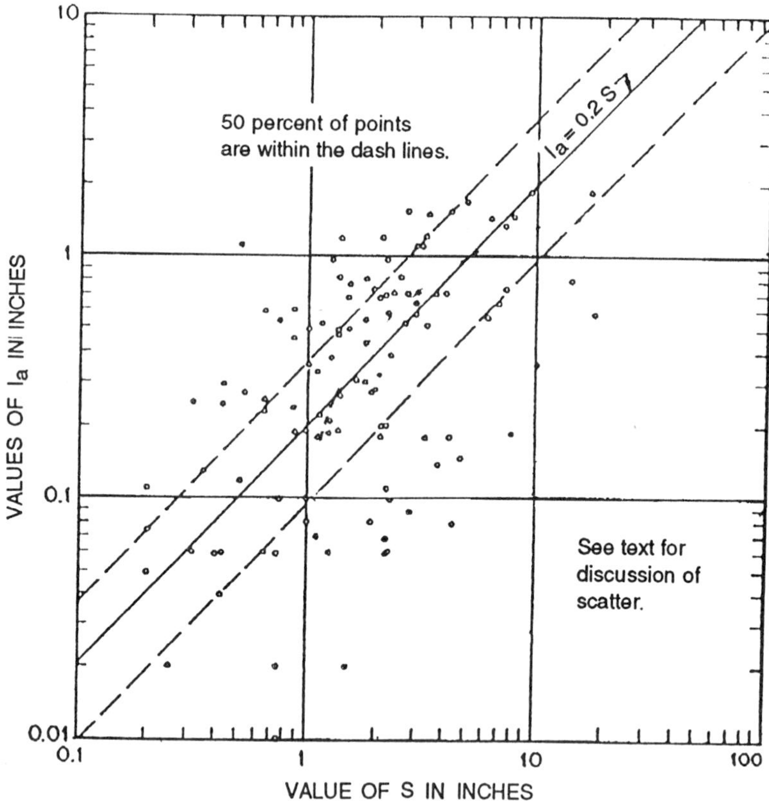

Fig. 2 Relationship between I_a and S (SCS, 1972).

precipitation in the preceding 5 days (Table 5). Definitions of AMC are given in Table 4. Depending on the AMC value, the curve number obtained from Table 1 is adjusted using Table 5.

2. *Peak Discharge*

Estimation of the peak discharge rate is important in evaluating the effects of management systems on engineering structures such as culverts or grassed waterways and in estimating the sediment transport capacity of the flow. The SCS graphical method (SCS, 1986) of computing the peak discharge from a homogenous area is based on nomographs of unit peak discharge versus time to concentration (also termed travel time) for selected values of I_a/R_d. The relationship for the four types of design storms used by the SCS (Fig. 3) is shown

Table 1 Runoff Curve Numbers for Hydrologic Soil-Cover Complexes Under Average Conditions of Antecedent Moisture

Land use	Cover treatment	Hydrologic condition	Hydrologic soil group			
			A	B	C	D
Fallow	Straight row		77	86	91	94
Row crops	Straight row	Poor	72	81	88	91
	Straight row	Good	67	78	85	89
	Contoured	Poor	70	79	84	88
	Contoured	Good	65	75	82	86
	Contoured and terraced	Poor	66	74	80	82
	Contoured and terraced	Good	62	71	78	81
Small grain	Straight row	Poor	65	76	84	88
	Straight row	Good	63	75	83	78
	Contoured	Poor	63	74	82	85
	Contoured	Good	61	73	81	84
	Contoured and terraced	Poor	61	72	79	82
	Contoured and terraced	Good	59	70	78	81
Close seeded legumes or rotation meadow	Straight row	Poor	66	77	85	89
	Straight row	Good	58	72	81	85
	Contoured	Poor	64	72	83	85
	Contoured	Good	55	69	78	83
	Contoured and terraced	Poor	63	73	80	83
	Contoured and terraced	Good	51	67	76	80
Pasture or range		Poor	68	79	86	89
		Fair	49	69	79	84
		Good	69	61	74	80
	Contoured	Poor	47	67	81	88
	Contoured	Fair	25	59	75	83
	Contoured	Good	6	35	70	79
Meadow		Good	30	58	71	78
Woods		Poor	45	66	77	83
		Fair	36	60	73	79
		Good	25	55	70	77
Farmsteads			59	74	70	77
Roads	Dirt		72	82	78	89
	Hard surface		74	84	90	92

Source: Soil Conservation Service, 1972.

Table 2 Classification of Soils by Their Hydrologic Properties

Hydrologic soil group	Permeability	Texture	Comments
A	High	Coarse-moderate	
B	Moderate	Fine-moderate	
C	Slow	Moderate-fine	May have impeding layer
D	Very slow	Very fine	Clays with high swelling potential

Table 3 Classification of Vegetation Covers

Vegetative cover	Condition	Vegetation or management	Cover (%)
Crop rotation	Poor	Row crops, small grains, and fallow	
	Good	Alfalfa and grasses	
Pasture or range	Poor	Heavy grazing	<50%
	Fair	Moderate grazing	50–75%
	Good	Light grazing	>75%

Table 4 Rainfall Limits for Estimating Antecedent Moisture Conditions

Antecedent moisture class	5 day total rainfall (in.)	
	Dormant season	Growing season
I	<0.5	< 1.4
II	0.5–1.1	1.4–2.1
III	>1.1	>2.1

Table 5 CN_I and CN_{III} corresponding to CN_{II}

CN_{II}	CN_I	CN_{III}
100	100	100
95	87	98
90	78	96
85	70	94
80	63	91
75	56	88
70	51	85
65	45	82
60	40	78
55	35	74
50	31	70
45	26	65
40	22	60
35	18	55
30	15	50
25	12	43
20	9	37
15	6	30
10	4	22
5	2	13

in Figs. 4a–d. The peak discharge relationship is

$$q_p = q_u A Q \tag{5}$$

where q_p = peak discharge (ft^3/s), q_u = unit peak discharge (smi/in.), A = drainage area (mi^2), and Q = runoff (in.). The time to concentration, T_c, needed to enter the nomographs is computed by two different methods depending on the flow length of the area. For flow lengths up to 300 ft, T_c (hr), is computed as

$$T_c = \frac{0.007(nL)^{0.8}}{P_2^{0.5}\, S_0^{0.4}} \tag{6}$$

where n = Manning's n ($s/ft^{1/3}$), L = flow length (ft), P_2 = 2-yr, 24-hr rainfall (in.), and S_0 = land slope (ft/ft). The flow or hydraulic length is the longest flow path from the upper end of the field to the outlet. It can be estimated from the field area in units of acres by (SCS, 1973)

$$L = 209A^{0.6} \tag{7}$$

The data used to derive Eq. (7) ranged in area from 10 to over 1000 acres and from 400 to over 20,000 feet in hydraulic length.

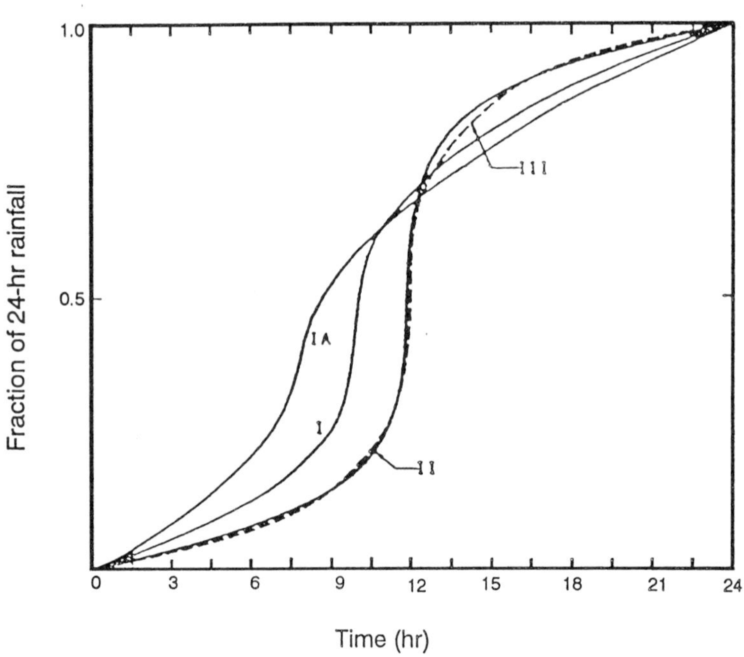

Fig. 3 SCS design storms.

For flow lengths greater than 300 ft, the time to concentration is the sum of Eq. (6) and

$$T_c = \frac{L}{3600u} \tag{8}$$

where u = average velocity (ft/s) of shallow concentrated flow. The average velocity is computed using the Manning equation for a design flow of 0.1 ft as

$$\begin{aligned} u &= 16.14S_0^{0.5} \quad \text{for unpaved surface} \\ u &= 20.33S_0^{0.5} \quad \text{for paved surface} \end{aligned} \tag{9}$$

The design storms used to enter the nomographs are illustrated in Fig. 3 and represent typical rainfall intensity distributions for regions of the United States. Type I is for the mediterranean-type climate of much of California while type IA is for the rest of the west coast of the United States. Both types are represented by wet winters and dry summers. Type III is for the coastal areas of the east and south which are subject to large tropical storms. Type II is for the rest of the country and is represented by short-duration, high-intensity rainfall.

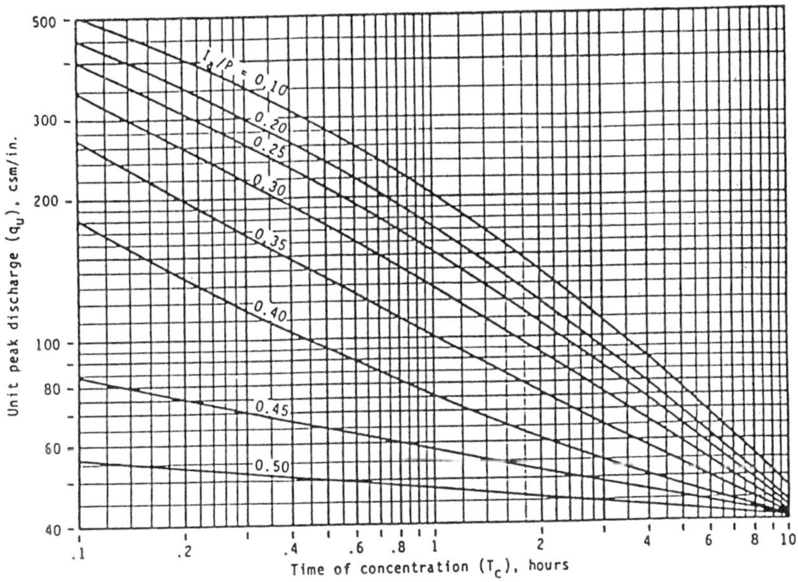

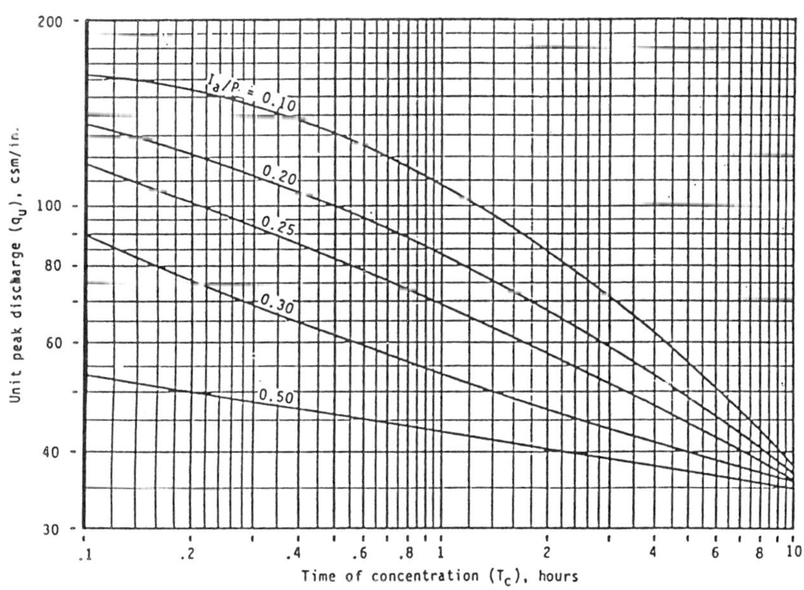

Fig. 4 SCS peak discharge nomographs.

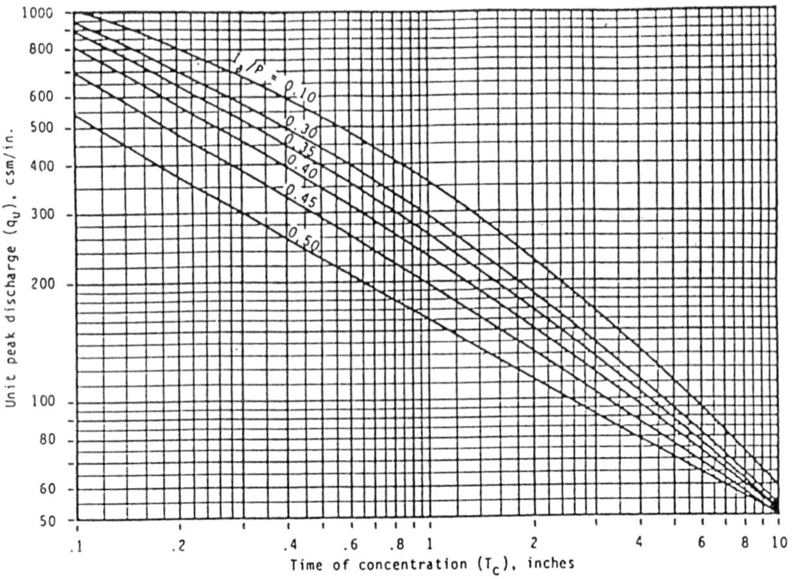

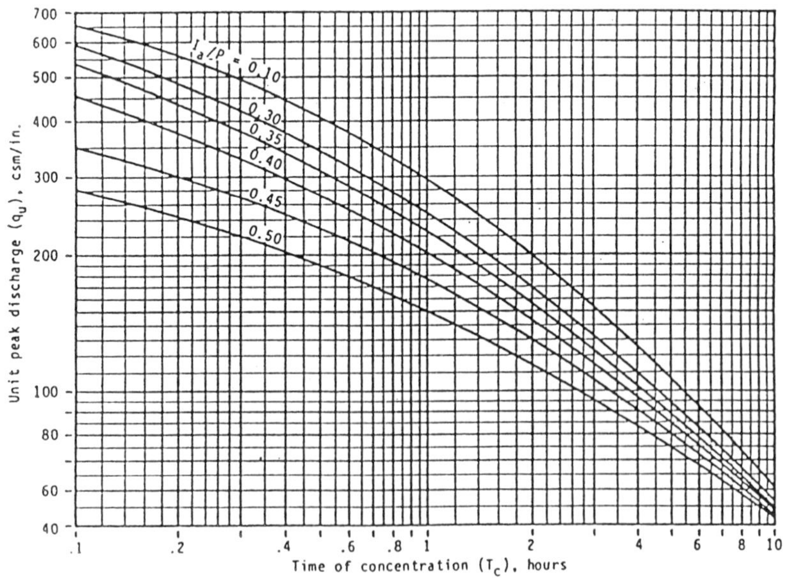

Fig. 4 Continued.

The limitations of the SCS graphical peak discharge method include (1) a homogeneous field which can be described by a single curve number, (2) a range of $0.1 < I_a/R < 0.5$, (3) a curve number > 40, and (4) a range of $0.1 < T_c < 100$ hr. If either I_a/R or T_c are found to be out of their ranges, it is suggested to use the closest limiting value of the range. Note that, for long flow paths, the computation of T_c is primarily a function of the physical dimensions of the field. For flow paths less than 300 ft, T_c is a function of the physical characteristics of the field and a characteristic rainfall amount.

B. Rainfall Excess Approach

In contrast to the storm total approach, the rainfall excess approach uses a time intensity rainfall distribution and an infiltration equation to compute a rainfall excess distribution. Rainfall excess is defined as all the rainfall which does not infiltrate into the soil but runs off the soil surface. The time distribution of rainfall excess is highly dependent on the rainfall intensity because, apart from the beginning of a runoff event, the infiltration approaches a steady rate which is not the case with the rainfall rate. For field scale applications, some form of the St. Vennant shallow water equations is used to route the rainfall excess over the flow surface. The computer runoff rate at the end of a field is dependent on the rainfall excess rate, the storage characteristics of the flow surface, the hydraulic characteristics of the soil, and the initial soil moisture conditions. The advantages of the approach are: (1) both the infiltration and the form of the surface water routing equations are conceptually physically based and their parameters have some physical significance, (2) because the parameter values have some physical significance, they can be used with more confidence than a more empirical approach when computing runoff under changing land use conditions, and (3) because the time distribution of runoff is highly dependent on the time distribution of rainfall excess, a potentially better representation of the runoff hydrograph can be obtained.

The steps involved in implementation of the rainfall excess approach are to compute infiltration, depression storage, rainfall excess, and then the runoff hydrograph.

1. *Infiltration*

Infiltration under rainfall is a two-phase process. During the first phase, the potential infiltration rate is grater than the rainfall rate. The actual infiltration rate is equal to the rainfall rate because the water can only enter the soil at the application rate. At a certain time, termed the *time to ponding*, the potential infiltration rate equals the rainfall rate and water begins to pond on the soil surface. The time to ponding depends on the rainfall intensity, soil surface characteristics, and the initial soil water content. After ponding, the infiltration rate decreases nonlinearly to a constant rate which is related to the saturated conductivity of the soil. For events in which ponding ceases and begins again, the initial soil moisture conditions change as the soil drains during the hiatus.

Methods to compute infiltration can be classified into empirical and conceptual models. Almost all of the models consider infiltration at a point and most assume ponded conditions at the start of the infiltration event. Although empirical infiltration models have suffered criticism because they are basically fitting equations with parameters which have little or no physical significance, empirical and conceptual infiltration models both have the same weakness—that of parameter estimation.

Conceptual Models. Conceptual models of infiltration are derived from some sort of formulation of the movement of water through a porous medium. As such, the parameters generally have physical significance and in theory can be measured. However, while infiltration during rainfall is influenced by porous media flow, it can also be significantly influenced by surface processes such as crusting or subsurface processes such as preferential flow paths. At the field scale, these surface and subsurface processes are highly variable and will affect the aggregate effect of infiltration on runoff.

The two conceptual models which have received the most interest in rainfall runoff modeling are the Philip (Philip, 1957) and Green-Ampt Mein-Larsen (GAML model) (Mein-Larsen, 1973) models, primarily because their parameters have the potential to be estimated from field data. Both the models have a formulation such that at the beginning of an event, the infiltration rate is a function of the saturated conductivity and a quantity which is a function of soil properties and initial soil moisture (sorptivity in the Philip equation or matric potential in the GAML model). For the Philip equation, as time increases, the infiltration rate approaches the saturated conductivity term. For the GAML model, infiltration approaches saturated conductivity as the cumulative infiltration increases. There have been numerous studies comparing the performance of the GAML and Philip models with each other and with other infiltration models with mixed resuults (see Singh, 1989 for a list of studies). In general, the studies have found that empirical models such as the Horton may fit data better, possibly because of more parameters, but that no one model is superior for all cases.

The GAML model as modified by Chu (1978) has been chosen by the Water Erosion Prediction Project (WEPP) (Nearing et al., 1989) and will be used in this chapter to illustrate the application of an infiltration model.

The rate form of the Green-Ampt equation (Green and Ampt, 1911) for the one-stage case of initially ponded conditions, assuming the ponded water depth is shallow, is

$$f = K_e \left(1 + \frac{\psi\theta_d}{F}\right) \tag{10}$$

where $f = dF/dt$ = infiltration capacity (m/s), K_e = effective saturated conductivity (m/s), ψ = average capillary potential (m), θ_d = soil moisture deficit (m/m), and F = cumulative infiltrated depth (m). The soil moisture deficit can be computed as

$$\theta_d = (1 - \theta_i)\eta_e \tag{11}$$

where θ_i = initial volumetric water content (m/m) and η_e = effective porosity (m/m). Equation (10) is a differential equation which is solved as

$$K_e t = F(t) - \psi\theta_d \ln\left[1 + \frac{F(t)}{\psi\theta_d}\right] \tag{12}$$

where t = time (s). Equation (12) can be solved for infiltrated depth for successive increments of time using a Newton-Raphson iteration and the solution used in Eq. (10) to obtain the instantaneous infiltration rate.

For the general case of time varying rainfall, Chu (1978) modified the Green-Ampt as modified by Mein and Larsen (1973) to account for multiple times to ponding assuming that, within any discrete time interval, the rainfall rate is constant. Chu computed an indicator, C_u (m), when ponding occurs within a given interval of rainfall intensity given that there was no ponding at the beginning of the interval as

$$C_u = R_i - V_i - \left(\frac{K_e\psi\theta_d}{r_{i-1} - K_e}\right) \tag{13}$$

where R is the cumulative rainfall depth (m), V is the cumulative rainfall excess depth (m), r is the rainfall rate (m/s), and i refers to the time step. The cumulative rainfall excess depth is computed as

$$V_i = R_i - F_i \tag{14}$$

If C_u is positive, ponding occurs before the end of the interval; if it is negative, no ponding occurs. The time to ponding, t_p (s), is computed as

$$t_p = \left(\frac{K_e\psi\theta_d}{r_{i-1} - K_e} - R_{i-1} + V_{i-1}\right)\frac{1}{r_{i-1}} + t_{i-1} \tag{15}$$

When there is ponding within a rainfall interval, the cumulative infiltration depth is computed using

$$K_e t_i = F_i - \psi\theta_d \ln\left(1 + \frac{F_i}{\psi\theta_d}\right) \tag{16}$$

in a Newton-Raphson iteration by computing the function g as

$$g = K_e t_i - F_i^j + \psi\theta_d \ln\left(1 + \frac{F_i^j}{\psi\theta_d}\right) \tag{17}$$

and its derivative with respect to F

$$dg = \frac{F_i^j}{F_i^j + \psi\theta_d} \tag{18}$$

where i refers to the time step and j refers to the iteration step. The iteration is stopped if the ratio

$$\frac{g}{dg} < \epsilon \tag{19}$$

where epsilon is an acceptable error. If the ratio is greater than epsilon, then the new value of F is

$$F_i^{j+1} = F_i^j + \frac{g}{dg} \tag{20}$$

The time, t_i, in Eq. (17) is corrected to account for the difference between instantaneous time to ponding and the actual time to ponding and is computed as

$$t_i = t + \frac{R_{tp} - V_{tp-1} - \psi\theta_d \ln(1 + R_{tp} - V_{tp-1}/\psi\theta_d)}{K_e} - t_p \tag{21}$$

where R_{tp} = amount of cumulative rainfall (m) at the time to ponding and V_{tp-1} = cumulative rainfall excess at the previous time step.

Similarly, Chu (1978) developed an indicator for the end of ponding C_p during an interval, assuming the surface was ponded at the beginning of the interval as

$$C_p = R_i - F_i - V_i \tag{22}$$

If C_p is positive, ponding continues; if it is negative ponding ceases within the interval. When there is no ponding within an interval, the cumulative infiltration is computed as

$$F_i = R_i - V_{i-1} \tag{23}$$

GAML Model Parameter Estimation. The most extensive analysis of GAML parameters has been done by Rawls et. al. (1982), who analyzed over 1000 soils. The average values derived from their analysis are given in Table 6. It should be noted that the effective saturated conductivity values reported in Table 6 were derived from soil core samples and may be higher than those observed

Table 6 Suggested Values for K_s (mm/hr), ψ (mm), and η_e (mm/mm)

Texture	K_s (mm/hr)	ψ (mm)	η_e (mm/mm)
Sand	90.0	49	.40
Loamy sand	30.0	63	.40
Sandy loam	11.0	90	.41
Loam	6.5	110	.43
Silt loam	3.4	173	.49
Silt	2.5	190	.42
Silty clay loam	1.5	214	.35
Clay loam	1.0	210	.31
Silty clay loam	0.9	253	.43
Sandy clay	0.6	260	.32
Silty clay	0.5	288	.42
Clay	0.4	310	.39

Source: Rawls et al., 1982.

in the field. To date, there are no well-developed methods to adjust these values for the influences of crusting, macropore development, or other physical or biological factors which may increase or decrease the conductivity of the soil.

2. *Rainfall Excess*

Rainfall excess is the portion of the rainfall which ponds on the surface during the period when the rainfall rate exceeds the infiltration rate. It is partitioned into depression storage and runoff which flows off the surface. The rainfall excess rate, $v(t)$ (L/T), can be conceptualized as

$$\begin{aligned} v(t) &= r(t) - f(t) && \text{for } h(t) > 0 \\ v(t) &= r(t) - \min[r(t), f(t)] && \text{for } h(t) \leq 0 \end{aligned} \tag{24}$$

where $h(t)$ = flow depth (L) and min = minimum. The top equation computes the rainfall excess rate when there is flow on the surface. The rainfall excess rate can be either positive or negative depending on whether the rainfall rate exceeds the infiltration capacity of the soil. Negative values of excess are during the recession of the hydrograph. The bottom equation computes the rainfall excess rate when there is no flow on the surface. If the value of rainfall excess is positive, then runoff begins. The conceptualization of rainfall excess in Eq. (24) is used when an infiltration equation is coupled with a routing method such as the kinematic wave model (see Woolhiser et al., 1991). Frequently, it is more convenient to define rainfall excess as

$$v(t) = r(t) - f(t) \qquad \text{for } r(t) > f(t)$$
$$v(t) = 0 \qquad \text{for } r(t) \le f(t) \tag{25}$$

The advantage of Eq. (25) is that the integral of $v(t)$ over the duration of the event is the total storm runoff and can be computed by an infiltration equation without having to rely on a rainfall excess routing procedure. The disadvantage is that the rainfall excess lost by infiltration during the recession of the hydrograph is not accounted for so that Eq. (25) will always overestimate the runoff volume.

3. *Depression Storage*

Depression storage is the portion of rainfall excess which is held in storage caused by microvariations in topography and which eventually infiltrates into the soil or evaporates. Depending on the degree of microrelief occurring on a surface, depression storage can greatly impact runoff amounts and rates. Gayle and Skaggs (1978) report values ranging from 0.6 to 7.5 mm of average storage depth measured on three agricultural soils with different tillage practices imposed. Onstad (1984) developed an equation which relates maximum depression storage, S_d (cm), to slope and random roughness from an analysis of over 1000 plots. The relationship is

$$S_d = 0.112r_r + 0.031r_r^2 - 0.012r_r S \tag{26}$$

where r_r is the random roughness (cm) and S is the slope of the flow surface (%). Onstad (1984) and others (Mitchell and Jones, 1976; Moore and Larsen, 1979) have suggested that runoff begins before depression storage is completely satisfied. Onstad developed an equation to compute the amount of rainfall excess, PR (cm), needed to satisfy depression storage while runoff is occurring. This equation is

$$PR = 0.329r_r + 0.073r_r^2 - 0.018r_r S \tag{27}$$

The amount of runoff which occurs during the period when depression storage is satisfied is found by subtracting Eq. (26) from Eq. (27).

Depression Storage Estimation. The rate at which depression storage is filled, even on the small plot scale, is difficult to measure directly. In addition, direct estimates of depression storage for different surface conditions are difficult because of the difficulty in quantifying the storage characteristics of different surfaces. Random roughness is an easier measurement and typical values as the result of various tillage implements were compiled by Zobeck and Onstad (1987) and are listed in Table 7.

Table 7 Field Operations and Random Roughness Values

Field operation	Random roughness (mm)
Harrow, spike	8
Drill, conventional and no-till	10
Planter, no-till and broadcast	
Rodweeder, plain	
Drill, semideep furrow	11
Rodweeder, shovel	
Sweeps	
Blades	13
Drill, deep furrow	
Fertilizer application	
Cultivator, field (sweeps) and row	15
Mulch treader	
Planter, row	
Cultivator, field (shovels)	18–19
Disk, tandem	
Chisel	23–28
Disk, 1-way (18–22 in. disks)	
Lister	
Disk, 1-way (22–26 in. disks)	31–33
Moldboard, 5–7 in. deep	
Disk, large offset	48–51
Moldboard, 8 in. deep	

Source: Zobeck and Onstad, 1987.

4. *Runoff Routing*

Dynamic infiltration-hydrograph models for overland flow consist of an infiltration function which computes the infiltration rate as it varies with time from an unsteady rainfall input and a routing function which transforms rainfall excess into flow depths on a flow surface. The choice of the infiltration function is somewhat arbitrary, but the routing function is generally some form of the St. Vennant shallow water equations. One such form, the kinematic wave model, has been shown (Woolhiser and Liggett, 1967) to be a valid approximation for most overland flow cases.

Kinematic Wave Model. The kinematic wave model is used to compute the flow depth and discharge rate at a specified distance down a flow surface during three time periods. The first time period is when the rise of the hydrograph occurs. During this period, water is being stored on the flow surface as a function of the rainfall excess rate and the flow surface characteristics of roughness, length, and slope. The end of this period is when a wave, termed the *kinematic wave*, originating at the top of the surface at the start of runoff, reaches the end of the surface. This time, termed the *time to equilibrium*, is equivalent to some definitions of the time of concentration. The second period is from the time to equilibrium to the time when the rainfall rate becomes less than the infiltration rate. For constant rainfall and infiltration, the hydrograph will be at steady state and the discharge rate will be equal to the rainfall excess rate. For variable input, the discharge rate will vary depending on the variation of the rainfall excess rate and the surface characteristics. The last period is when the recession of the hydrograph occurs. During this period, water held in storage drains off the surface. For the case of an infiltrating surface, the drying front where the flow depth goes to zero travels from the top to the bottom of the flow surface.

The kinematic equations for flow on a plane are the continuity equation

$$\frac{\partial h}{\partial t} + \frac{\partial q}{\partial x} = v \tag{28}$$

and a depth-discharge relationship

$$q = \alpha h^b \tag{29}$$

where h = depth of flow (m), q = discharge per unit width of the plane (m^3/m-s), α = depth-discharge coefficient, b = depth-discharge exponent, and x = distance from top of plane (m). When the Chezy relationship is used, $\alpha = CS^{1/2}$ where C = Chezy coefficient ($m^{1/2}$/s) and $b = 3/2$. When the Manning relationship is used, $\alpha = S^{1/2}/m$ where n = Manning coefficient (s/$m^{1/3}$) and $b = 5/3$. If the initial and boundary conditions are

$$h(x,0) = h(0,t) = 0 \tag{30}$$

which means that there is zero inflow at the top of the flow surface being considered and zero runoff at the beginning of the computations, then Eqs. (28) and (29) can be solved by the method of characteristics. The method involves rewriting Eqs. (28) and (29) as simple ordinary differential equations in terms of the flow depth at a distance on the plane. These equations are termed the characteristic equations. The equations for depth and distance along a characteristic $c(t,x)$ at a given time are (see Eagleson, 1970, for a derivation of the characteristic equations)

$$\frac{dh}{dt} = v(t) \tag{31}$$

and

$$c(t,x) = \frac{dx}{dt} = \alpha b h(t)^{b-1} \tag{32}$$

The characteristic, Eq. (32), defines a locus of points in the time–space plane on which the flow depth is computed by Eq. (31). The characteristic equations are integrated to get

$$h = h_1 + \int_{t_1}^{t_2} v(w)\, dw \tag{33}$$

$$x = x_1 + \alpha b \int_{t_1}^{t_2} h(w)^{b-1}\, dw \tag{34}$$

where x_1 = distance down the plane where the depth is equal to h_1 (m), h_1 = depth at time t_1 (m), t_1 and t_2 = limits of integration (s), and w = dummy variable of integration.

The general solution procedure is to solve Eq. (33) for the flow depth at a time and then solve Eq. (34) for the distance from the top of the plane that the depth occurs. Because it is generally the hydrograph at the end of the plane which is of interest, the distance solved for in Eq. (34) is the length of the plane. The discharge rate is computed by solving Eq. (29) given the depth found by Eq. (33).

Although the method of characteristic solution is relatively easy to implement as a computer algorithm, it is difficult to illustrate for the general case of unsteady rainfall excess. Therefore, it is useful to consider the more restrictive example of constant rainfall excess which allows for an analytical solution of Eqs. (31) and (32).

The analytic solution can be shown to consist of four distinct zones in the characteristic space of the t-x plane (Fig. 5). The x-axis represents the initial

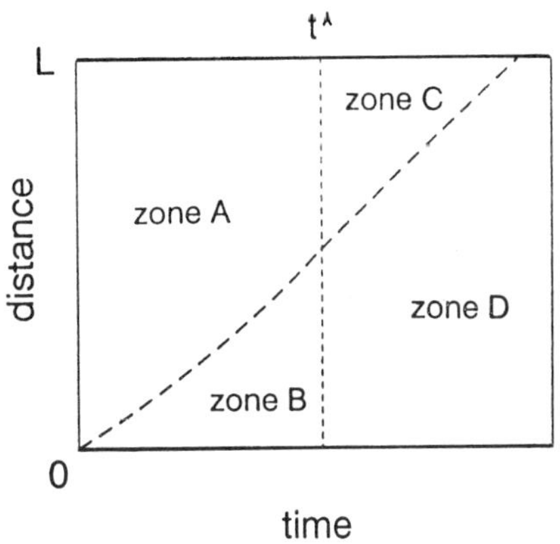

Fig. 5 Four zones in the t-x plane.

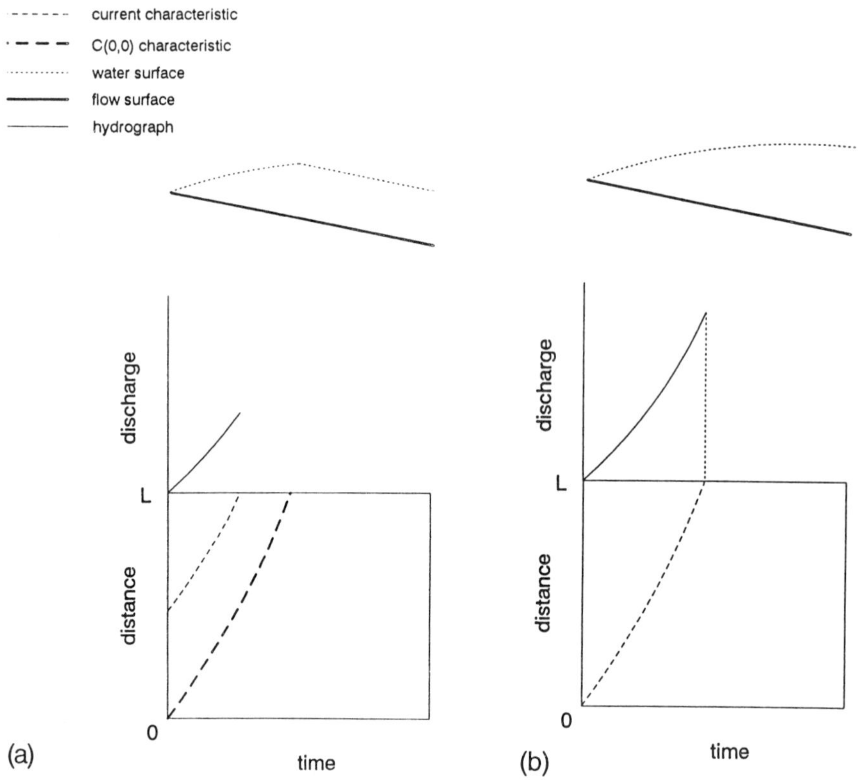

Fig. 6 Characteristics and hydrographs for (a) zone A, (b) zone B, (c) zone C, and (d) zone D.

condition of zero flow at time zero, while the t-axis represents the boundary condition of zero inflow at the top of the flow surface. Because the flow depth is zero on these two axes, they are used as starting points for the solution of Eqs. (31) and (32). The general procedure is to integrate all the characteristics which originate on the x-axis at $t = 0$ until the characteristic from $x = 0$ and $t = 0$ reaches the end of the flow surface. Then all the characteristics which originate at $x = 0$ for all times $t > 0$ are integrated until the end of the event.

1. Zone A—Rise of the Hydrograph (Fig. 6a):

$$\int_0^t dh = v \int_0^t dt$$
$$h = vt \tag{35}$$

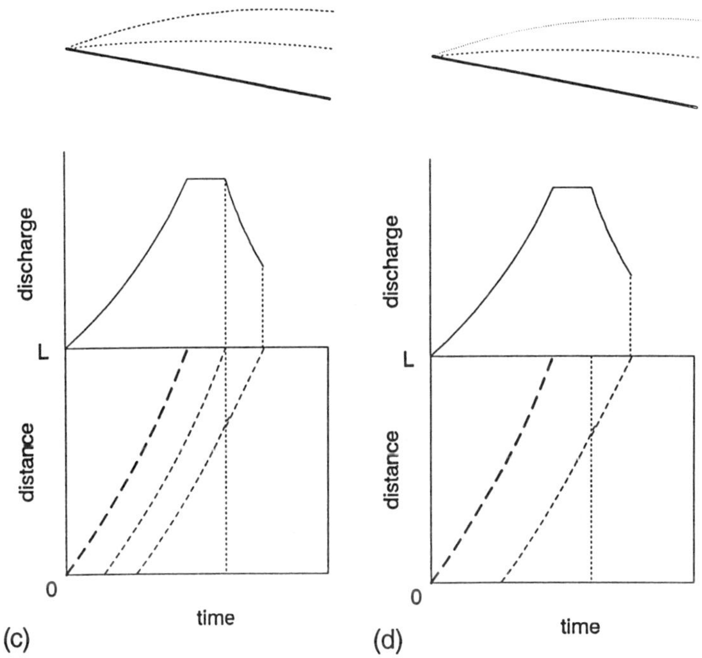

Fig. 6 Continued.

$$\int_0^t dx = \alpha v^{b-1} \int_0^t t^{b-1}\, dt$$
$$x = \alpha v^{b-1}\, t^b \tag{36}$$

Equations (35) and (36) are solved at time zero by selecting successive times to compute the increase in depth on the flow surface until the computed value of x is greater than or equal to the length of the flow surface. When $x = L$, the flow surface is at equilibrium and the discharge rate is equal to the rainfall excess rate. This time, termed the time to kinematic equilibrium, t_e (s), is computed as

$$t_e = \left(\frac{L}{\alpha v^{b-1}}\right)^{1/b} \tag{37}$$

The characteristic which starts at the origin of the t-x plane and reaches $x = L$ at t_e is termed the limiting characteristic and separates the zone A and B solutions.

2. Zone B—Equilibrium Hydrograph (Fig. 6b):

$$\int_{t_0}^{t} dh = v \int_{t_0}^{t} dt$$
$$h = v(t - t_0) \tag{38}$$

$$\int_{t_0}^{t} dx = \alpha v^{b-1} \int_{t_0}^{t} (t - t_0)^{b-1}\, dt$$
$$x = \alpha v^{b-1}(t - t_0)^b \tag{39}$$

After the time to equilibrium, the solution is computed using Eqs. (38) and (39). The time t_0 is the time when the characteristic which is being integrated begins at the top of the flow surface at some time on the t-axis. Equation (39) is computing the time it takes a characteristic to travel from the top of the flow surface to the end. For constant rainfall excess, this time is equal to t_e and because the flow is at steady state within this zone, the discharge rate is simply

$$q = vL \tag{40}$$

3. Zone C—Recession (Fig. 6c):

$$\int_{t_0}^{D_v} dh + \int_{D_v}^{t} dt = v \int_{t_0}^{D_v} dt + v \int_{D_v}^{t} dt$$
$$h = v(D_v - t_0) \equiv h_v \tag{41}$$

$$\int_{t_0}^{D_v} dx + \int_{D_v}^{t} dx = \alpha\, b v^{b-1} \int_{t_0}^{D_v} (D_v - t_0)^{b-1}\, dt$$
$$+ \alpha b\, h_v^{b-1} \int_{D_v}^{t} dt \tag{42}$$
$$x = x_v + \alpha b\, h_v^{b-1}(t - D_v)$$

where x_v is the distance the characteristic has traveled at time D_v. It is computed as zone C solutions occur after rainfall excess ceases. The characteristic is integrated

$$x_v = \alpha v^{b-1}\, (D_v - t_0)^b \tag{43}$$

in two steps: the first is during the period between the time the characteristic originates on the t-axis, t_0, and the end of rainfall excess, D_v; the second is during the period between D_v and the time the characteristic reaches the end of the plane during which time the rainfall excess is zero. As can be seen by Eqs. (41) and (42), the characteristics within this zone are straight lines and as the depth approaches zero, the time the characteristic takes to reach the end of the flow surface approaches infinity.

4. Zone D—Partial Equilibrium (Fig. 6d):

$$\int_0^{D_v} dh + \int_{D_v}^{t} dt = v \int_0^{D_v} dt + v \int_{D_v}^{t} dt \tag{44}$$

$$h = v\,D_v \equiv h_v$$

$$\int_0^{D_v} dx + \int_{D_v}^{t} dx = \alpha b\, v^{b-1} \int_0^{D_v} (t - t_0)^{b-1}\, dt + \alpha b\, h_v^{b-1} \int_{D_v}^{t} dt \tag{45}$$

$$x = \alpha v^{b-1}\, D_v^b + \alpha b\, h_v^{b-1}(t - D_v)$$

Zone D solutions occur when the duration of rainfall excess is less than t_e and the discharge rate is always less than the rainfall excess rate. The solution is similar to that of zone C, except that the flow depth and discharge rate are constant from the time the rainfall excess ends to the time the characteristic from $t = 0$ and $x = 0$ reaches the end of the surface.

The *kinematic wave parameter* to be estimated is the roughness coefficient in the depth-discharge coefficient. Engman (1989) used data from large plot rainfall simulator experiments and computed the Manning's n values shown in Table 8.

III. SOME APPROXIMATIONS FOR RUNOFF ESTIMATION

For some cases, simple approximations of more complex representations of the rainfall runoff process are sufficient. As with the choice of a total storm runoff or a rainfall excess based approach, using an approximation will depend on the objective of the study and initial conditions of the area to be studied.

A. Infiltration

Li et al. (1976) used the nondimensional terms

$$t_* = \frac{tK_e}{\psi\theta_d} \tag{46}$$

$$F_*(t_*) = \frac{F(t)}{\psi\theta_d} \tag{47}$$

Table 8 Manning's n (s/m$^{1/3}$) Values for Rainfall Simulator Plots

Cover/treatment	Residue (T/ac)	Manning's n (s/m$^{1/3}$)
Bare/fallow	<1.4	0.045
Grass/sod		0.530
Chisel	<1.4	0.075
	1.4–1	0.180
	1–3	0.340
	>3	0.450
Range/natural		0.130
Disk/harrow	<1.4	0.078
	1.4–1	0.170
	1–3	0.270
	>3	0.310
Notill	<1.4	0.053
	1.4–1	0.083
	1–3	0.350
Plow (fall)	<1.4	0.055
Coulter	<1.4	0.110

Source: Engman, 1989.

to write the Green and Ampt equation as

$$t_* = F_*(t_*) - \ln(1 + F_*(t_*)) \tag{48}$$

By using the first term of a power series expansion of the natural log term in Eq. (48), they derived the following quadratic approximation of infiltrated depth, F_{q*}, for the case of initial ponded conditions as

$$F_{q*}(t_*) = \frac{1}{2}\left(t_* + \sqrt{t_*^2 + 8t_*}\right) \tag{49}$$

If the infiltration rate is nondimensionalized as

$$f_*(t_*) = \frac{f(t)}{K_e} \tag{50}$$

then Li et al.'s infiltration rate, f_{q*}, is computed as

$$f_{q*}(t_*) = 1 + \frac{1}{F_{q*}(t_*)} \tag{51}$$

B. Peak Discharge

As mentioned in the introduction for the kinematic wave equation, the basic problem in computing the runoff hydrograph is determining the dynamics of how water is stored on the flow surface. Given a constant rainfall excess rate, eventually the flow rate off the surface will equal the rainfall excess rate. We can use this to approximate the peak discharge from small areas by studying the relationship among the time and rate variables of rainfall excess and routed runoff. For the case of constant rainfall excess, the flow depth and discharge rate increase during the period $t < t_e$ and are constant for $t \geq t_e$. If the duration of the rainfall excess is less than t_e, then the maximum flow depth, h_p (m), is

$$h_p = v_c D_v \tag{52}$$

Substituting Eq. (52) into Eq. (29), using the definition of t_e, and simplifying, the peak discharge, q_p (m/s), is

$$q_p = v_c \left(\frac{D_v}{t_e}\right)^b \qquad \text{for } D_v < t_e \tag{53}$$

When the duration of rainfall excess is greater than the time to kinematic equilibrium (i.e., equilibrium), then the peak flow rate is simply

$$q_p = v_e \qquad \text{for } D_v > t_e \tag{54}$$

Equations (53) and (54) can be generalized by defining the following quantities and rewriting as

$$q_* = \frac{q_p}{v_c} \qquad t_* = \frac{t_e}{D_v} \tag{55}$$

$$q_* = t_*^{-b} \qquad \text{for } t_* \geq 1 \tag{56}$$

$$q_* = 1 \qquad \text{for } t_* < 1 \tag{57}$$

Equations (56) and (57) are illustrated in Fig. 7.

For variable rainfall excess, the definition of t_e [Eq. (37)] is not exactly true. In addition, for times greater than when the characteristic from time and distance zero reaches the end of the flow surface, the discharge rate approaches but never exactly equals the rainfall excess rate. Using Eqs. (55) and (56) as a starting

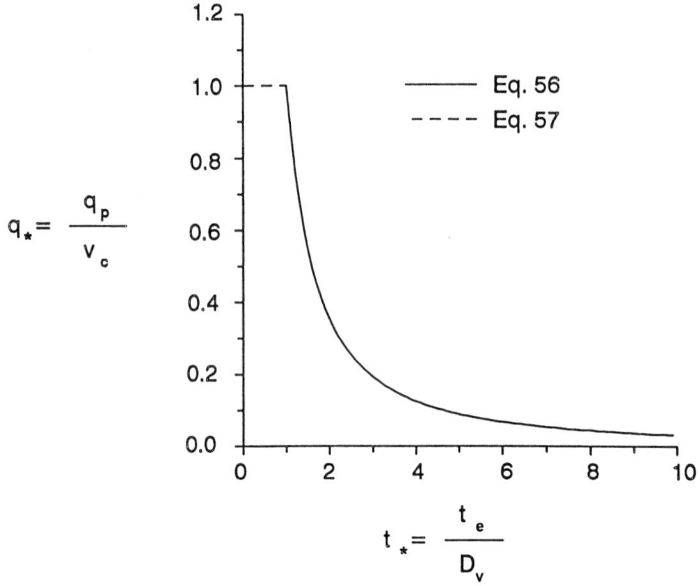

Fig. 7 Peak discharge relationship for constant rainfall excess.

point, we redefine the nondimensional quantities in Eq. (54)

$$q_* = \frac{q_p}{v_a} \qquad t_* = \frac{t_a}{D_v} \tag{58}$$

and define a nondimensional rainfall excess rate as

$$v_* = \frac{v_a}{v_p} \tag{59}$$

The average rainfall excess rate, v_a (m/s), is computed as

$$v_a = \frac{V}{D_v} \tag{60}$$

where t_a is computed using v_a as

$$t_a = \left(\frac{L}{\alpha \, v_a^{b-1}} \right)^{1/b} \tag{61}$$

A plot of the results of a number of simulations of the kinematic wave model for a range of values of q_*, t_*, and v_* suggested the following relationships

$$q_* = t_*^{-b} \qquad \text{for } t_* > 1 \tag{62}$$

$$q_* = \frac{1}{t_*} \qquad \text{for } 1 > t_* > t_{**} \tag{63}$$

$$q_* = \frac{1}{v_*} - \frac{m}{m+1}\,\frac{1-v_*}{v_*}\,t_* \qquad \text{for } t_{**} > t_* > 0 \tag{64}$$

The intersection of Eqs. (63) and (64), t_{**}, is found by combining the equations, substituting t_{**} for t_*, and solving for t_{**} using the quadratic formula as

$$t_{**} = \frac{1 - (1 - 2.4\,(v_* - v_*^2))^{1/2}}{1.2\,(1 - v_*)} \tag{65}$$

The relationships of Eqs. (62), (63), and (64) are plotted in Fig. 8. The average error using the range of conditions for the simulation was 1%, 10%, and 5% for Eqs. (62), (63), and (64) respectively, or a total combined error of 6.6%

C. Recession Infiltration

For the case of partial equilibrium described in Section II.B.1, the runoff volume can be significantly less than the rainfall excess volume. Stone et al. (1993)

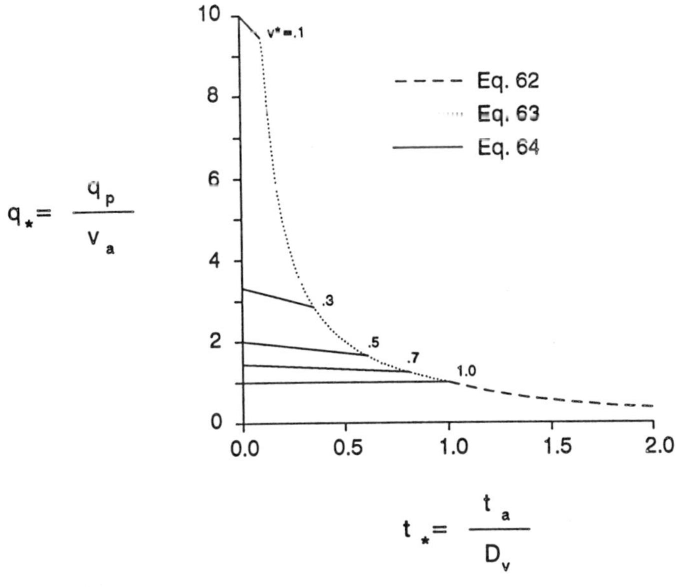

Fig. 8 Approximate peak discharge relationship for variable rainfall excess.

developed a simple relationship to compute the amount of reduction of rainfall excess which occurs during the recession of the hydrograph. By defining the following nondimensional quantities

$$Q_* = \frac{Q_v}{V_v} \qquad f_* = \frac{f_f}{v_a} \tag{66}$$

where Q_v = adjusted runoff volume (m), V_v = rainfall excess volume (m), and f_f = final infiltration rate (m/s) at the last time of nonzero rainfall excess rate, the reduction is computed as

$$Q_* = \frac{1}{b+1}\frac{f_*+1}{f_*}t_*^{-b} \qquad \text{for } t_* \geq \left(\frac{f_*+1}{f_*}\right)^{1/b} \tag{67}$$

and

$$Q_* = 1 - \frac{b}{b+1}\left(\frac{f_*}{f_*+1}\right)^{1/b} t_* \qquad \text{for } t_* < \left(\frac{f_*+1}{f_*}\right)^{1/b} \tag{68}$$

Equations (67) and (68) are illustrated in Fig. 9 for the depth-discharge coefficient computed using the Chezy coefficient.

1. *Tolerable Error in Runoff Volume Estimation*

Equation (67) can be used to compute a threshold in terms of t_* above which recession infiltration should be considered by rewriting Eq. (67) in the form of an inequality

$$t_* \leq (1 - Q_*)\frac{b+1}{b}\left(\frac{f_*+1}{f_*}\right)^{1/b} \tag{69}$$

IV. SIMULATION MODELS

The curve number method and the coupled infiltration-kinematic wave equations described in this chapter form the basis for many of the simulation models used to estimate field scale erosion. In terms of model application, most fall into two broad categories: event-based models used to compute runoff and erosion for a single rainfall event and continuous simulation models used to compute runoff and erosion for an extended period of time, typically using a daily time step. The selection of either type of model will depend on input parameter data availability and the objectives of the application of the model. In general, continuous simulation models require more input data than the event-based models because they require parameter values for model components which compute initial conditions or update other parameter values. For example, the event model

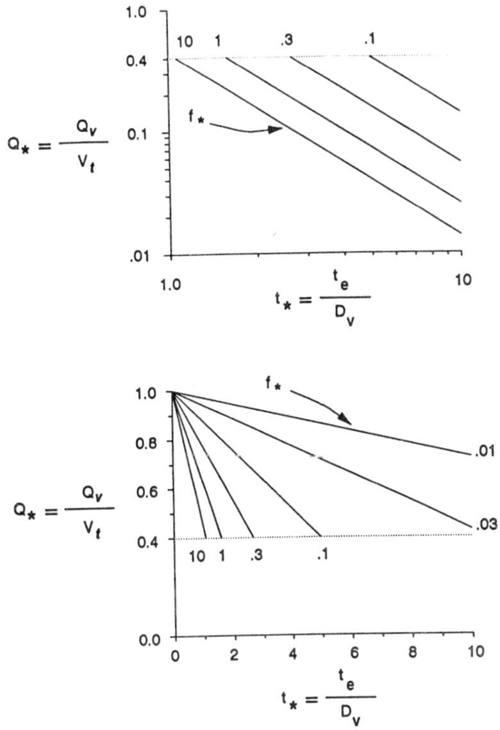

Fig. 9 Approximate recession infiltration relationship.

KINEROS requires four parameters to compute infiltration including initial soil moisture. In contrast, the continuous version of WEPP requires not only the same analogous parameters for infiltration, but also requires additional parameters for the computations of the water balance, crop growth, residue decomposition, and the effects of tillage on infiltration. Event-based models are best suited for design purposes in which the input rainfall and initial conditions can be specified for a specific return period or acceptable failure rate of a structure such as a culvert or terrace. Continuous simulation models can also be used for design purposes but are better suited for evaluating the long term effects of management systems on runoff and erosion.

Although the curve number method has been used for event models, the coupled infiltration-kinematic wave model approach has gained increasing popularity in recent years partly because the approach is based on the fundamentals of conservation of mass and momentum, and partly because of the increase of computer speed. The model KINEROS (Woolhiser et al., 1991) is an example of this coupled approach. Infiltration is computed using the Smith-Parlange infiltration equation which is quite similar to the Green-Ampt model described in Section II.B.1. The computed rainfall excess is routed interactively using a four-point implicit numerical scheme so that infiltration is computed during the recession of the hydrograph. The input variable is a time depth distribution of rainfall and the input parameters include physical characteristics of the flow plane, a roughness coefficient (either Chezy or Manning), Smith-Parlange infiltration parameters, and erosion parameters. Because the model uses a numerical routing procedure, it can be applied to cases where the infiltration characteristics change in the downslope direction as would be the case with strip cropping or changes in soil texture.

Both the curve number method and the coupled infiltration-kinematic wave approach have been used in runoff-erosion continuous simulation models. One of the first of these type of models, CREAMS (Chemicals, Runoff, and Erosion from Agricultural Management Systems, Knisel, 1980), has both options although the curve number option has been more widely used in applications. The implementation of the curve number method in CREAMS refined the selection of a condition I, II, or III curve number from one based on an antecedent moisture class to one which is updated based on a depth weighted average of the soil moisture within the root zone. The soil moisture is computed using the water balance equation on a daily time step. Similar implementations of the curve number and water balance equation have been used by several other runoff erosion models including EPIC (Williams et al., 1983), SWRRB (Williams et al., 1985), and SPUR (Wight et al., 1987). The CREAMS approach for computing evapotranspiration component of the water balance includes input of temperature and solar radiation, depth of bare soil evaporation, and a distribution of leaf area index for the growing season. The EPIC and field scale SPUR models compute daily crop growth which is then translated into incremental changes in leaf area index. Downward movement of soil water in all the above models is computed using a simple layered storage model in which water moves to the next layer when the storage capacity is exceeded in the current year. The infiltration-kinematic wave option of CREAMS is similar in concept to that of the WEPP model described in this chapter with the major difference in the method the kinematic wave peak discharge is approximated. However, the approach, as with the WEPP model, requires a time-intensity distribution of rainfall. Observed rainfall data of this type is typically not available for most applications and at the time the CREAMS model was developed, there was no large data base of rainfall statistics to use in a rainfall simulation model. The WEPP model, the hydrology computations of which were described in this chapter, was developed to use a more process-based approach than the then existing runoff erosion models. With respect to the hydrologic calculations, two features distinguish it from the CREAMS approach. First, a data base comprised of climate statistics for over 1000 locations in the United States was developed to be used in the climate generator, CLIGEN (Nicks and Lane, 1989). The CLIGEN model uses a rainfall disaggregation model based on a double exponential distribution to compute a rainfall intensity distribution from four rainfall characteristics; depth, duration, time to peak intensity, and peak intensity. Second, the values of infiltration and runoff parameters are adjusted on a daily basis to account for the effects of management and climate. For example, a soil-model computes the changes in soil bulk density and random roughness immediately following a tillage operation and adjusts the infiltration parameters and the Chezy coefficient to reflect these changes. As time increases and the soil becomes consolidated, the soil submodel computes the increase in bulk density and decrease in random roughness which is then used to update the infiltration and runoff parameters.

At present, there are no objective criteria for the selection of a particular simulation model or methodology for a given situation. Key issues in model selection are the complexity of the model, input variable and parameter uncertainty, and systematic model errors. According to Lane and Nichols (1993) model complexity is defined as a function of the number of input variables, input parameters, and number of model runs required to do a noninteractive sensitivity analysis; variable and parameter uncertainty is defined as a function of the coefficient of variation of the variables and parameters; and model error is defined as a function of the number of basic physical laws contained within the model. They found using these definitions that the curve number method has low complexity and parameter uncertainty and high systematic model error while the Green-Ampt approach has high complexity and uncertainty but low systematic error. Their approach is a step in providing objective criteria to merge model selection with complexity, uncertainty, and error.

V. SUMMARY AND DISCUSSION

We have shown that runoff is a primary driving variable in the water induced erosion process. Although the impact of rain drops detaches soil particles, it is primarily flowing water that transports detached particles to areas of concentrated flow, and depending on the sediment load and the transport capacity of the flow, detaches or re-entrains previously detached particles or deposits particles being transported. The relative importance of runoff in erosion and sedimentation processes increases with increasing scale. This is why many of the conservation practices employed to reduce erosion at the field scale are in fact practices to reduce the volumes and rate of runoff. Accurate estimation of the effects of an agricultural management system or individual practice on the volume and rate of runoff are essential in the evaluation and planning associated with soil and water conservation.

Methodologies to estimate runoff derived at point scales and watershed scales have been applied to field scale runoff processes. Two approaches for modeling runoff, the U.S. Soil Conservation Service Curve Number method and the infiltration-kinematic wave method, are described in this chapter. Examples of the first approach describe the SCS method for computing direct storm runoff and the SCS graphical method for computing peak discharge. Examples of the second approach describe the Green-Ampt Mein-Larsen model for computing rates and volumes of infiltration and rainfall excess, a method for computing depression storage, and the use of the kinematic wave model for computing the runoff hydrograph and peak discharge. Approximate methods commonly used in the second approach include the Li approximation for computing total infiltration amount and rainfall excess volume, an approximate method for computing peak discharge, and a method for estimating recession infiltration.

The type of approach taken will depend on the data available and the objectives of the analysis. For example, a study to screen management systems to replace a current system may only need an estimate of annual average runoff differences among the systems while a study to design the capacity of a grassed waterway may need detailed information of the flow profile down the channel length.

Factors influencing runoff include rainfall characteristics, soil properties, antecedent moisture, as well as land use and management practices. Of critical importance for future research on improved runoff estimation methodologies are the interactions and feedback mechanisms between these factors, and how they vary with time, space, and process intensity scales. Knowledge of these factors and their scale properties are necessary to extend knowledge of point processes to processes operating at the field scale.

REFERENCES

Chu, S. T. 1978. Infiltration during an unsteady rain. *Water Resources Research* 14(3): 461–466.

Eagleson, P. S. 1970. *Dynamic Hydrology*. McGraw-Hill, New York, 462 pp.

Engman, E. T. 1989. The applicability of Manning's n values for shallow overland flow. Proceedings of the International Conference for Centennial of Manning's Formula and Kuichling's Rational Formula, B. C. Yen, Ed. 22–26 May 1989, University of Virginia. pp. 299–308.

Gayle, G. A. and R. W. Skaggs. 1978. Surface storage on bedded cultivated lands. *Transactions of the ASAE* 21(1): 101–104, 109.

Green, W. H. and G. A. Ampt. 1911. Studies in soil physics. I. The flow of air and water through soils. *Journal of Agricultural Science* 4: 1–24.

Horton, R. E. 1940. An approach toward a physical interpretation of infiltration capacity. *Soil Science Society of America Proceedings* 5: 399–417.

Knisel, W. G. 1980. CREAMS: A Field-Scale Model for Chemicals, Runoff, and Erosion from Agricultural Management Systems. USDA, Conservation Res. Rep. No. 26, 640 p.

Lane, L. J. and M. H. Nichols. 1993. Complexity, uncertainty, and systematic error in hydrologic models. In GIS-Hydrologic Modeling-DSS Technical Workshop. Occasional Paper No. 11/94. Land and Water Resources Research and Development Corporation, Canberra ACT, Australia. 51 p.

Li, R. M., M. A. Stevens, and D. B. Simons 1976. Solutions to Green-Ampt infiltration equation. *ASCE Journal of Irrigation and Drainage* 102(2): 239–248.

Mein, R. G. and C. L. Larson. 1973. Modeling infiltration during a steady rain. *Water Resources Research* 9(2): 384–394.

Mitchell, J. K. and B. A. Jones, Jr. 1976. Micro-relief surface depression storage: analysis of models to describe the depth-storage function. *Water Resources Bulletin* 12(6): 1205–1222.

Moore, I. D. and C. L. Larson. 1979. Estimating micro-relief surface storage from point data. *Transactions of the ASAE* 20(5): 1073–1077.

Nearing, M. A., G. R. Foster, L. J. Lane, and S. C. Finkner. 1989. A process-based soil erosion model for USDA-Water Erosion Prediction Project technology. Transactions of the ASAE 32(5): 1587–1593.

Nicks, A. D. and L. J. Lane. 1989. Weather Generator. In *USDA-Water Erosion Prediction Project: Hillslope Profile Model Documentation.* L. J. Lane and M. A. Nearing (editors). NSERL Report No. 2 USDA-ARS National Soil Erosion Research Laboratory, West Lafayette, In.

Onstad, C. A. 1984. Depression storage on tilled soil surfaces. *Transactions of the ASAE* 27(3): 729–732.

Philip, J. R. 1957. The theory of infiltration: 4. Sorptivity and algebraic infiltration equations. *Soil Science* 84: 257–264.

Rawls, W. J., D. L. Brakensiek, and K. E. Saxton. 1982. Estimation of soil water properties. *Transactions of the ASAE, Special Edition, Soil and Water* 25: 1316–1320

Singh, V. J. 1989. *Hydrologic Systems: Watershed Modeling,* Vol. II. Prentice-Hall, Englewood Cliffs, N.J.

Soil Conservation Service (SCS). 1971. National Engineering Handbook. Section 4. Hydrology. USDA-SCS. Washington, D.C.

Soil Conservation Service (SCS). 1973. A method for estimating volume and rate of runoff in small watersheds. USDA-SCS. TP-149. Washington, D.C.

Soil Conservation Service (SCS). 1986. Urban Hydrology for Small Watersheds. USDA-SCS Engineering Division. Technical Release 55. Washington, D.C.

Stone, J. J., L. J. Lane, and E. D. Shirley. 1992. Infiltration and runoff simulation on a plane. *Transactions of the ASAE* 35(1): 161–170.

Stone, J. J., E. D. Shirley, and L. J. Lane. 1993. Impact of recession infiltration on runoff volume computed by the kinematic wave model. *Transactions of the ASAE* 36(5): 1353–1361.

Wight, J. R. and J. W. Skiles eds. 1987. SPUR. Simulation of Production and Utilization of Rangelands. Documentation and User Guide. USDA, ARS-63. 372 p.

Williams, J. R., A. D. Nicks, and J. G. Arnold. 1985. Simulator for water resources in rural basins. *ASCE Journal of Hydraulic Engineering.* 111(6): 970–986.

Williams, J. R., K. G. Renard, and P. T. Dyke. 1983. A new method for assessing the effect of erosion on productivity-The EPIC model. *Journal of Soil Water Conservation*. 38: 381–383.

Wischmeier, W. H. and D. D. Smith. 1978. Predicting rainfall erosion losses—a guide to conservation planning. Agr. Handbook No. 537. USDA, Washington.

Woolhiser, D. A. and J. A. Liggett. 1967. Unsteady, one-dimensional flow over a plane—the rising hydrograph. *Water Resources Research*, 3(3): 753–771.

Woolhiser, D. A., R. E. Smith, and D. C. Goodrich. 1990. KINEROS, a Kinematic Runoff and Erosion Model; Documentation and User Manual. ARS Publication No. 77.

Zobeck, T. M. and C. A. Onstad. 1987. Tillage and rainfall effects on random roughness. A review. *Soil Tillage Research* 9: 1–20.

11

Common Soil and Water Conservation Practices

Paul W. Unger
Conservation and Production Research Laboratory, Agricultural Research Service, U.S. Department of Agriculture, Bushland, Texas

I. INTRODUCTION

Conservation of the world's soil and water resources is essential for sustaining food and fiber production for an ever-increasing human population and for preserving the environment. The basic factors, processes, and forms of soil erosion by water have been discussed in previous chapters. Briefly, erosion is possible wherever raindrops strike or water flows across unprotected soil surfaces. Both actions may detach particles from the soil mass and transport them across the soil surface. Hence, the primary means of reducing or preventing erosion by water are to reduce or prevent particle detachment and transport of the particles in runoff (overland water flow) across the surface. Reducing runoff is also important for conserving water. While runoff may occur on virtually any land, the emphasis of this chapter will be on soil and water conservation practices for lands that are suited for cultivated crop production, namely, land-use capability classes I, II, III, and IV (SCSA, 1982). Practices for other lands that affect soil and water conservation on cropland will be discussed where appropriate.

II. PARTICLE DETACHMENT AND TRANSPORT

Energy for particle detachment and transport is derived from raindrops that strike the soil surface and from runoff across the soil surface. Thus, to reduce detachment and transport, the energy must be dissipated by other means. The kinetic energy of rainfall is effectively dissipated by a soil surface cover of crop plants or residues. For maximum effectiveness, the entire soil surface must be covered, but a partial cover may reduce particle detachment and soil losses to tolerable levels.

Runoff detaches soil particles due to the energy provided, with detachment usually increasing with increases in runoff velocity. Hence, runoff velocity must be decreased to reduce particle detachment. While surface cover decreases particle detachment by raindrops, it has no direct effect on particle detachment by runoff. For reducing runoff velocity, plant stem density is the controlling factor. The stems must be capable of withstanding the force and thereby dissipating the energy of the runoff.

Particle detachment may result also from soil wetting that causes dispersion of unstable aggregates. When aggregates disperse, surface sealing usually occurs (FAO, 1965), which increases runoff with a concomitant potential increase in further particle detachment and soil loss. Neither surface cover nor plant density directly affects dispersion of unstable aggregates due to soil wetting.

III. APPROACHES TO SOIL AND WATER CONSERVATION

Runoff is the primary carrier of soil particles from cropland and, hence, the ultimate cause of soil erosion by water. Thus, to control erosion, runoff must be eliminated or reduced to a water flow rate that cannot transport detached soil particles, if detached particles are present. Because water lost as runoff also is of no direct benefit for crop production on land where the runoff occurred, controlling runoff is essential also for water conservation purposes.

Soil and water conservation practices are divided into three approaches: (1) reducing runoff—practices included are designed to prevent initiation of runoff; (2) retaining runoff—accepting initiation of runoff, yet reducing distance of runoff flow and retaining it in the field; (3) controlling runoff—practices accept initiation of runoff and flow out of the field, yet with minimum erosion hazard.

A. Reducing Runoff

Reduced sediment transport from fields and increased water conservation is achieved by reducing the amount and velocity of runoff. Reducing the amount lessens the quantity of water available for sediment transport, whereas reducing the velocity provides time for sediments to settle from water before they are transported from the field. It also provides more time for water to infiltrate the soil, thus improving water conservation. Practices that reduce runoff and, hence, reduce sediment transport and improve water conservation include conservation tillage, mulches, cover crops, and chemical additives. These practices are directed toward preventing runoff initiation by intercepting raindrops (mulches, covers) and increasing soil aggregate stability (soil conditioners). Both methods maintain relatively high infiltration rates and, therefore, minimize runoff and the ensuing erosion.

1. *Conservation Tillage*

Conservation tillage, which, by definition (CTIC, 1990), is a crop residue management system, has been widely acclaimed for its erosion control benefits. Where water erosion is of primary concern, at least 30% of the soil surface must be covered by residues after planting for a system to qualify as conservation tillage. In practice, however, 30% surface coverage by residues may not be adequate on some soils; other soils may not need as much surface cover to protect them against erosion (Kemper and Schertz, 1992).

Conservation tillage reduces stormwater runoff primarily because the surface residues intercept falling raindrops, thus reducing soil surface sealing and maintaining favorable water infiltration rates. Conservation tillage, however, also reduces runoff due to increased hydraulic roughness resulting from crop residues on the soil surface. As hydraulic roughness increases, runoff velocity is reduced because of the increasing tortuosity of the runoff path. In addition to the increased hydraulic roughness caused by residues alone, residues also add to the existing roughness of the soil itself (Gilley, Kottwitz, and Wieman, 1991). Residues cause water ponding on the surface and obstruct or divert water flow across the surface. The reductions in runoff flow velocity provide more time for water infiltration. Reduced runoff and increased infiltration are not indicative of increased surface water storage per se. However, for greater infiltration to occur, water must be retained on the surface for a longer time in most cases. Surface residues with conservation tillage provide this benefit with respect to water infiltration. Provided adequate residues are available, some types of conservation tillage are mulch tillage, reduced tillage, and no tillage.

Mulch tillage. Mulch tillage is the practice of tilling a soil in a manner that retains most crop residues on the soil surface. Mulch tillage is also called mulch farming, trash farming, stubble mulch tillage, subsurface tillage, and plowless farming (SSSA, 1987). Tillage implements that retain most residues on the soil surface include (a) sweeps—60 cm or wider, (b) rodweeders—without or with semichisels or small sweeps, (c) straight-blade machines, (d) chisel plows, and (e) one-way plows (can be used when large amounts of residue are present) (Unger, 1984). The goal of mulch tillage is to control weeds and prepare a seedbed for the next crop, yet retain adequate residues on the surface to control erosion and improve water conservation. Typical amounts of the original residues that remain after each operation with various implements are given in Table 1.

The mulch-tillage farming system was developed to combat wind erosion similar to that which plagued portions of the U.S. Great Plains states and the Canadian prairie provinces during a major drought in the 1930s. Although developed for wind erosion control, the value of mulch tillage for controlling water erosion and conserving water was soon recognized (McCalla and Army, 1961). Mulch tillage aids water erosion control and water conservation because the surface residues dissipate the energy of raindrops, maintain relatively high infiltration rates, and reduce runoff, provided the residues remain anchored in the soil to avoid being carried away by the flowing water.

With 2.2 Mg ha^{-1} of wheat (*Triticum aestivum* L.) residues on the surface, annual soil losses with mulch tillage were 2.8 Mg ha^{-1} in a maize (*Zea mays* L.)–oats (*Avena sativa* L.)–wheat rotation at Lincoln, Nebraska (USA), and 8.1 Mg ha^{-1} in a wheat-fallow system at Pullman, Washington (USA). With moldboard plowing, which eliminated most surface residues, annual soil losses at the two locations were 13.4 and 40.2 Mg ha^{-1}, respectively (Zingg and Whitfield, 1957).

When surface residue amounts are low, mulch tillage may reduce runoff by disrupting soil surface crusts and creating surface depressions, but sediment loss may not follow the same trend. On Pullman clay loam (fine, mixed, thermic Torrertic Paleustoll) at Bushland, Texas (USA), runoff from 1984 to 1991 averaged 27 and 43 mm annually from mulch-tillage and no-tillage watersheds, respectively, that were cropped to dryland winter wheat and grain sorghum [*Sor-*

Table 1 Effect of Tillage Machines on Surface Residue Remaining after Each Operation

Tillage machine	Approximate residue maintained (%)
Surface cultivators	
wide-blade cultivator, rodweeder	90
Mixing-type cultivators	
heavy-duty cultivator, chisel, other similar-type machines	75
Mixing and inverting disk machines	
one-way flexible disk harrow, one-way disk, tandem disk, offset disk	50
Inverting machines	
moldboard, disk plow	10

Source: Anderson (1968).

ghum bicolor (L.) Moench] in a three-year rotation. The greater runoff from the no-tillage watershed occurred mainly during fallow after grain sorghum harvest when residue cover was limited and major surface sealing occurred on the exposed bare soil. However, even though runoff was greater with no tillage, average soil-water contents at planting time for wheat and grain sorghum were greater on the no-tillage than on the mulch-tillage watershed. The greater water content was attributed to reduced evaporation with no tillage.

Soils in semiarid regions often dry to the depth of tillage after each operation, and that water must be replaced before water storage can occur at greater depths. In some years, dryland crops may not produce adequate residues for either tillage method to qualify as conservation tillage, based on the 30% surface cover requirement. Therefore, soil and water conservation may be limited. However, annual soil losses on Pullman clay loam were 0.40 Mg ha^{-1} with mulch tillage and 0.19 Mg ha^{-1} with no tillage (Jones, 1992). The greatest annual soil losses were 7.3 and 2.3 Mg ha^{-1} from mulch-tillage and no tillage watersheds, respectively (personal communication, O. R. Jones, Bushland, Texas, 1992). The *T*-value (tolerable annual soil loss rate) for this soil is 11.2 Mg ha^{-1} (5 tons $acre^{-1}$).

For a simulated rainfall study on Pullman clay loam in the field at Bushland, Texas, soil losses for dry and wet runs after dryland grain sorghum were greater with moldboard plowing than for mulch (sweep) tillage, for which residues were either removed or retained in place. After dryland wheat, losses tended to be lower with mulch tillage, but were not significantly lower. Losses were, or tended to be, even lower with no tillage (Table 2). Duration of water application was similar on moldboard-plowed and mulch-tillage plots, which points to the advantage of surface residues for controlling soil losses. Both tillage methods thoroughly loosened the soil to similar depths (about 15 cm).

Reduced tillage. Controlling weeds is a major reason for tilling a soil. Hence, tillage can be reduced if weeds can be controlled by other means, such as herbicides, which control weeds during at least part of the crop production cycle. The reduction in tillage may be in intensity or frequency of operation. As for mulch tillage, major goals are erosion control and water conservation, which are achieved by retaining residues on the surface as long as possible and especially during major erosive periods. Some major reduced-tillage systems, along with some variations of those systems, are

- Fall (autumn) plow, field cultivate
- Spring plow, spring cultivate
- Fall (autumn) chisel, field cultivate
  - Chisel, plant
  - Chisel, disk + harrow
  - Chisel, field cultivate + harrow

Table 2 Sediment Losses during Dry or Wet Runs of Simulated Rainfall on Pullman Clay Loam as Affected by Previous Crop and Tillage Method, Bushland, Texas

Previous crop and measurement	Tillage treatments[a] ($Mg\ ha^{-1}$)					
	Plow	Roto	Sweep-RR	Sweep-RL	Notill-RR	Notill-RL
Wheat						
dry run	4.8a[b]	2.5b	0.21b	0.26b	0.14b	0.09b
wet run	4.3a	1.9bc	0.16bc	0.21b	0.12bc	0.09c
Sorghum						
dry run	4.4a	3.1ab	0.29ab	0.18bc	0.13c	0.10c
wet run	2.7a	2.1ab	0.24a	0.14ab	0.11ab	0.06b

[a]Tillage treatments were plow—moldboard plowing plus disking; roto—rotary tillage; sweep-RR—sweep tillage with residues removed; sweep-RL—sweep tillage with residues left in place; notill-RR—no tillage with residues removed; notill-RL—no tillage with residues left in place.
[b]Row values for a given run followed by the same letter or letters are not significantly different at the 0.05 probability level (Duncan's multiple range test).
Source: Unger (1992a).

Combination coulter or disk-chisel, and
Alternate chisel-moldboard plow
Disk and plant
Till-plant (for example, lister or sweep tillage)
Strip tillage (for example, rotary tillage)
Tillage-herbicide combinations

Variations of the above systems are numerous, depending on equipment available, producer preferences, soil conditions, and crops grown. As compared with clean tillage, which eliminated most surface residues, erosion (sediment transport) generally was reduced as tillage intensity or frequency was reduced because of the greater amounts of crop residues retained on the soil surface (Unger, 1984).

No tillage. No tillage (also known as zero tillage, slot planting, ecofallow, sod planting, chemical fallow, and direct drilling) is the method of planting crops with no seedbed preparation or surface disturbance other than opening the soil for the purpose of placing the seed at the desired depth. Seed placement can be achieved by cutting a small slit or punching a hole in the soil. With no tillage, weeds normally are controlled with herbicides and crops are not cultivated. To facilitate planting, up to 25% of the soil surface can be disturbed or tilled in a no-tillage system (Lessiter, 1982).

The tremendous value of no-tillage crop production for reducing sediment losses was demonstrated for a severe rainstorm in Ohio (USA) (Table 3). For

Table 3 Runoff and Sediment Yield from Maize Watersheds at Coshocton, Ohio (USA), during a Severe Rainstorm

Tillage	Slope (%)	Rainfall (%)	Runoff (%)	Sediment yield (Mg ha^{-1})
Plowed, clean-tilled sloping rows	6.6	140	112	50.7
Plowed, clean-tilled contour rows	5.8	140	58	7.2
No-tillage, contour rows	20.7	129	64	0.07

Source: Harrold and Edwards (1972).

that storm, sediment loss from the no-tillage watershed was negligible compared with that from other watersheds, even though the slope of the no-tillage watershed was much greater. No tillage is especially effective for reducing runoff when surface residue amounts are adequate and there are no inherent or induced soil characteristics that hinder infiltration. The literature contains numerous other reports showing the value of no tillage for reducing runoff or soil losses; examples are shown in Tables 4 to 9. Under most conditions, runoff or soil losses were lowest with no tillage, with the reduction in soil losses being greater than the reduction in runoff. Factors contributing to a smaller runoff reduction than soil loss reduction, with no tillage, include the fact that the soil may be already saturated with water, thus infiltration is reduced, and when the soil may have a restrictive layer than hinders infiltration. In contrast, surface residues with no

Table 4 Runoff and Soil Loss from Bedford Silt Loam (9% Slope) during 1-h Simulated Rainstorms, 1972[a]

	Runoff (mm)		Soil loss (Mg ha^{-1})	
Tillage system	1st Hour	2nd Hour	1st Hour	2nd Hour
Spring plow, disk, plant	45	56	23.3	27.1
Spring chisel, field cultivate, plant	11	38	0.7	2.5
Till-plant	37	50	7.4	8.5
No-tillage	26	39	3.6	3.8

[a]Storms of 65 mm h^{-1} were applied within four weeks after planting maize in rows that ran across the slope.
Source: Griffith et al. (1977).

Table 5 Comparisons of Surface Cover, Soil Erosion, and Runoff from a Rainfall Simulation Study for Bare-Fallow, Stubble-Mulch, and No-Tillage Wheat-Fallow System

Period	System	Cover (%)	Runoff (%)	Soil loss (Mg ha^{-1})
Fallow after havest (October)	Bare fallow	62	9	0.66
	Stubble mulch	91	15	0.80
	No tillage	91	11	0.72
Fallow after tillage (May)	Bare fallow	4	36	9.40
	Stubble mulch	92	9	0.21
	No tillage	96	1	0.17
Wheat 10-cm tall (October)	Bare fallow	26	35	7.25
	Stubble mulch	38	24	2.56
	No tillage	85	5	0.55
Wheat 45-cm tall (May)	Bare fallow	78	43	2.09
	Stubble mulch	83	29	0.84
	No tillage	88	16	0.34

Source: Dickey et al. (1983).

tillage reduce runoff flow velocity, thus providing more time for sediments to settle from the sediment-laden runoff water. Of course, the residues also minimize soil particle detachment, both by raindrop impact and by flowing water.

Dickey et al. (1983) compared surface cover, runoff, and soil losses under simulated rainfall conditions for a wheat-fallow system where different tillage methods were used (Table 5). Runoff and soil losses were similar at the first sampling when cover amounts were relatively high with all treatments. At subsequent samplings, cover was always greatest and runoff least with the no-tillage system. In all cases, soil losses were <1.0 Mg ha^{-1} with the no-tillage treatment. Similar reductions in runoff were reported for a study involving maize on different soil slopes in Nigeria (Table 8). In this study, soil losses were negligible with no-tillage on all slopes.

For a study on a silt loam soil, Gilley et al. (1986a) uniformly spread maize residues on the surface at rates of 0, 1.12, 3.36, 6.73, or 13.45 Mg ha^{-1} after the soil was tilled. The respective surface covers were 0%, 10%, 31%, 51%, and 83%. Simulated rainfall was applied at a 28-mm h^{-1} rate on days 1, 2, and 3 of the study. Average equilibrium runoff rates with the different surface covers were 15.6, 10.7, 6.0, 1.8, and 0 mm h^{-1}, respectively.

The effect of surface cover provided by grain sorghum and soybean (*Glycine max* L.) on runoff of simulated rainfall from a silty clay soil was studied by Gilley, Finkner, and Varvel (1986b). Water was applied at a rate of 48 mm h^{-1}.

Table 6 Soil Losses under Different Tillage Systems and Crops Using a Rainfall Simulator

Crop and treatments	Soil loss[a] 1975/76 (Mg ha^{-1})	Soil loss[a] 1975/76 (%)	Soil loss[a] 1978/79 (Mg ha^{-1})	Soil loss[a] 1978/79 (%)
Soybean				
Conventional tillage (previous wheat crop also conventional tilled)	7.4	100	7.5	100
Conventional tillage (fallowed previous to soybean)	—	—	1.3	17.3
Direct drilled (previous wheat crop, direct drilled)	2.1	28.4	1.1	14.7
Wheat				
Conventional tillage (previous soybean crop conventional tillage)	4.7	100	3.0	100
Direct drilled (previous soybean crop direct drilled)	3.3	70.2	2.0	66.7
Bare soil	103.0		97.4	

[a]Soil losses are the total erosion losses during the whole vegetative period and after harvest, including the following growing stages: I—sowing to 30 days; II—30–60 days; III—60–90 days, and IV—after harvest.

Source: Sidiras et al. (1983).

The equilibrium runoff rate decreased by 73% as sorghum residue cover increased from 0% to 44%. It decreased 68% as soybean residue cover increased from 0% to 56%.

For most field studies, runoff responses indicate the combined effect of tillage methods on residue and soil properties. Residues are either retained on the surface or incorporated with soil. Some soil properties affected are porosity and random and oriented roughness. An increase in these properties due to tillage usually causes water to infiltrate the soil more rapidly or to be retained in surface depressions and then infiltrate. Residue-related hydraulic roughness is maximum when residues are retained on the surface. In contrast, tillage increases soil-related hydraulic roughness and decreases residue-related hydraulic roughness when residues are incorporated. However, residue incorporation may cause soil-related hydraulic roughness to decrease more rapidly with time (rainfall) than where residues are retained on the soil surface. Surface residues stabilize soil

Table 7 Effect of Tillage Practices on Soil Loss (Mg ha^{-1}) in 1976 and 1977

Year	Treatments	1st season	2nd season	Total
1976	No tillage	0.04	0.01	0.05
	Manual	3.41	1.20	4.61
	Plow	5.01	2.75	7.76
	Conventional	5.27	3.19	8.46
	Bare fallow	10.22	7.41	17.63
	LSD (P = 0.05)	3.18	2.53	4.50
1977	No tillage	0.06	0.02	0.08
	Manual	4.10	2.50	6.60
	Plow	5.90	2.95	8.85
	Conventional	6.20	3.50	9.80
	Bare fallow	11.56	7.96	19.52
	LSD (P = 0.05)[a]	3.21	2.40	4.63

[a]Least significant difference.
Source: Osuji et al. (1980).

surfaces, whereas bare surfaces are subject to the impact of falling raindrops, which may cause aggregate dispersion, surface sealing, and increased runoff.

In general, runoff is greater where conservation tillage systems are used than where practices such as contouring and furrow diking are used (providing depressions for retaining water on the surface). However, water storage in soil over an extended period such as a fallow period between crops may be similar or even greater with conservation tillage than with contouring or furrow diking. Russel (1939) showed that basin listing (furrow diking) prevented runoff, but

Table 8 Effect of Tillage on Runoff and Soil Losses from Land Cropped to Maize in Nigeria[a]

	Bare fallow		Plowed		No tillage	
Slope (%)	Runoff (%)	Soil loss (Mg ha^{-1})	Runoff (%)	Soil loss (Mg ha^{-1})	Runoff (%)	Soil loss (Mg ha^{-1})
1	18.8	0.2	8.3	0.04	1.2	0.0007
5	20.2	3.6	8.8	2.16	1.8	0.0007
10	17.5	12.5	9.2	0.39[b]	2.1	0.0047
15	21.5	16.0	13.3	3.92	2.2	0.0015

[a]Rainfall was 44.2 mm.
[b]Probably an error.
Source: Rockwood and Lal (1974).

Table 9 Effects of Plowing and Maize Stover on Runoff and Soil Loss from Guelph Loam (8% Slope) at Guelph, Ontario (Canada)

	Mean annual losses			
	May to October, nine-year period		November to April, six-year period	
Treatment	Runoff (mm)	Soil (Mg ha^{-1})	Runoff (mm)	Soil (Mg ha^{-1})
Stover left on field				
Not plowed	11	31	31	2
Fall plowed	25	246	82	83
Stover removed				
Not plowed	37	381	106	76
Fall plowed	25	403	83	218

Source: Ketcheson (1977).

resulted in the greatest evaporative losses of water. Consequently, soil-water storage during the study period was less than with some surface-residue treatments (Table 10). Unger (1992b) compared the effects of open- and diked-furrow treatments on runoff and soil-water storage on fields used for a wheat-sorghum-fallow rotation. No tillage was used after wheat, reduced tillage after sorghum. Furrow diking treatments did not significantly affect average runoff and water storage, again indicating the value of surface residues for reducing runoff and conserving water.

Although conservation tillage systems have been widely promoted because of their soil and water conservation benefits, there are also some real or perceived disadvantages associated with these systems. Often mentioned is the increased use of chemicals (herbicides, insecticides, fungicides, and fertilizers), which may increase crop production costs and introduce hazardous substances into the environment. In addition, conservation tillage, especially no tillage, is not well adapted to poorly drained soils, soils in cold climates, or to soils in a degraded condition. Under such conditions, crop yields often are lower with conservation tillage than with conventional (clean) tillage.

2. *Mulches*

Crop residues are a type of mulch, and they have been discussed in previous sections with respect to reducing runoff, sediment transport or losses, and water conservation. Other mulching materials that affect runoff and, hence, sediment transport and water conservation include stones, gravel, paper, coal, and bitu-

Table 10 Water Storage, Runoff, and Evaporation from Field Plots at Lincoln, Nebraska, 10 April to 27 September 1939

Treatment	Storage (mm)	Runoff (mm)	Evaporation loss (mm)	Evaporation loss (%)[a]
Straw, 2.2 Mg ha^{-1}, normal subtillage	30	26	265	83
Straw, 4.5 Mg ha^{-1}, normal subtillage	29	10	282	88
Straw, 4.5 Mg ha^{-1}, extra-loose subtillage	54	5	262	82
Straw, 9.0 Mg ha^{-1}, normal subtillage	87	Trace	234	73
Straw, 17.9 Mg ha^{-1}, no tillage	139	0	182	57
Straw, 4.5 Mg ha^{-1}, disked in	27	28	266	83
No straw, disked	7	60	254	79
Contour basin listing	34	0	287	89

[a]Based on total precipitation, which was 321 mm for the period.
Source: Russel (1939).

men. Use of these materials usually is limited to high-value crops and where labor is abundant (Unger, 1984).

3. *Cover Crops*

Cover crops are close-growing crops such as grasses, legumes, or small grains that are grown primarily to provide seasonal protection against erosion and for soil improvement. They usually remain on the land for less than one year (SCS, 1977). In practice, cover crops affect runoff, sediment transport, and water conservation in a manner similar to that resulting from other growing crops and crop residues. As plant density and cover increase, the potential for runoff decreases (Sidiras et al., 1985; Roth et al., 1987). A major disadvantage of using cover crops, especially in subhumid and semiarid regions, is that these crops use water that could be used by the next crop. Hence, timely precipitation or irrigation is essential for achieving good crop yields where cover crops are used.

4. *Chemical Additives*

A major factor contributing to high runoff is surface sealing that results from raindrop impact or water flowing across soils having low-stability surface aggregates, especially when few or no crop residues are on the surface. Under such conditions, application of materials that stabilize the surface could reduce aggregate breakdown and surface sealing. As a result, infiltration may remain at favorable rates, which could improve soil and water conservation.

A material that has shown promise to conserve soil and water on low-stability soils is phosphogypsum (PG). When PG was surface applied to a ridged sandy soil in a field in Israel at a rate of 10 Mg ha^{-1}, runoff was reduced 6-fold and soil movement from ridge tops to furrow bottoms was reduced 20-fold. Under laboratory conditions using a rainfall simulator on the same soil, increasing the slope from 5% to 25% resulted in a 2-fold increase in soil loss when PG was applied and a 12-fold increase for the control treatment. The PG was highly effective for preventing erosion on the steeply sloping soil (Agassi et al., 1989).

Application of PG at 3.0 Mg ha^{-1} to a clay loam in Texas reduced runoff compared to a bare soil treatment, but the reduction was not as large as that obtained with 2.2 Mg ha^{-1} of wheat straw on the surface. In addition to the reduced runoff, soil drying was slowest with the PG treatment, which resulted in greater sorghum seedling emergence than the bare soil treatment. Final crust strength, however, was greatest where PG was applied (Benyamini and Unger, 1984).

Lentz et al. (1992) injected anionic polymers [polyacrylamide (PAM) or starch copolymer solutions] into irrigation water for the first 1 to 2 h of an 8-h irrigation of a silt loam. Sediment loss from treated furrows was significantly reduced as compared with that from untreated furrows for the first (treated water) and second (nontreated water) irrigations, but not for the fourth (nontreated water) irrigation. The PAM treatment resulted in better erosion control than the starch treatment. For the initial irrigation at low flow rates, the mean sediment load in runoff from furrows treated with water having a 10-mg kg^{-1} PAM concentration was reduced 97% compared with that from furrows irrigated with untreated water. Sediment losses in treated furrows were reduced about 50% for subsequent irrigations with untreated water.

B. Retaining Runoff (in the field)

Practices that retain runoff on cropland include contour tillage, furrow dikes, level terraces, and land leveling. These involve some type of soil surface manipulation and their effectiveness, in general, increases in the order listed. These practices prevent runoff under small-storm conditions, but water from large storms may overtop and wash out the earthen structures.

1. Contour Tillage

Contouring involves performing cultural operations such as plowing, planting, and cultivating across the slope of the land so that elevations along rows are as near to level as practical. When ridge-forming tillage, for example, lister tillage, is used on the contour and the resultant furrows are blocked at the ends, runoff and sediment transport are prevented for storms of low to moderate intensity and duration. However, such tillage may provide little or no protection against runoff and sediment transport for severe storms that may cause extensive overtopping and breaking of the contoured ridges.

The degree of protection afforded by a given contour tillage system increases as size of ridges increases and is influenced also by soil type and slope. The effectiveness and persistence of ridges for preventing runoff are greater on stable soils than on soils that settle extensively when wetted. In general, fine-textured soils are more stable than coarse-textured soils in this regard.

The potential for runoff and, hence, erosion on contoured land generally increases with increases in land slope (Table 11). The potential is lowest for slopes of 2% to 7%. Below this range, the land slope approaches equality with the contour row slope and the soil loss ratio (*P*-value) approaches 1.0. At slopes greater than 2% to 7%, contour row capacity decreases and the *P*-value again approaches 1.0 (Wischmeier and Smith, 1978).

Lister tillage on the contour under some conditions results in each ridge serving as a miniature level terrace, thus promoting uniform water storage over the entire field and reducing the potential for runoff and sediment transport. At

Table 11 Values of Support-Practice Factor, *P*

	Land slope (%)				
Practice	1.1–2	2.1–7	7.1–12	12.1–18	18.1–24
Contouring	0.60	0.50	0.60	0.80	0.90
Contour strip cropping					
R-R-M-M[a]	0.30	0.25	0.30	0.40	0.45
R-W-M-M	0.30	0.25	0.30	0.40	0.45
R-R-W-M	0.45	0.38	0.45	0.60	0.68
R-W	0.52	0.44	0.52	0.70	0.90
R-O	0.60	0.50	0.60	0.80	0.90
Contour listing or ridge planting	0.30	0.25	0.30	0.40	0.45
Contour terracing[b]	$0.6/\sqrt{n}$[c]	$0.5/\sqrt{n}$	$0.6/\sqrt{n}$	$0.8/\sqrt{n}$	$0.9/\sqrt{n}$
Non support practice	1.0	1.0	1.0	1.0	1.0

[a]R = row crop, W = autumn-seeded grain, O = spring-seeded grain, M = meadow. The crops are grown in rotation and so arranged on the field that row-crop strips are always separated by a meadow or winter-grain strip.

[b]These values estimate the amount of soil eroded to the terrace channels and are used for conservation planning. For prediction of off-field sediment, these values are multiplied by 0.2.

[c]*n* = number of approximately equal-length intervals into which the field slope is divided by the terraces. Tillage operations must be parallel to the terraces.

Source: Stewart et al. (1975).

Spur, Texas (USA), runoff from 1927 to 1952 averaged 50 mm yr^{-1} with contour rows and 70 mm yr^{-1} with sloping rows. Cotton (*Gossypium hirsutum* L.) lint yields averaged 211 and 131 kg ha^{-1} annually on the respective areas (Fisher and Burnett, 1953). In the southern Great Plains (USA), flat tillage on the contour (sweep or one-way-disk plowing) as compared with such tillage without regard to land slope increased wheat grain yields by 10% (Finnell, 1944), which was much less than the increase for cotton planted on contour ridges. Flat tillage on the contour has limited value for preventing runoff and sediment transport and for improving water conservation.

In India, with 1300-mm annual precipitation and 25% soil slope, runoff averaged 52 and 29 mm from sloping-row and contoured watersheds, respectively. Soil losses averaged 88 and 33 Mg ha^{-1} and potato (*Solanum* spp.) yields averaged 12.6 and 13.4 Mg ha^{-1} on the respective areas. On land with 2% slope, runoff was 38% and 13%, soil loss was 14.4 and 4.4 Mg ha^{-1}, and maize yields were 1.3 and 1.9 Mg ha^{-1} on sloping-row and contoured fields, respectively (Singh, 1974).

For a 140-mm rainstorm in Ohio (USA), runoff on a clean-tillage watershed with contour furrows was about one-half as great as from a clean-tillage watershed with sloping rows (up and down the slope), and sediment loss was 50.7 Mg ha^{-1} with sloping rows but only 7.2 Mg ha^{-1} with contour rows. Sediment loss was negligible with no tillage on contour rows, even though the slope was much greater and runoff was greater on the no-tillage watershed (Table 3). Maize was grown on all watersheds.

2. *Furrow Diking*

Furrow diking (also called tied ridging, furrow damming or blocking, basin listing) results in the formation of small earthen dikes or dams across the furrows of a ridge-furrow system such as those resulting from lister tillage. The objective of furrow diking is to capture and hold rainfall in place until it infiltrates into the soil; some may be lost by evaporation. Furrow diking can be accomplished at low cost with simple equipment, and it has been widely adopted in some regions, mainly to prevent runoff for improved water conservation (Jones and Stewart, 1990). If runoff is prevented, then particle transport is prevented also.

Furrow diking is not a new practice (Hudson, 1981). In the United States it was first used in 1931, but slow operating speeds, poor weed control, erratic yield responses, and seedbed preparation and subsequent tillage difficulties caused abandonment of the practice by 1950. In the 1970s, interest in furrow diking was revived in the United States of America because of the development of improved equipment and herbicides (Jones and Clark, 1987). The principle of furrow diking is simple, but successful application of the practice to a field situation is complex. Factors, to be considered include rainfall characteristics, soil properties, and the cropping system used.

To estimate the potential of furrow diking to reduce or prevent runoff, Jones and Clark (1987) analyzed 27 years of runoff data from a wheat–grain sorghum–fallow (WSF) rotation at Bushland, Texas. They found that from 465-mm average annual precipitation, there was a 50% probability of 30-mm runoff from row-planted sorghum, 25 mm from fallow, and 12 mm from drilled wheat. They also found that storms with more than 51 mm of rainfall accounted for only 9% of the total precipitation at Bushland but for 40% of the runoff from the WSF rotation. Based on these results, they recommended that furrow diking should provide at least 60 mm of depression storage capacity to minimize the potential for runoff from the furrow-diked land. Other means of determining the required depression storage capacity for a given location include analyses based on soil properties and historical rainfall data using computer simulation modeling (Jones and Stewart, 1990).

Morin and Benyamini (1988) and Morin et al. (1984) calculated and predicted potential runoff based on soil properties and on probability analyses of historical rainfall data. Based on this information, they determined the amount of surface detention storage needed for successful performance of a furrow-diking system. Steps involved in their analyses were to (1) determine soil infiltration characteristics under a particular management regime using a rainfall simulator, (2) analyze the long-term rainfall records of the region to determine storm intensity distribution, (3) calculate runoff rates and amounts for rainstorms using a method that combines the soil infiltration function and storm intensity pattern, and (4) predict synthetic long-term runoff values using the rainfall probability distribution.

Computer models to determine effects of furrow diking on stormwater runoff were developed by Krishna and Arkin (1988), Krishna et al. (1987), Krishna and Gerik (1988), and Williams et al. (1988) for various cropping conditions in Texas. These models are useful for determining such factors as detention storage needed, when to install furrow dikes, and when dikes can be removed. When combined with crop growth models, potential effects of furrow diking on crop yields can be determined. In general, furrow diking in Texas most likely will be effective where annual rainfall ranges between about 500 and 800 mm (Krishna and Gerik, 1988). With average rainfall much below 500 mm, the probability for capturing sufficient water is low and, hence, the probability of consistent yield increases is low also. Where average rainfall exceeds 800 mm, the potential for good crop yields is relatively high without the aid of runoff water conserved by furrow diking. Of course, prevention of runoff under higher rainfall conditions does provide protection against soil erosion, even though crop yields may not be affected.

Several practices have been developed that involve the use of diked furrows to reduce runoff where crops are irrigated. In Washington (USA), Aarstad and Miller (1973) minimized runoff and generally increased hay, potato, and sugarbeet (*Beta vulgaris* L.) yields when furrow diking was used on sprinkler-irrigated land. furrow diking is essential for elimination of runoff for the LEPA (low-energy precision application) sprinkler irrigation system developed by Lyle and Bordovsky (1981). The LEPA irrigation equipment consists of a lateral-move or center-pivot overhead sprinkler system that delivers water at low pressure through drop tubes to within a few centimeters of the soil surface. Water is discharged as gentle rainfall, but at a high rate to a small area, to diked furrows that have sufficient capacity to retain all water. Retaining water in place results in highly uniform distribution of the applied water. The diked furrows result in water application efficiencies exceeding 90% for the LEPA system. Kincaid et al. (1990) obtained lower runoff and slightly greater potato yields when they used furrow dikes in conjunction with low-pressure center-pivot irrigation.

To improve rainfall and irrigation water retention on the soil surface under furrow-irrigated conditions, Stewart, Dusek, and Musick (1981) developed the LID (limited irrigation–dryland) system for regions where both precipitation and irrigation water supplies are limited and the soil is slowly permeable. The system is based on the premise that a producer has a given, but limited, amount of water available for irrigating a crop. The water is applied on a set schedule, based on irrigation well yields and delivery-system capacity. The producer also recognizes that the supply is insufficient to adequately irrigate the entire field. Furrow dikes, which are installed from end to end in the field, control the distance that irrigation water advances in the furrows. With the LID system, the upslope portion of a field (about one-half) may be adequately irrigated, the third quarter receives runoff from the irrigated portion, and the lower quarter is dry-farmed. Seeding rates and fertilizer applications are adjusted according to anticipated water supplies and yield levels for a given portion of the field (fully irrigated, partially irrigated, or dryland). All furrows are diked, and water is applied to all or alternate furrows. Water fills successive basins as it overtops a dike and advances downslope. Because the amount applied is designed to adequately irrigate a given portion of the field, rainfall influences the distance that irrigation water advances in the furrows. The advantages of this system are that it prevents runoff, uses irrigation water more efficiently, irrigates more land in wet seasons and less in dry seasons, and allows irrigation sets to be moved at regular intervals without regard to the distance that water has advanced in the furrows.

Furrow diking is highly effective for preventing runoff and, hence, sediment transport on nearly level fields if the amount of water that can be stored, plus infiltration during the storm, is greater than the volume of water from the largest storm that is likely to occur (Hudson, 1981). In Israel, soil losses were 10 times greater from plowed and disked plots and 15 times greater from ridged plots than from furrow-diked plots (Rawitz et al., 1983). In other studies in Israel, furrow diking prevented runoff (Agassi et al., 1989), with water conserved by furrow diking resulting in potato yield increases of 8% to 18% (Agassi and Levy, 1993). Soil losses from ridged land were 17.0 Mg ha^{-1} without dikes and 5.1 Mg ha^{-1} with dikes in alternate furrows in Nigeria (Kowal, 1970). For the latter case, the loss was slightly greater than that from land with flat tillage.

Furrow diking may result in accelerated soil losses on sloping land if ridge overtopping occurs. To reduce the potential for such losses, a common practice is to make the dikes lower than the ridges so that, if dike failure occurs, runoff will be between the ridges (Jones and Stewart, 1990). However, in Israel, dikes are made higher than ridges so that runoff will occur along ridge tops, thus avoiding dike failure (Morin and Benyamini, 1988). To minimize the potential for dike failure resulting from exceptionally large storms, supporting practices such as terracing and contour tillage have been recommended. Furrow diking is not recommended for regions having a high probability of large storms, where adequate surface depression storage cannot be achieved (Jones and Stewart, 1990), and where an impermeable soil layer that would reduce water infiltration is near the surface.

3. *Soil Pitting/Chain Diking*

Soil pitting on cropland involves the formation of small depressions at closely spaced intervals to retain runoff from rainstorms. The pits are made with equipment similar to that used for furrow diking. As the equipment is pulled across the land surface, paddles or blades on a tripping drag or a rotating mechanism having some resistance to turning make depressions in the soil. The system is well adapted for use on drilled or broadcast-planted grain crops.

Chain diking results in surface conditions similar to those obtained by soil pitting, but different equipment is used. A chain diker has specially shaped blades welded to a large ship-anchor chain to make indentations in the soil surface as the diker is pulled over loose, flat-tilled land. As the chain rotates while being pulled, a broadcast pattern of diamond-shaped basins about 10 cm deep is formed on the surface. The basins on the soil surface reduce runoff, thus providing more water for crops and increasing crop yields. For a three-year period, a chain-diking treatment reduced runoff 46% compared with a nondiked treatment on a fine sandy loam (unpublished data, H. T. Wiedemann, Vernon, Texas, 1992). The basins do not appear to hinder subsequent crop production operations (Wiedemann and Smallacombe, 1989).

4. *Level Terraces*

Contour tillage and furrow diking involve only a relatively small amount of surface soil manipulation and are applicable to land with relatively slight slopes. In contrast, terracing usually involves major surface soil manipulation, and it has been applied to a wide range of slopes, depending on the type of terrace constructed.

Terraces that prevent runoff are level (on the contour of the land) and have blocked outlets. They should be of sufficient size to retain all stormwater on the surface until it infiltrates the soil. Although intended mainly to conserve water, these terraces are highly effective for preventing sediment transport from fields because runoff is prevented, but erosion may occur between the terraces. Because of the possibility of overtopping, level terraces usually are restricted to regions having a low probabilities of major rainstorms. To minimize the potential for damage, level terraces should have provisions for safe removal of excess water during major storms.

Level terraces may have broad or narrow bases. The most common type on gently sloping land is the broad-base terrace for which the ridge and channel are cropped, usually the same as for the interval between terraces. Steep backslope terraces with a relatively narrow base are commonly used on more steeply sloping land (Wischmeier and Smith, 1978), but narrow-base terraces with steep side slopes are sometimes used on gently sloping land. Such narrow terraces minimize soil movement required for terrace construction.

A variation of the level terrace is the conservation bench terrace (CBT) for which the adjacent upslope portion of the terrace interval is leveled. Runoff from the remaining unleveled area (watershed) of the terrace interval is spread over the leveled area, which has a greater storage capacity than that of level terraces. Consequently, potential runoff water from larger storms can be retained on a portion of the land. While sediment transport may occur within the field, sediment transport from the field is prevented because runoff is prevented (Zingg and Hauser, 1959).

Ratios of unleveled to leveled areas of CBTs have ranged from 1:1 to 6:1 (Unger, 1984), with the appropriate ratio being influenced by soil and precipitation characteristics. A major constraint to widespread use of CBT systems is the high cost of terrace construction and leveling of a portion of the land. Use of narrow CBT systems reduces construction costs and has resulted in similar water and soil conservation benefits as the wider systems (Jones, 1981; Unger, 1984).

5. *Land Leveling*

Level bench terraces, for which the entire interval between terraces is leveled, are widely used on steep slopes where land for crop production is limited and where precipitation is relatively high as in Peru, Nepal, Indonesia, Malaya, China, Japan, The Philippines, and other countries. Irrigation terraces have a raised outer lip that retains applied water uniformly on the entire bench. On steep slopes, bench terraces separate the slope into a series of steps, with horizontal or nearly horizontal ledges separated from each other by vertical or nearly vertical walls. The walls may be constructed of brick, stone, or timber. If the soil is sufficiently stable, the wall may be held in place by vegetation (Hudson, 1981).

Level bench terraces have been used also where precipitation is limited, mainly as a water conservation practice, but also to prevent sediment losses. When narrow bench terraces (10 m wide) were constructed on gently sloping land (about 1%), only a small amount of soil had to moved, thus greatly reducing the cost of land leveling (Jones, 1981). By using laser-controlled soil-moving equipment, land leveling can be achieved at a relatively low cost because one person (the tractor operator) can perform the entire operation.

C. Controlling Runoff

Under some conditions, it may not be practical nor desirable to prevent or greatly reduce runoff. However, to control erosion under those conditions, the runoff must occur at controlled, nonerosive rates. Runoff at low rates also provides time for water infiltration, thus enhancing the potential for satisfactory water conservation. Practices that result in runoff at controlled rates include land smoothing, stripcropping, graded furrows, graded terraces, variations of bench terraces, discontinuous parallel terraces, land imprinting, and tillage per se. The objective of these practices is to safely convey excess water from croplands to nearby waterways and streams.

1. Land Smoothing

Land smoothing is the practice of mechanically shifting soil from high to low points in fields. By eliminating low points, water is retained on the field more uniformly, thus reducing the potential for concentrated flow that could cause accelerated sediment transport and uneven water storage in the field. While important for reducing concentrated water flow, land smoothing should not eliminate microdepressions and surface roughness that are important for temporary water storage on the surface and, thus, for reducing runoff, sediment transport, and water conservation (Unger, 1984).

2. Stripcropping

Stripcropping is the practice of growing protective crops and row crops in alternating strips on the contour. Stripcropping is more effective than contouring alone for controlling erosion (Table 11). The protective and row crop strips usually are of equal width. Sod and winter small-grain crops tend to be more effective than spring grain crops for controlling erosion. With stripcropping, sediments from the cropped areas are filtered out as the runoff is slowed by the protective strips. Stripcropping reduces sediment transport from fields, but does not necessarily prevent transport between strips (Wischmeier and Smith, 1978). When runoff is slowed, the potential for water conservation is increased because there is more time for infiltration to occur.

A variation of stripcropping involving equal-width protective and row-crop strips is buffer stripcropping. It involves using narrow protective strips alternated with wider-cropped strips. The location of strips is determined by the width and arrangement of adjoining strips in a rotation or the location of steep, severely eroded (or highly erosive) areas on slopes. The buffer or protective strips usually occupy irregular areas on sloping land and are seeded to perennial grasses and legumes (Wischmeier and Smith, 1978), which may provide feed for livestock in addition to their value for reducing sediment transport. Narrow strips of plants that form a dense barrier are highly effective for trapping sediment as water flows through the barrier. Vetiver grass (*Vetiveria zizanioides*) has been found to have desirable characteristics for trapping sediments, thus naturally tending to reduce the land slope between the barrier strips with time (Greenfield, 1988; Erskine, 1992). Reducing the slope reduces the potential for subsequent sediment movement in the interval between the barriers. Other plants that provide a dense barrier should be similarly effective for reducing sediment movement within fields. A potential problem with buffer strips is gullying and rilling below the strips due to concentrated flow on lands having irregular topography.

3. *Graded Furrows*

In contrast to contour furrows that retain water and sediment, graded furrows are designed to convey excess water from fields at low flow rates. Flow rate reduction minimizes sediment transport by providing more time for sediment settling. It also reduces the energy of flowing water, thus reducing its capacity to detach and transport sediments. When furrow gradients varied from near 0% at the upper end to about 1% at the lower end of a 300-m-long field in Texas, soil loss was comparable to that from a terraced field (Richardson, 1973). Each graded furrow served as a miniature graded terrace, thus avoiding the more concentrated flow that occurred in the graded terrace. The graded furrows also provided for more uniform water storage on the entire watershed. As a result, runoff for a 32-month period totaled 187 mm from the graded-furrow watershed and 236 mm from the terraced watershed without graded furrows.

4. *Graded Terraces*

As for graded furrows, graded terraces are designed to convey excess water from fields at nonerosive flow rates. Their design is based on soil conditions, and the gradient within channels may be variable to improve terrace alignment, especially if underground drains are available to remove some water from fields (SCS, 1977). Because of low water flow rates, graded terraces reduce sediment transport by providing time for sediments to settle and for infiltration to occur. Sediment transport is reduced also within the field (between terraces) because the terraces reduce the potential for concentrated downslope water flow that could detach and transport sediments.

5. *Variations of Bench Terraces*

Level bench terraces were discussed relative to retaining runoff in the field and reducing sediment transport (Sec. III.B.4). Other bench terraces that are not level, but which reduce runoff and sediment transport and improve water conservation, include outward-sloping, inward-sloping (or reverse-slope), and step terraces. The outward- and inward-sloping bench terraces are widely used for cultivated crops in some countries, whereas the step terraces are used for tree crops where regular cultivation is not required (Hudson, 1981).

6. *Intermittent Terraces*

Bench terraces may not be practical if the soil is too shallow or the land slope is too steep. In such cases, intermittent terraces, sometimes called orchard terraces, may be developed for tree crops. Orchard terraces are leveled or have reverse slopes, and each is designed for a single row of trees.

Other types of intermittent terraces, or variations of them, are hillside ditches, lock-and-spill hillside ditches, and ''Fanya Juu'' terraces. The term ''hillside ditches'' is applied to bench terraces having several shapes, with a common feature being a ditch dug on the contour to capture soil and water.

The lock-and-spill hillside ditches are variations of the hillside ditches. Rather than on the contour, lock-and-spill ditches are constructed at a slight grade with low crosswalls remaining in the ditch to divide the larger basins into smaller units. The crosswalls encourage infiltration, yet allow excess water to flow across the walls to protected outlets during runoff events. Soil trapped in the basins is periodically spread over the adjacent land.

''Fanya Juu'' terraces are intermittent terraces used in Kenya, where they are constructed by throwing the excavated soil uphill to build an embankment above the ditch. The intent of this practice is for the embankment to trap sediments, thus reducing the slope of the land between the terraces.

When using intermittent terraces on steep erosion-prone slopes, an important requirement is that the interval between terraces be protected by a vigorous cover crop (Hudson, 1981). In general, these terraces are not suited for mechanized crop production, but are used where adequate labor is available.

7. *Discontinuous Parallel Terraces*

Severely dissected landscapes having highly undulating surfaces are difficult to farm when conventional terrace systems are installed on them. One method of controlling sediment losses on such landscapes is to install discontinuous parallel terraces using riser inlets and underground pipes with outlets to convey excess water from the fields, yet trap sediments in basins behind the terraces (Laflen et al., 1972). Constructing the terraces parallel to existing field boundaries provides straight rows for easy performance of farming operations. With clean-cultivated maize, more than 97% of the sediment was trapped near the point of origin by the basins. An 86-mm storm transported about 40 Mg ha^{-1} and a 50-mm storm transported about 17 Mg ha^{-1} of sediment into the basins (Mielke, 1985).

8. *Land Imprinting*

The land-imprinting system was developed to improve vegetation on overgrazed and shrub-infested arid-to-semiarid rangelands while protecting them from accelerated runoff and erosion. When operated across the slope of the land, the imprinter forms an interconnected system of water-shedding and water-absorbing furrows that constitute a miniature rainfed irrigation system. Teeth of the imprinter crush and cut aboveground plant materials, partially embed them in soil, and deposit the remainder on the surface as a mulch (Dixon, 1981).

Land imprinting involves using a massive steel roller faced with angular steel teeth to form relatively stable depressions (imprints) on the soil surface. As the imprinter is pulled across the land surface by a tractor, previously spread seed is pressed into the soil. When rains occur, seed, topsoil, and plant litter are concentrated in the imprints, thus enhancing the potential for seed germination and successful seedling establishment. Although designed primarily for rangeland improvement, the system has potential also for water and soil conservation on cropland, especially where residues from a previous crop are present on land in semiarid regions that is to be cropped to small grains. To enhance the probability of success, imprinting should coincide with periods of anticipated rainfall, and high-quality seed of suitable species should be used for the conditions under which the plants are to be grown (Dixon, 1981).

9. *Tillage Per Se*

Runoff usually is greater with clean tillage than where residues are retained on the surface or where other conservation measures are applied. However, runoff and, hence, sediment transport and water conservation are affected also by the tillage method used when residues are not present and other conservation measures are not applied. The differences result from various factors, including tillage depth, resultant surface-water storage capacity, surface-soil stability (affects surface sealing and infiltration), and soil porosity.

The effects of tillage-induced plow layer porosity and surface roughness on simulated rainfall infiltration (hence, runoff) are shown in Table 12. Cumulative infiltration approached the total pore space and surface-roughness retention volumes before runoff started and exceeded the total volumes before 25 mm of runoff occurred for the plow treatment. The retention volumes were not filled for the other treatments, even though 50 mm of runoff occurred. Smoother surfaces with treatments other than plowing apparently resulted in more rapid surface-soil dispersion and surface sealing, which reduced infiltration into the tillage-loosened soil. Although sediment transport was not reported, it undoubtedly would have followed trends inverse to those for infiltration.

10. Other Practices

Numerous other practices, which will not be discussed, provide some water retention and erosion control benefits. The following is a list of some such practices:

Vertical mulching
Slot mulching
Deep tillage
Profile modification
Contour strip rainfall harvesting

Table 12 Effect of Tillage-Induced Plow Layer Porosity and Surface Roughness on Cumulative Infiltration of Simulated Rainfall

	Surface conditions		Cumulative infiltration[a]		
Tillage treatment[b]	Pore space[c] (mm)	Roughness (mm)	To initial runoff (mm)	To 25-mm runoff (mm)	To 50-mm runoff (mm)
Untilled	81	8	9	21	24
Plow	137	50	171	217	230
Plow-disk-harrow	124	25	53	73	84
Cultivated	97	29	57	83	91
Rotovated	117	15	24	38	41

[a]Water applied at a 12.7-cm h^{-1} rate.
[b]Plowing and rotovating performed to a 15-cm depth; cultivating to a 7.5-cm depth on otherwise untilled soil.
[c]Measured to the tillage depth.
Source: Burwell et al. (1966).

Rainfall collection in microcatchments
Fish-scale terraces

REFERENCES

Aarstad, J. S., and Miller, D. E. (1973). Soil management to reduce runoff under center-pivot sprinkler systems. *J. Soil Water Conserv. 28*: 171–173.

Agassi, M., and Levy, G. J. (1993). Effect of the diked furrow technique on potato yield. *Potato Res. 36*: 247–251.

Agassi, M., Shainberg, I., Warrington, D., and Ben-Hur, M. (1989). Runoff and erosion control in potato fields. *Soil Sci. 148*: 149–154.

Anderson, D. T. (1968). Field equipment needs in conservation tillage. In *Conservation Tillage in the Great Plains* (Proc. Workshop, Lincoln, NE, 1968), Great Plains Agric. Counc. Publ. 32, pp. 83–91.

Benyamini, Y., and Unger, P. W. (1984). Crust development under simulated rainfall on four soils. *Agron. Abstr.* pp. 243–244.

Burwell, R. E., Allmaras, R. R., and Sloneker, L. L. (1966). Structural alteration of soil surfaces by tillage and rainfall. *J. Soil Water Conserv. 21*: 61–63.

CTIC (Conservation Technology Information Center). (1990). Tillage definitions. *Conserv. Impact 8(10)*: 7.

Dickey, E. C., Fenster, C. R., Laflen, J. M., and Mickelson, R. H. (1983). Effects of tillage on soil erosion in a wheat-fallow rotation. *Trans. Am. Soc. Agric. Eng. 26*: 814–820.

Dixon, R. M. (1981). Imprinting vegetation system: How it works and things that can go wrong. (Unpublished mimeo. report.)

Erskine, J. M. (1992). Vetiver grass: Its potential use in soil and moisture conservation in southern Africa. *South Afr. J. Sci. 88*: 298–299

FAO (Food and Agricultural Organization of the United Nations) (1965). *Soil Erosion by Water*. FAO, Rome.

Finnell, H. H. (1944). Water conservation in Great Plains wheat production. *Texas Agric. Exp. Stn. Bull.* B-655, College Station.

Fisher, C. E., and Burnett, E. (1953). Conservation and utilization of soil moisture. *Texas Agric. Exp. Stn. Bull.* B-767, College Station.

Gilley, J. E., Finkner, S. C., Spomer, R. G., and Mielke, L. N. (1986a). Runoff and erosion as affected by corn residue. Part II: Rill and interrill components. *Trans. Am. Soc. Agric. Eng. 29*: 161–164.

Gilley, J. E., Finkner, S. C., and Varvel, G. E. (1986b). Runoff and erosion as affected by sorghum and soybean residue. *Trans. Am. Soc. Agric. Eng. 29*: 1605–1610.

Gilley, J. E., Kottwitz, E. R., and Wieman, G. A. (1991). Roughness coefficients for selected residue materials. *J. Irrig. Drainage Eng. 117*: 503–514.

Greenfield, J. C. (1988). *Vetiver grass (Vetiveria zizanioides), A Method of Vegetative Soil and Moisture Conservation*, 2nd ed. World Bank, New Delhi.

Griffith, D. R., Mannering, J. V., and Moldenhauer. (1977). Conservation tillage in the eastern Corn Belt. *J. Soil Water Conserv. 32*: 20–28.

Harrold, L. L., and Edwards, W. M. (1972). A severe rainstorm test of no-till corn. *J. Soil Water Conserv. 27*: 30.

Hudson, N. (1981). *Soil Conservation*, 2nd ed. Cornell Univ. Press, Ithaca, New York.

Jones, O. R. (1981). Land forming effects on dryland sorghum production in the southern Great Plains. *Soil Sci. Soc. Am. J. 45*: 606–611.

Jones, O. R. (1992). Water conservation practices in the southern High Plains. In *Proc. Fourth Annual Conf., Colorado Conserv. Tillage Assn.* (S. E. Hinkle, ed.), Sterling, CO, pp. 22–25.

Jones, O. R., and Clark, R. N. (1987). Effects of furrow dikes on water conservation and dryland crop yields. *Soil Sci. Soc. Am. J. 51*: 1307–1314.

Jones, O. R., and Stewart, B. A. (1990). Basin tillage. *Soil Tillage Res. 18*: 249–265.

Kemper, W. D., and Schertz, D. (1992). Purpose of the workshop. Presentation at Residue Management Workshop, Kansas City, MO.

Ketcheson, J. (1977). Conservation tillage in eastern Canada. *J. Soil Water Conserv. 32*: 57–60.

Kincaid, D. C., McCann, I., Busch, J. R., and Hasheminia, M. (1990). Low pressure center pivot irrigation and reservoir tillage. In *Visions of the Future* (Proc. Third Nat. Irrig. Symp., Phoenix, AZ), Am. Soc. Agric. Eng. Publ. 04-90, St. Joseph, MI.

Kowal, J. (1970). The hydrology of a small catchment basin at Samuru, Nigeria. IV: Assessment of soil erosion. *Niger. Agric. 7*: 134–147.

Krishna, J. H., and Arkin, G. F. (1988). Furrow diking technology for agricultural water conservation and its impact on crop yields in Texas. *Texas Water Resources Inst. Publ.* TR-140, Texas A&M Univ., College Station.

Krishna, J. H., Arkin, G. F., Williams, J. R., and Mulkey, J. R. (1987). Simulating furrow-dike impacts on runoff and sorghum yields. *Trans. Am. Soc. Agric. Eng. 30*: 143–147.

Krishna, J. H., and Gerik, T. J. (1988). Furrow-diking technology for dryland agriculture. In *Challenges in Dryland Agriculture—A Global Perspective* (Proc. Int. Conf. on Dryland Farming, Amarillo/Bushland, TX) (P. W. Unger, T. V. Sneed, W. R. Jordan, and R. Jensen, eds.), Texas Agric. Exp. Stn., College Station, pp. 258–260.

Laflen, J. M., Johnson, H. P., and Reeve, R. C. (1972). Soil loss from tile-outlet terraces. *J. Soil Water Conserv. 27*: 74–77.

Lentz, R. D., Shainberg, I., Sojka, R. E., and Carter, D. L. (1992). Reducing furrow erosion with polymer amended irrigation waters. *Agron. Abstr.*, p. 330.

Lessiter, F. (1982). No-tillage defined. *No-Till Farmer 10(6)*: 2.

Lyle, W. M., and Bordovsky, J. P. (1981). Low energy precision application (LEPA) irrigation system. *Trans. Am. Soc. Agric. Eng. 24*: 1241–1245.

McCalla, T. M., and Army, J. T. (1961). Stubble mulch farming. *Adv. Agron. 12*: 125–196.

Mielke, L. N. (1985). Performance of water and sediment control basins in northeastern Colorado. *J. Soil Water Conserv. 40*: 524–528.

Morin, J., and Benyamini, Y. (1988). Tillage method selection based on runoff modeling. In *Challenges in Dryland Agriculture—A Global Perspective* (Proc. Int. Conf. on Dryland Farming, Amarillo/Bushland, TX) (P. W. Unger, T. V. Sneed, W. R. Jordan, and R. Jensen, eds.), Texas Agric. Exp. Stn., College Station, pp. 251–254.

Morin, J. Rawitz, E., Hoogmoed, W. B., and Benyamini, Y. (1984). Tillage practices for soil and water conservation in the semi-arid zone. III: Runoff modeling as a tool for conservation tillage design. *Soil Tillage Res. 4*: 215–224.

Osuji, G. E., Babalola, O., and Aboaba, F. O. (1980). Rainfall erosivity and tillage practices affecting soil and water loss on a tropical soil in Nigeria. *J. Environ. Manage. 10*: 207–217.

Rawitz, E., Morin, J., Hoogmoed, W. B., Margolin, M., and Etkin, H. (1983). Tillage practices for soil and water conservation in the semi-arid zone. 1: Management of fallow during the rainy season preceding cotton. *Soil Tillage Res. 3*: 211–232.

Richardson, C. W. (1973). Runoff, erosion, and tillage efficiency on graded-furrow and terraced watersheds. *J. Soil Water Conserv. 28*: 162–164.

Rockwood, W. G., and Lal, R. (1974). Mulch tillage: A technique for soil and water conservation in the tropics. *Span 17*: 77–79.

Roth, C. H., Vieira, M. J., Derpsch, R., Meyer, B., and Frede, H. G. (1987). Infiltrability of an Oxisol in Parana, Brazil as influenced by different crop rotations. *J. Agron. Crop. Sci. 159*: 186–191.

Russel, J. C. (1939). The effect of surface cover on soil moisture losses by evaporation. *Soil Sci. Soc. Am. Proc. 4*: 65–70.

SCS (Soil Conservation Service). (1977). *National Handbook of Conservation Practices.* U.S. Dept. Agric., Soil Conserv. Serv., U.S. Gov. Print. Off., Washington, DC.

SCSA (Soil Conservation Society of America). (1982). *Resource Conservation Glossary.* Soil Conserv. Soc. Am. (now Soil and Water Conserv. Soc.), Ankeny, IA.

Sidiras, N., Derpsch, R., and Mondardo, A. (1983). Effect of tillage systems on water capacity, available moisture, erosion, and soybean yield in Parana, Brazil. In *No-Tillage Crop Production in the Tropics* (Proc. Symp., Monrovia, Liberia) (I. O. Akobunda and A. E. Deutsch, eds.), Int. Plant Prot. Ctr., Oregon St. Univ., Corvallis.

Sidiras, N., Heinzmann, F. X., Kahnt, G., Roth, C. H., and Derpsch, R. (1985). The importance of winter crops for controlling water erosion, and for the summer crops on two oxisols in Parana, Brazil. *J. Agron. Crop Sci. 155*: 205–214.

Singh, G. (1974). The role of soil and water conservation practices in raising crop yields in dry farming areas of tropical India. In *UNDP International Expert Consultation on the Use of Improved Technology for Food Production in Rainfed Areas of Tropical Asia.* FAO, Rome, Mimeo. Rpt.

SSSA (Soil Science Society of America). (1987). *Glossary of Soil Science Terms.* Soil Sci. Soc. Am., Madison, WI.

Stewart, B. A., Dusek, D. A., and Musick, J. T. (1981). A management system for the conjunctive use of rainfall and limited irrigation of graded furrows. *Soil Sci. Soc. Am. J. 45*: 413–419.

Stewart, B. A., Woolhiser, D. A., Wischmeier, W. H., Caro, J. H., and Frere, M. H. (1975). *Control of Water Pollution from Cropland*, Vol. 1, *A Manual for Guideline Development.* U.S. Dept. Agric. Rpt. No. ARS-H-5-1. U.S. Gov. Print. Off., Washington, DC.

Unger, P. W. (1984). *Tillage Systems for Soil and Water Conservation.* FAO Soils Bull. 54, FAO, Rome.

Unger, P. W. (1992a). Infiltration of simulated rainfall: Tillage system and crop residue effects. *Soil Sci. Soc. Am. J. 56*: 283–289.

Unger, P. W. (1992b). Ridge height and furrow blocking effects on water use and grain yield. *Soil Sci. Soc. Am. J. 56*: 1609–1614.

Wiedemann, H. T., and Smallacombe, B. A. (1989). Chain Diker—A new tool to reduce runoff. *Agric. Eng. 70*: 12–15.

Williams, J. R., Wistrand, G. L., Benson, V. W., and Krishna, J. H. (1988). A model for simulating furrow dike management and use. In *Challenges in Dryland Agriculture —A Global Perspective* (Proc. Int. Conf. on Dryland Farming, Amarillo/Bushland, TX) (P. W. Unger, T. V. Sneed, W. R. Jordan, and R. Jensen, eds.), Texas Agric. Exp. Stn., College Station, pp. 255–257.

Wischmeier, W. H., and Smith, D. D. (1978). *Predicting Rainfall-Erosion Losses from Cropland East of the Rocky Mountains.* U.S. Dept. Agric. Handbook 282. U.S. Gov. Print. Off., Washington, DC.

Zingg, A. W., and Hauser, V. L. (1959). Terrace benching to save potential runoff for semiarid land. *Agron. J. 51*: 289–292.

Zingg, A. W., and Whitfield, C. J. (1957). A summary of research experience with stubble-mulch farming in the western states. U.S. Dept. Agric. Tech. Bull. 1166. U.S. Gov. Print. Off., Washington, DC.

12

Soil Stabilizers

G. J. Levy
Institute of Soils and Water, The Volcani Center, Agricultural Research Organization, Bet-Dagan, Israel

I. INTRODUCTION

Erosion was recognized as a problem in early civilizations, and various attempts were made to deal with it. Water erosion is the product of detachment of soil by hydraulic shear stress in concentrated confined flow (rill erosion), and/or detachment and transport by raindrop impact and shallow overland flow (interrill erosion). Conservation practices must therefore limit soil detachment and runoff volume and velocity. Traditional strategies for erosion control are based mainly on measures that (1) protect the soil from raindrop impact to decrease detachment, (2) increase surface depression storage and soil roughness to reduce runoff volume and velocity, and (3) alter slope-length gradient and direction of runoff flow.

An alternative approach to soil conservation, which has recently gained much attention, is one that advocates the modification of some soil properties responsible for the susceptibility of the soil to erosion. Increasing aggregate stability at the soil surface and preventing clay dispersion are known to control seal formation, increase the infiltration rate, and reduce runoff in cultivated soils. In addition, stable aggregates at the soil surface are less susceptible to detachment by raindrop impact and to transportation by runoff water. Improving aggregate stability and preventing clay dispersion can be obtained by applying soil amendments to the soil. The current understanding that seal formation, runoff, and soil erosion are soil surface phenomena gave rise to the concept that it is only necessary to treat and modify the properties of the soil surface rather than those of the entire cultivated layer. Consequently, small amounts of soil amendments are needed, which make their use for agricultural purposes economically feasible.

This chapter discusses the potential of two types of soil amendments, namely gypsum and synthetic organic polymers, in controlling seal formation, runoff, and water erosion. For the latter, the discussion focuses mainly on interrill erosion from cultivated soils.

II. GYPSUM

A. Introduction

Gypsum is a relatively common mineral that is widely available in agricultural areas. Its specialized agronomic uses and effects have been presented recently in an extensive review by Shainberg et al. (1989). Gypsum is by far the most-used amendment for sodic soils reclamation because of its low cost, availability, and ease of handling. The beneficial effect of gypsum in preventing clay dispersion and maintaining soil permeability was recognized at the beginning of the century (Hilgard, 1906). A great many studies have followed and established the role of gypsum in controlling soil sealing, infiltration rate, and soil erosion not only in sodic soils from semiarid regions but also in dispersive low-Na soils from humid and temperate regions.

B. Sources of Gypsum and Its Aqueous Chemistry

Gypsum ($CaSO_4{*}2H_2O$) and its dehydration products, the hemihydrate (plaster of Paris, $CaSO_4{*}1/2H_2O$) and anhydrite ($CaSO_4$), are extensively distributed minerals found worldwide in sedimentary evaporite deposits (Hurlbut and Klein, 1971). Nongeological sources of gypsum stem from industrial processes producing $CaSO_4$ by products. The most common and important one is the production of phosphoric acid from rock phosphate (apatite) by wet-process acidulation:

$$Ca_{10}(PO_4)_6F_2(s) + 10H_2SO_4 + 20H_2O \longrightarrow 10CaSo_4{*}2H_2O(s) + 6H_3PO_4 + 2HF$$

Phosphoric acid is used to manufacture high-analysis P fertilizers, while the by-product gypsum, termed phosphogypsum, is collected as a waste product. Another potentially growing source of by-product gypsum is the capture of SO_2 from stack gases produced by fossil-fuel-fired electric-power-generating plants.

The various sources of gypsum are slightly soluble salts in aqueous solutions, dissolving to an extent of approximately 2.5 g L^{-1}, or 15 mM. This level of solubility is a substantial contribution to the ionic strength of most soil solutions, yet it is low enough to allow continued release of salt to the soil over a considerable time period. Other Ca salts are either less soluble ($CaCO_3$) or more soluble ($CaCl_2$, $Ca(NO_3)_2$).

In addition to the effect of some chemical properties of the solution such as ionic strength and ion pairing, gypsum dissolution is also significantly affected by the surface area of the gypsum particles. Keren and Shainberg (1981) found that phosphogypsum (PG) was more rapidly soluble than mined gypsum (Fig. 1). The finer particles of the PG compared with the mined gypsum, and hence their larger surface area was considered the main cause for the higher dissolution rate of the PG (Keren and Shainberg, 1981).

C. Effect of Gypsum on Seal Formation and Infiltration Rate

1. *Soil Dispersivity*

Many soils in semiarid-to-humid regions have an unstable structure which makes them susceptible to erosion and thus difficult to manage because of their tendency to disperse and develop compacted structure particularly at or near the soil surface. The colloidal clay is the soil fraction which predominantly affects the physical behavior of soils since it possesses the greatest specific surface area and is thus most active in physicochemical processes such as swelling and dispersion. In smectitic soils clay dispersion depends on the type of exchangeable cation and on the composition and concentration of the soil solution (Quirk and Schofield, 1955). In kaolinitic soils clay dispersion depends mainly on the pH of the soil solution (Schofield and Samson, 1954) and the presence of dispersive anions (Shanmuganathan and Oades, 1983). In field soils, preventing clay dispersion (i.e., keeping the clay in a flocculated state) is a necessary condition for the formation and stabilization of soil structure.

2. *Seal Formation and Infiltration Rate*

Soil infiltration rate (IR) is defined as the volume flux of water flowing into the profile per unit surface area, and has the dimensions of velocity. In general, soil IR is high at first, particularly when the soil is initially dry, but it tends to decrease monotonically to approach asymptotically a constant rate—the final IR. When water delivery rate exceeds soil infiltration capacity, the latter determines the actual IR and the process becomes surface controlled.

Decrease in soil infiltration capacity often results from gradual deterioration of soil structure at the soil surface and the formation of a surface seal. When a seal with a very low hydraulic conductivity is formed, its reduced permeability

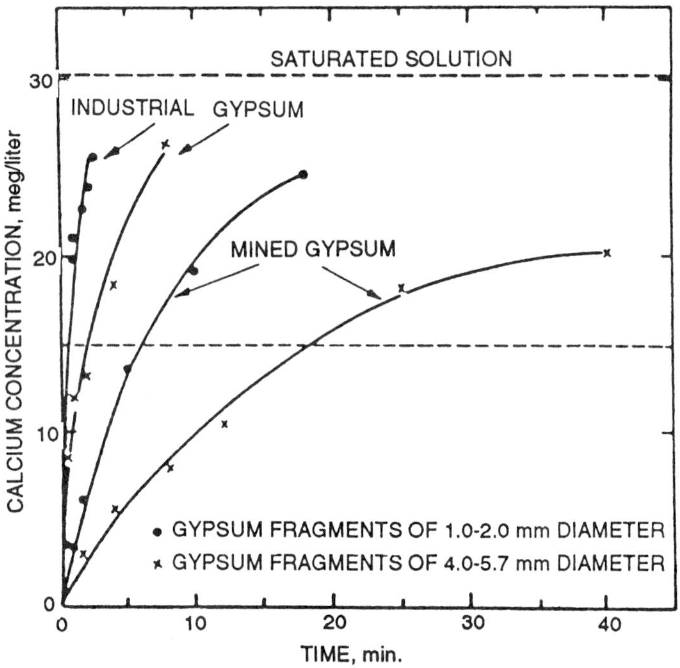

Fig. 1 Calcium concentration in water vs. time plots for industrial and mined gypsum samples of two fragment sizes. (From Keren and Shainberg, 1981.)

determines the IR of the soil (Baver, Gardner, and Gardner, 1972; Morin and Benyamini, 1977; Hillel, 1980).

Seal formation in soils exposed to drops' impact is due to two mechanisms (McIntyre, 1958; Agassi et al., 1981): (1) physical disintegration of soil aggregates and their compaction; and (2) a physicochemical dispersion and movement of clay particles into a region of 0.1–0.5 mm depth, where they lodge and clog the conducting pores. The two mechanisms act simultaneously as the first enhances the latter. The IR of the seal is significantly affected by the electrolyte concentration of the soil solution at the soil surface (i.e., that of the applied water) and the exchangeable sodium percentage (ESP) of the soil (Agassi et al., 1981; Kazman et al., 1983).

3. Controlling Surface Sealing with Phosphogypsum

The chemical dispersion of the surface particles during rain, can be prevented by maintaining the electrolyte concentration of the soil solution at the soil surface above a critical concentration (termed flocculation value), which ensures that the clay particles are in a coagulated state. Spreading PG (or any readily available electrolyte source) at the soil surface is an effective measure for preventing clay dispersion.

The effect of source, amount, and fragment size of gypsum on the infiltration rate of a loess with ESP 30 was studied by Keren and Shainberg (1981). Without gypsum application the IR decreased sharply to a constant value of 2 mm h^{-1} (Fig. 2). When 3.4 Mg ha^{-1} of powdered PG and mined gypsum were spread on the soil surface, the final IRs were 7.5 and 5.5 mm H^{-1}, respectively. When coarser particles of gypsum were used (4–5.7 mm), the mined gypsum had no effect on the IR irrespective of the amount added (Fig. 2). Conversely, coarse fragments of PG were effective in maintaining high IR. The latter increased with the increase in the amount of PG applied (Fig. 2).

The optimal amounts and application methods of PG were studied by Agassi, Morin, and Shainberg (1982) on three smectitic soils from Israel. They found

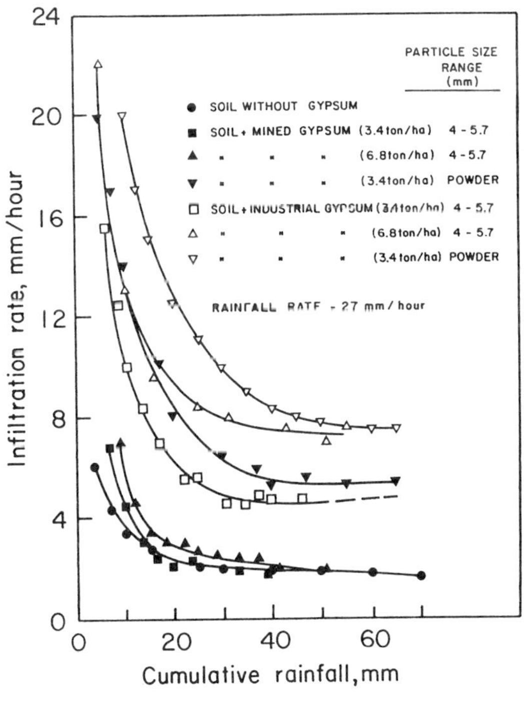

Fig. 2 The effect of industrial and mined gypsum (at amounts and particle sizes as indicated) on infiltration rate of loess soil as a function of the cumulative infiltration under laboratory conditions. (From Keren and Shainberg, 1981.)

that rates of 3, 5, or 10 Mg ha^{-1} of PG spread over the soil surface had almost similar effects on the IR (Fig. 3A). Spreading PG over the soil surface was more effective than mixing it to a depth of 5 mm, because when mixed only 20% was available at the upper 1 mm where the seal is formed (Fig. 3B). Thus, Agassi et al. (1982) concluded that in soils susceptible to seal formation, incorporating the PG with the soil by disk or plow should be discouraged. Moreover, efficiency of PG in maintaining high IR (especially if spread over) was still evident after six consecutive storms of 35 mm each with three to five days intervals between them (Fig. 3C). In addition, Agassi et al. (1982) investigated the effect of aggregate size at the soil surface (0–3 mm and 0–10 mm) on PG efficiency. They noted that PG efficiency in controlling seal formation was higher in the small aggregate size than in the large size, which was opposite to expectations and to the trend observed for the control treatment. Agassi et al. (1982) suggested that washing away by erosion of the PG from the large aggregates' surfaces to the intervening depressions enhanced seal formation on these surfaces, and the effect of PG is evident only in the depressions. The relative inefficiency of PG in maintaining high IR when large aggregates are present at the soil surface was verified by Agassi et al. (1985b). They observed that in the case of a coarse relief an amount of 10 Mg ha^{-1} of PG is needed in order to maintain high IR, as opposed to the 5 Mg ha^{-1} of PG recommended by Agassi et al. (1982) for smooth reliefs.

The efficacy of PG application on seal formation and the IR of calcareous and noncalcareous smectitic soils having different ESP levels was studied by Kazman et al. (1983). The IR curves presented in Fig. 4 clearly show that as the ESP of the soil increased, the depth of rain required to reach the final IR decreased and at several occasions a lower final IR was obtained. Because the raindrop energy was kept the same throughout the experiments, the differences in the IR curves were the result of chemical dispersion caused by soil sodicity. When 5 Mg ha^{-1} of PG was applied to the soil (spread over) the rate of drop in the IR became more gradual compared with the nontreated soils and the final IR maintained values >8 mm h^{-1} (Fig. 4). PG was effective also in soils with ESP as low as 1.8, thus indicating that some chemical dispersion took place even when exchangeable Na occupies just 2% of the soil exchange capacity.

The advantageous effect of PG in maintaining high IR and reducing runoff was also observed in the field under natural rainfall conditions. Keren et al. (1983) constructed plots (25 $\times$ 6 m) in two bare calcareous Haploxeralf with ESP 4.6 and 19.3 to which PG was added at an equivalent rate of 5 and 10 Mg ha^{-1}, respectively. In a winter with a total of 182 mm of rain in seven storms, PG reduced runoff from 22.8 to 3.8 mm and from 64.2 to 18.7 mm for the ESP 4.6 and 19.3 soils, respectively. The results of this study also confirmed those obtained in the laboratory with respect to the method of PG application. Spreading PG over the soil surface was more effective in controlling runoff than mixing

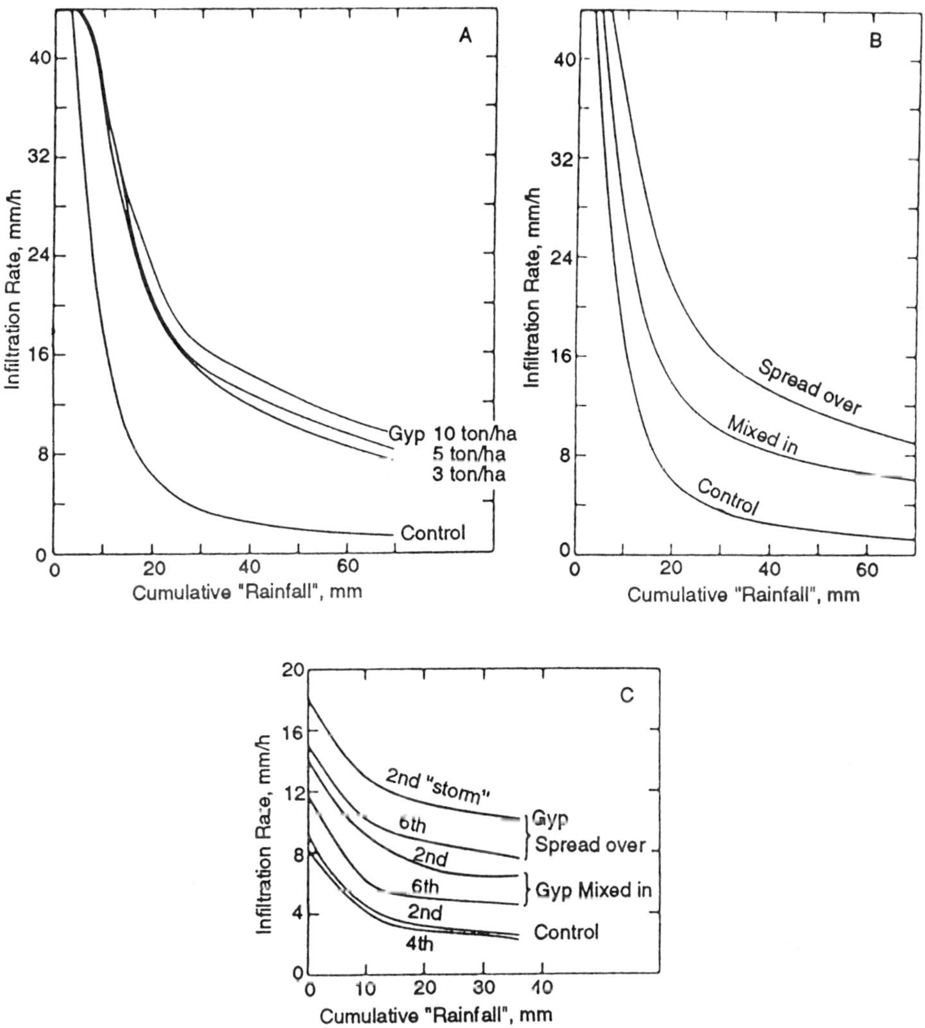

Fig. 3 Infiltration rates of loess (Nahal-Oz) soil as a function of cumulative simulated rainfall: (A) The effect of quantities of applied gypsum; (B) the effect of method of application; (C) the effect of consecutive "rainfalls" and gypsum application. (From Agassi et al., 1982.)

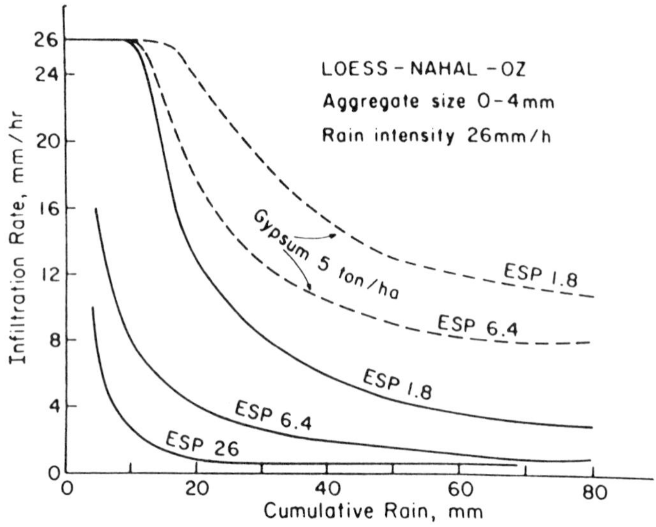

Fig. 4 The infiltration rate of Nahal-Oz soil as a function of cumulative rain. The effect of soil ESP and phosphogypsum application. (From Kazman et al., 1983.)

the PG with the upper 100 mm of the soil. The two methods of application resulted in 10.3% and 16.5% runoff, respectively (Keren et al., 1983). A different field trial tested the effect of PG on runoff/rain relation in three commercial winter wheat fields in Israel (Agassi et al., 1985b). During three consecutive winters, PG reduced runoff to 10–30% of that of the untreated soil.

Controlling seal formation and maintaining high IR by PG application was also studied on kaolinitic soils from the southeastern United States. Miller (1987) studied the effect of PG on the IR of three kaolinitic Ultisols from Georgia (USA). In the absence of PG, the IR stabilized at 10 mm h^{-1} with sealing apparent at the soil surface. Surface application of PG at a rate of 5 Mg ha^{-1} resulted in final IR of 25 mm h^{-1}. Miller and Scifres (1988) found that PG was also efficient in preventing a major deterioration in the permeability of a kaolinitic soil surface treated with a Na-containing fertilizer ($NaNO_3$). PG application together with the fertilizer or after its addition delayed runoff initiation until 45 and 25 mm of rain, respectively, had accumulated. It also resulted in improved final IR (Miller and Scifres, 1988). In a recent study, Norton et al. (1993) compared the efficiency of a number of gypsiferous materials on the IR of three humid soils of the eastern United States. These investigators found that two sulfitic materials from coal desulfurization were less effective than PG in controlling runoff; and one by-product consisting of almost pure gypsum was more effective than PG. Norton et al. (1993) concluded that the greater the solubility of the amended material the greater its efficacy in controlling seal formation and maintaining high IR. Stern et al. (1991a) studied the effect of PG on 19 predominantly kaolinitic soils from South Africa. The susceptibility of the nontreated soils to sealing depended upon the presence of smectitic impurities in their clay fraction. Soils containing these impurities were classified as dispersive, and soils having no detectable amounts of smectite were classified as stable (Stern et al., 1991a). The results showed that for both stable and dispersive soils addition of 5 Mg ha^{-1} of PG increased the final IR at the end of an 80 mm storm to values >18 and >8 mm h^{-1}, respectively (Fig. 5). In a further study Stern et al. (1991b) investigated the efficacy of PG in controlling runoff from small runoff plots (1.5 m^{-1}) under natural rainfall conditions in two kaolinitic and one illitic soil. Annual runoff from the control plots ranged between 33% and 73% of the rain. PG reduced runoff to 0.15–0.82 of that of the untreated plots. The efficiency of PG in reducing runoff was inversely correlate with rain intensity. The effect of PG lasted throughout the first rainy season (320–510 mm of rain). However, once cumulative precipitation exceeded 700 mm during the consecutive season, the effect of PG diminished (Stern et al., 1991b).

It is evident that smectitic soils from a semiarid region (Israel) and kaolinitic soils from humid (southeastern United States) and semiarid (South Africa) regions respond, when exposed to rain, in a similar manner to surface application of PG. This suggests that in many of the cultivated soils in the world, PG can be considered as a beneficial soil amendment in combating seal formation and preventing runoff.

4. *Mechanisms of Phosphogypsum Effect on Soil Sealing*

The main mechanism by which PG affects the IR of soils exposed to a low-electrolyte source of water is by dissolution and release of electrolytes into the soil-surface solution. Agassi et al. (1986) compared the IR of a loess soil treated with 5 Mg ha^{-1} powdered PG and exposed to distilled water rain to that of the same soil exposed to saline "rainwater" of 0.01 M $CaCl_2$ or saturated PG solution. Although the salinity of the percolating water was similar (= 2 dS m^{-1}) in all three treatments, the IR curves obtained with the saline solutions were lower than that obtained with the spread over powdered PG. It was evident that PG affects the IR of the soil not only through its effect on the electrolyte concentration of the percolating water (Agassi et al., 1986). These researchers postulated that powdered PG at the soil surface may affect seal formation by two additional physical mechanisms: (1) by interfering mechanically with the structure of the seal and thus disturbing the formation of a continuous seal of low permeability; (2) by serving as a mulch and partially protecting the soil surface from the beating action of the raindrops.

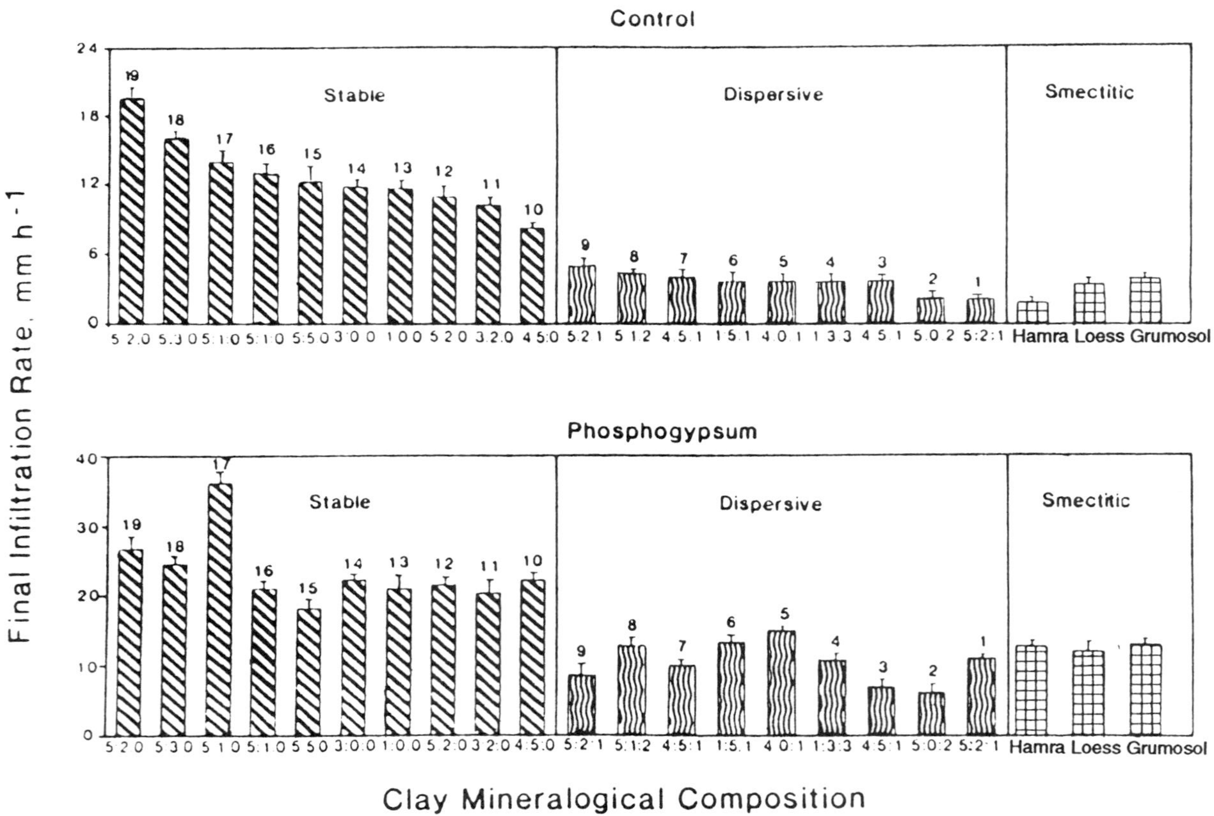

Fig. 5 Final IRs of the control and PG-treated for the stable, dispersive, and smectitic soils. The numbers above the columns represent the soil number. The numbers below the columns represent the mineral quantity (5—high, 1—traces) for kaolinite, illite, and smectite. Bars represent standard deviation. (From Stern et al., 1991a.)

D. Effect of Gypsum on Soil Erosion

1. Phosphogypsum and Soil Loss from Gentle Slopes (<10%)

The effect of PG on soil erosion from gently sloped kaolinitic (Miller, 1987) and smectitic soils (Warrington et al., 1989; Levin et al., 1991) studied in laboratory rainfall simulators is presented in Table 1. Rain energy, intensity, and depth varied for the three studies, so a quantitative comparison is impossible. Nevertheless, all soils developed seals with low final IR (<10 mm h^{-1}), runoff accounted for 40–82% of the rain applied in the nontreated soils. PG addition at the rate of 5 Mg ha^{-1} resulted in a decrease in runoff of 0.3–2.5 times and roughly decreased soil loss by 50% compared with the control (Table 1). Under natural rain conditions, PG efficiency in controlling soil erosion was reported to be even higher. Results showed that soil losses from PG-treated gently sloped Wischmeier plots (25 m long) were 3–5 times lower than that from the control plots (Agassi et al., 1985b).

Water erosion involves the detachment of soil by hydraulic shear stress in concentrated confined flow (rill erosion), and soil detachment and transport by raindrop impact and shallow interrill flow (interrill erosion) (Lane et al., 1987). Interrill soil erosion is the predominant type of erosion at gently sloping soils (Walker et al., 1977). The effect of PG on soil detachment by raindrop impact is relatively small compared with untreated soils. That is so because in both instances soil detachment is restricted, despite the fact that an opposing process is involved in each treatment. The formation of a seal with high shear strength may limit soil detachment by raindrops in untreated soil (Bradford et al., 1987).

Table 1 Effect of Phosphogypsum (PG) on Infiltration Rate (IR), Percent Runoff, and Soil Loss from Soils with Gentle Slope

Soil classification	Dominant clay minerals[a]	Final IR (mm h^{-1})		Runoff (%)		Soil loss (Mg ha^{-1})	
		Control	PG	Control	PG	Control	PG
Cecil[b]	K	10.0	25	42	12.7	0.266	0.096
Wedowee[b]	K	5.0	17	72	31.7	1.135	0.442
Worsham[b]	K	4.0	12	74	55.7	1.315	0.732
Rhodoxeralf[c]	St, K	2.9	8	66	32.3	9.350	4.140
Haploxeralf[d]	St, I	4.0	7	82	61.9	10.200	6.670
Chromoxerert[d]	St, I	4.3	15	76	47.6	4.800	2.200

[a]I = illite, K = kaolinite, St = smectite.
[b]Georgia, U.S. soils (Miller, 1987).
[c]Israeli soil (Warrington et al., 1989).
[d]Israeli soils (Levin et al., 1991).

In surface PG-amended soils, the surface shear strength is lower than that of an untreated soil (Warrington et al., 1989), but the aggregates at the soil surface are stabilized and rain detachment is also significantly reduced.

The capacity of runoff water to cause interrill erosion is smaller than that of raindrop impact because of its smaller detachment capacity (Young and Wiersma, 1973). Runoff flow detachment capacity depends on runoff volume and velocity. Runoff velocity, in turn, is determined by the slope angle and the roughness of the soil surface. PG application reduces the volume of runoff to approximately half of that obtained in untreated soil and increases the roughness of the soil surface and the tortuosity of the flow paths, thereby decreasing runoff velocity.

PG increases the electrolyte concentration in both the runoff and the percolating water and thus prevents clay dispersion (Agassi et al., 1981) and results in bigger aggregates at the soil surface. The latter, in turn, are less susceptible to erosion by runoff flow. The higher electrolyte concentration in runoff water from PG-amended soil also affects the sediment concentration in the runoff water. Sediment deposition owing to sediment settling out under gravity occurs continuously during erosion (Rose, 1985). Miller (1987) found that no clay-size particles were present in the runoff water of PG-treated soils, compared with 15–30% in untreated soils. Thus, PG treatment increases the size of the suspended material leading to enhanced sediment deposition, which consequently results in smaller amounts of soil loss.

2. Phosphogypsum and Soil Loss from Steep Slopes

The fact that both empirically based models for erosion prediction such as the universal soil loss equation (USLE), and process-based ones like the water erosion prediction project (WEPP), include slope as a factor in their prediction equations indicates the significance of slope for water erosion. For soils susceptible to sealing, the slope has an intricate effect on IR, runoff, and erosion (Poesen, 1987). As the slope increases, runoff velocity and, subsequently, erosion increase. However, since the seal formed may be eroded, it is possible that, with an increase in slope and seal erosion, the IR may increase and runoff may decrease. Poesen (1987) found that increasing the slope led to higher IR, less runoff, and a decrease in the shear strength of the seal. The effects of increasing the slope were attributed to the following (Poesen, 1987): (1) more splash and sheet erosion occurred, which removed the seal continuously; (2) the number of raindrop impacts per unit surface area decreased; (3) the normal component of drop impact force decreased; and (4) the occurrence of a thin film of water, through which the compactive force of the raindrop is enhanced (Palmer, 1964), is reduced.

Warrington et al. (1989) investigated the effect of PG on erosion from a smectitic sandy loam soil at slopes ranging between 5% and 30%. The final IR increased for both treated and untreated soil with increase in slope (Fig. 6). The results for the latter are similar to those obtained by Poesen (1987) and support the explanation that the IR increases due to removal of the seal by erosion. However, erosion is not so pronounced in PG-treated soils; thus a different explanation for the increase in IR with increased slope is in order. In the presence of PG clay dispersion is prevented and the seal formed results predominantly from drop impact. With increase in slope, both the number of drops impacting a unit surface area and the normal component of drop impact decrease. Consequently, a less developed seal is formed and the IR increases (Agassi et al., 1985a).

The marked effect of PG at various slopes on soil loss is presented in Fig. 7. Increasing the slope from 5% to 25% increased soil loss by 668% and 190% for the untreated and PG-treated soil, respectively (Warrington et al, 1989). Large increases in soil loss in the control soil occurred only at slopes >10% (Fig. 7). At gentle slopes (<10%), runoff velocities are low and the main cause for soil loss is soil detachment by raindrop impact. As raindrop detachment is insensitive to slope, there is only a slight increase in erosion with changes in the slope at slopes below 10%. Once the slope is greater than 10% the velocity of the runoff water is high enough to produce a hydraulic shear stress capable of detaching soil particles from the soil surface and thus enhance erosion. At such slopes (>10%), Warrington et al. (1989) observed the formation of rills in the untreated soil that reached a depth of 7.5 mm and a width of 20 mm at the 30% slope. Once rills start forming, soil loss increases at an accelerated rate. In the PG-treated soil, on the other hand, no rills were observed and the soil surface kept a certain degree of roughness at all slope angles (Warrington et al., 1989). The latter restricted runoff flow velocity and prevented high soil loss in steep slopes.

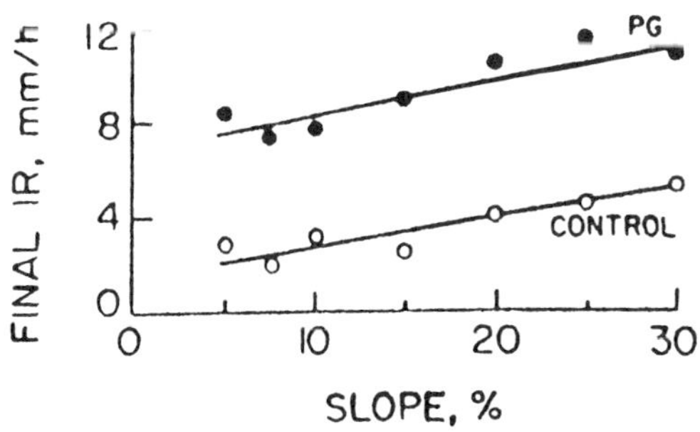

Fig. 6 Effect of slope and phosphogypsum treatment (PG) on the final infiltration of sandy loam soil under simulated rainfall. (From Warrington et al., 1989.)

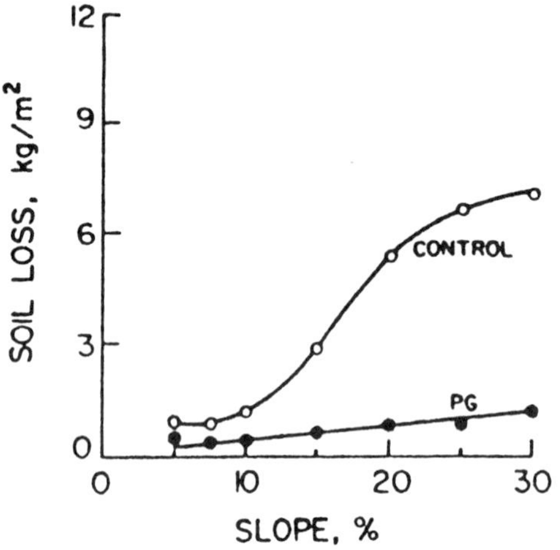

Fig. 7 Effect of slope and phosphogypsum treatment (PG) on soil loss from a sandy loam soil exposed to simulated rain. (From Warrington et al., 1989.)

The effect of PG on runoff and erosion from a Chromoxerert in Israel was measured by Agassi et al. (1990) in small field plots (1 × 1.5 m) exposed to natural rainstorms. Agassi et al. (1990) observed that at the gentle slope, PG reduced soil losses to 6–10% of that in the control, whereas in the steeper slopes PG reduced erosion to 1–3% of that in the control. Addition of PG reduced runoff from 60% in the control to 15% in the treated soil and was only slightly influenced by the slope. Agassi et al. (1990) concluded that the beneficial effect of PG on reducing soil loss is due not only to the decrease in runoff but also to decreased sediment concentration. Their observations were similar to those made in laboratory studies using rainfall simulators (Warrington et al., 1989).

In an additional field experiment in a Typic Rhodoxeralf, Agassi and Ben-Hur (1991) studied the effect of PG on runoff and erosion from short (1.5 m) and long (10 m) plots constructed at a 48% slope. PG reduced runoff by 23% irrespective of plot length. Conversely, erosion was 6.4 times higher in the long plots compared with the small plots, and PG efficiency in reducing erosion was smaller in the long plots than in the small ones. It was concluded that in the small plots erosion is determined by interrill erosion processes, whereas in the long plots rill erosion is the predominate mechanism (Agassi and Ben-Hur, 1991). Thus, once the slope factor (slope angel × length) is large enough, it enables rill formation and accelerates erosion which diminishes the beneficial effect of PG.

The effect of PG in controlling soil erosion from steep slopes was also studied in kaolinitic soils. Stern (1990) studied the effect of 5 Mg ha^{-1} surface-spread

PG on soil loss from five kaolinitic soils from South Africa placed at 5% and 30% slopes in a laboratory rainfall simulator. Runoff was not affected by slope angle, and, excluding the Rhodustalf(2) soil, PG reduced runoff by 25–50% (Table 2). These results were similar to those obtained in smectitic soils (Warrington et al., 1989; Agassi et al., 1990). With respect to soil loss, Stern (1990) separated the soils into stable (no swelling minerals in the clay fraction) and unstable (some impurities of swelling minerals) soils. The latter are more erodible (Table 2) because they are more dispersive and formed a less permeable seal that generated higher levels of runoff. PG addition significantly reduced soil loss from both the stable and unstable soils. However, the effect of PG was more pronounced in the more erodible soils, especially in the steep slopes (Table 2). Stern (1990) attributed this phenomenon to the susceptibility of the unstable soils to chemical clay dispersion, which in the PG-amended soils was diminished. Under less erosive conditions, such as gentle slopes or in chemically stable soils, PG efficacy in controlling erosion is less dramatic, though still important (Stern, 1990).

3. *Mechanism of PG Effect on Soil Erosion*

Surface-spread PG, by releasing electrolytes at the soil surface to the percolating and runoff water, decreases soil erosion by the following mechanisms:

Table 2 Effect of Phosphogypsum (PG) on Percent Runoff and Soil Loss from Kaolinitic Soils at Two Extreme Slopes

Soil classification	Dominant clay minerals[a]	Slope (%)	Runoff (%)		Soil loss ($Mg\ ha^{-1}$)	
			Control	PG	Control	PG
Paleudalf(1)	K, I, Is	5	87.6	53.0	1.31	0.47
		30	81.3	55.7	6.77	2.12
Rhodudalf	K, I, Is	5	76.5	45.6	1.60	0.58
		30	73.2	46.7	7.15	1.18
Rhodustalf(1)	K, I, Is	5	62.7	47.5	1.18	0.84
		30	68.0	49.0	3.39	1.16
Paleudalf(2)	K, I	5	40.7	33.0	0.38	0.22
		30	40.2	35.0	0.42	0.25
Rhodustalf(2)	K, I	5	44.5	8.5	0.32	0.14
		30	47.3	32.9	0.84	0.45

[a]I = Illite, Is = interstratified materials (swelling minerals, K = kaolinite.
Source: Stern (1990).

1. The fraction of rain that penetrated the soil was increased, and hence runoff depth decreased.
2. The stability of surface aggregates was increased and fewer particles were detached by either raindrop impact or overland flow.
3. Velocity of runoff flow was decreased by maintaining rough surface at the soil surface.
4. Flocculation of suspended clay size particles and the resultant deposition thereof was enhanced.

The efficiency of PG in reducing soil erosion from various slopes extends to soils of varying mineralogy. Thus, PG can be considered as a universal agent in stabilizing cultivated soils of gentle slopes and natural or artificial steep slopes.

III. POLYMERS

A. Introduction

Interest in organic polymers as soil conditioners and stabilizers was enhanced in the early 1950s when the Monsanto company developed Krilium, which was a trade name for a couple of different types of polymers such as vinyl acetate, malic acid, and hydrolyzed polyacrylonitrile (Chepil, 1954). Enthusiasm at the time was high because of the urgent need for improving the physical properties of soils. However, it became apparent after a few years that the methodology used was too costly for field agriculture, and hence commercial use of polymers was abandoned in the early 1960s. There has recently been a renewed interest in organic polymers as soil conditioners in general (*Soil Sci. Special Issue 141* (5), 1986) and particularly in improving aggregate stability, controlling seal formation and runoff, and preventing soil erosion (e.g., Shaviv et al., 1986; Helalia and Letey, 1988; Shainberg et al., 1990; Smith et al., 1990). This was due, in part, to the development of more effective and less expensive polymers. Additionally, the current understanding of seal formation phenomenon allows the concept of treating the soil surface only. Hence, the volume of soil needed for treatment has become very small, resulting in lesser quantities of polymers required and making their use for agricultural purposes more economically attractive.

B. Polymers and Their Properties

Polymers are small recognized repeating units (monomers) coupled together to form extended chains. Their chain lengths in solution range between a few thousand and 3×10^5 μm and average diameters are 0.5–1.0 μm (Schamp et al., 1975). Besides being long, polymer chains are flexible, multisegmented, and polyfunctional. Polymers are characterized mainly by the following parameters: (1) molecular weight, (2) molecular conformation (coiled or stretched), (3) type of charge, and (4) charge density.

Polysaccharides are the largest group of polymers in soil and constitute 10–30% of soil organic matter. Soil polysaccharides are strongly implicated in soil aggregate stabilization (Hepper, 1975), but are effective for only a short time, from weeks to a few months. Hence, when aiming at improving soil physical conditions, synthetic organic polymers have been preferred over natural polymers.

Polyacrylamide (PAM) and polysaccharide (PSD) are two synthetic organic polymers that have recently been extensively investigated with respect to their efficacy as soil conditioners. PAM is a by-product of petrochemical processes (Azzam, 1980), and its molecular weight varies in the range of 10×10^3 Dalton (medium weight) to 15×10^6 Dalton (high weight). PAM can be either cationic, nonionic, or anionic, with the latter form being the most commonly used. The charge density of the anionic PAM depends on the degree of hydrolysis, i.e., the number of amine groups substituted with OH^- (Stutzmann and Siffert, 1977). The PSD is a guar derivation, with the latter being classified as a galactomannan consisting of mannose and galactose units (Aly and Letey, 1988). The PSDs have low-to-medium molecular weight ($0.2–2.0 \times 10^6$ Dalton). Their charge density is determined by the number of substitutions of nonionic *cis*-hydroxyl groups with quaternary ammonium, hydroxpropyl, and carboxyl groups in order to get a cationic, nonionic, and anionic PSD, respectively (Aly and Letey, 1988). Cationic PSD is the one most commonly used.

C. Polymer-Solid Interactions

1. Polymer-Reference Clays Interactions

Polymer adsorption onto the clay holds the key to the formation and properties of clay-polymer complexes, which in turn enhance our understanding of polymer behavior in soil. Generally, adsorption isotherms of polymers fit the Langmuir equation (Schamp and Huylebroeck, 1973), and, with few exceptions, the adsorption is irreversible (Stutzmann and Siffert, 1977). When flat surfaces are available (e.g., peptized clay) adsorption of the polymer is rapid. Conversely, in a porous structure (flocculated clay), adsorption is slow and decreases with increasing molecular weight (Schamp and Huylebroeck, 1973). Of the various polymer properties (molecular weight and conformation, type and density of charge, temperature, acidity, etc.), type of charge largely determines the mechanisms controlling polymer adsorption and thus received much attention (e.g., Theng, 1979, 1982).

The adsorption of nonionic flexible linear polymer onto a clay surface increases with the molecular weight of the polymer and generally leads to the desorption of numerous water molecules. Thus the adsorption of such polymers is usually an entropy-driven process. On average, 40% of the total number of segments in the polymer chain can be attached to the clay surface (Theng, 1982), and the adsorption is almost "irreversible."

The interaction of charged (anionic and cationic) polymers with clays is more complex than those involving nonionic species (Hesselnik, 1977). Because of their charge the polymers can undergo changes in surface charge and conformation (stretching-coiling transformation) in response to changes in the pH and ionic strength of the ambient solution.

The adsorption of cationic polymers by clays occurs through electrostatic (Coulombic) interactions between the cationic groups on the polymer and the negatively charged sites on the clay surface. Polycations compete with exchangeable and electrolyte cations for exchange sites on the clay. Hence, adsorption of polycations by clay increases with a decrease in the valency of the exchangeable cation (Gu and Doner, 1992) and decreases with an increase in the electrolyte concentration of the solution (Aly and Letey, 1988). In addition, Gu and Doner (1992) found that increasing the pH of the solution increased the adsorption of a polycation by illite.

Negatively charged polymers tend to be repelled from the clay surface, and little adsorption occurs. In addition, unlike nonionic and cationic polymers, anionic polymers do not enter the interlayer space of expanding minerals (e.g., Ruehrwein and Ward, 1952). Adsorption is promoted by the presence of polyvalent cations which act as "bridges" between the anionic groups on the polymer and the negatively charged sites on the clay (Mortensen, 1962). Increasing the ionic strength of the solution reduces the electrostatic repulsion between the polyanion and the clay surfaces (Ruehrwein and Ward, 1952; Mortensen, 1959), and may also lead to decreased charge and size of the polymer (Mortensen, 1959), both of which enhance adsorption of polyanions (e.g., Aly and Letey, 1988; Gu and Doner, 1992). Acidic conditions also favor polyanion adsorption (Theng, 1982).

The amount of polymer adsorbed by clay depends on the type of charge. Gu and Doner (1990) reported that the amount of PSD adsorbed by Silver Hill Na-illite was 100, 35, and 8 g polymer per kilogram clay for cationic, nonionic, and anionic PSD, respectively (Fig. 8). Conversely, the amount of a nonionic PSD adsorbed by Wyoming Na-montmorillonite (80 g kg^{-1}) was similar to that of a cationic PSD (Aly and Letey, 1988). With respect to PAM, in its anionic form only small quantities are adsorbed, 2–3 g kg^{-1} (Stutzmann and Siffert, 1977; Aly and Letey, 1988). By comparison, Wyoming Na-montmorillonite adsorbs 45 g kg^{-1} of nonionic PAM (Aly and Letey, 1988). Stutzmann and Siffert (1977) postulated that the small quantities of PAM adsorbed on the clay result from the fact that PAM is adsorbed only on the external surfaces of the clay

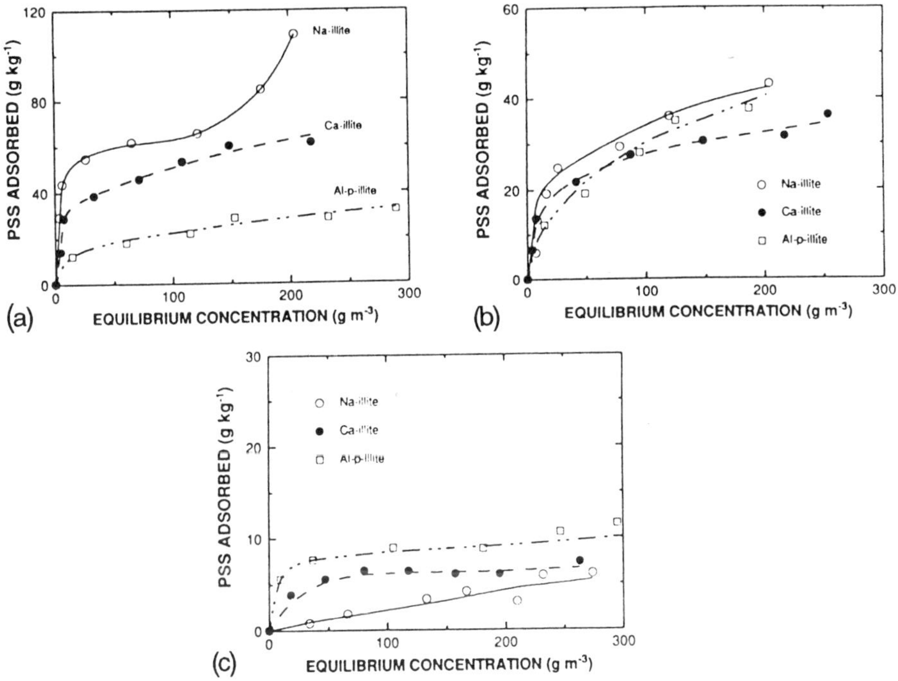

Fig. 8 Adsorption isotherms of (a) cationic, (b) nonionic, and (c) anionic polysaccharide by Na-, Ca-, and Al-p-illite. (From Gu and Doner, 1990.)

particles, and hence that PAM adsorption depends on the external cation-exchange capacity of the mineral.

External charge density of the adsorbing clay is important in polymer adsorption. Ben-Hur et al. (1992b) observed a considerably larger adsorption by illite than by montmorillonite, irrespective of the type of the polymer (i.e., cationic or anionic). The higher charge density of illite compared with montmorillonite (when external surfaces are considered) was considered the reason for the higher adsorption on illite (Ben-Hur et al., 1992b).

2. *Polymer-Soil Interactions*

Adsorption of polymers by soils has not been extensively studied, perhaps because it was implicitly assumed that the interaction of polymers with soil clays should not differ significantly from their interaction with reference clays. Knowledge of the adsorptive behavior of polymers on soil aggregates is essential for predicting their effect on soil stability and some other relevant physical properties. Polymers with a wide range of properties are currently available so that, with the understanding of polymer-soil interactions, the optimal polymer treatment can potentially be obtained.

Nadler and Letey (1989) studied the adsorption of three anionic polymers by a coarse loamy soil. Similar to anionic polymer-clay systems (Aly and Letey, 1988; Gu and Doner, 1992), Ca soils and initial high pH resulted in increased polymer adsorption. However, the increase was very small. The adsorption levels (Fig. 9) were, though, 2–3 orders of magnitude lower than those obtained by Aly and Letey (1988) for the same polymers on montmorillonite. It was postulated that (1) the higher specific surface area and the amount of charge available for adsorption associated with the clay-size fraction of the montmorillonite, and (2) the smaller accessible and active surfaces in the soil because of the presence of organic matter and aggregation are responsible for the lower adsorption of the polymers by the soil (Nadler and Letey, 1989).

Malik and Letey (1991) extended the study of Nadler and Letey (1989) by including additional soils, washed quartz sand, and cationic polymers. Polymer adsorption on soil was in the following decreasing order: high-charge cationic PSD>low-charge anionic PAM>low-charge cationic PSD>high-charge anionic PAM>no-charge PAM. Adsorption on sand was only little less than on soil. Conversely, adsorption on clay was in the order low-charge cationic PSD>low-charge anionic PAM (Malik and Letey, 1991). Thus, Malik and Letey (1991)

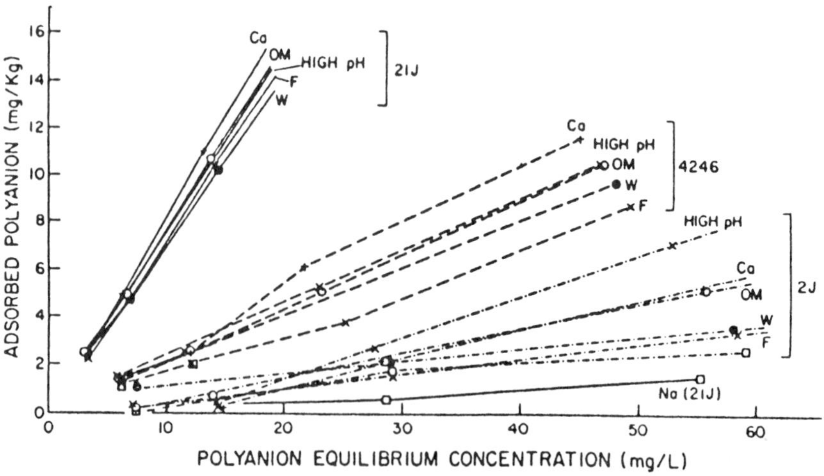

Fig. 9 Adsorption isotherms of three anionic polysaccharides, 21J, 4246, and 2J. Ca = Ca saturated soil, OM = organic matter depleted soil, F = fine size fraction of soil, W = nonpretreated soil. (From Nadler and Letey, 1989.)

concluded the following: (1) adsorption by soils is mostly on external surfaces and the polymers did not penetrate the aggregates; (2) charge density of the polymer determines its adsorption on charged surfaces (e.g., clays); and (3) molecular size and conformation of the polymer determine its adsorption on weakly- and noncharged surfaces (e.g., soils and sand), with increasing molecular size and chain extension of the polymer leading to increased adsorption.

The solid properties that primarily affect polymer adsorption are

1. The surface area on which the polymer can be adsorbed; since, in the case of soils, the polymer hardly penetrates the aggregates, it is the external surface area of the aggregates that is significant for the adsorption process. Consequently, adsorption onto soil material is significantly smaller than onto the clay-size fraction.
2. Surface charge density and type which determine the magnitude of the force by which the polymer is attracted to or repelled from the solid surfaces.

On the other hand, the polymer properties that affect its adsorption are mainly its charge density and type that influences electrostatic attraction or repulsion to the solid, and its molecular weight and conformation, especially in the case of weakly- or noncharged surfaces.

D. Controlling Seal Formation and Infiltration Rate with Polymers

It is preferable to add the polymer in a solution form to the soil surface rather than to add it as a dry powder or granules for controlling seal formation and maintaining high infiltration rates (Wallace and Wallace, 1986). Because seal formation and resulting low IR values may occur in soils exposed to both natural rain and to overhead sprinkler irrigation, the method of polymer application may vary accordingly. In the case of natural rains, polymers must be added to the soil surface, generally by spraying concentrated polymer solutions (in the range of 500 to 5000 g m^{-3}), prior to the rainy season. In case of overhead sprinkler irrigation, polymer can be added to the soil surface as for rain, or it can be dissolved in the irrigation water to form very dilute solutions (10–20 g m^{-3}) whenever the need arises. For simplicity the discussion on polymer efficiency in controlling seal formation is divided into two sections according to the method of polymer application.

1. *Polymer Addition to the Soil Surface*

The effect of polymers on stabilizing surface particles during rainstorms and, hence, on preventing seal formation has been studied extensively both under laboratory and field conditions. In the early 1970s, Gabriels et al. (1973) showed that surface application of a small amount (38 kg ha^{-1}) of an anionic polyacrylamide was highly successful in maintaining high IR values and preventing runoff.

Shainberg et al. (1990) used a laboratory rainfall simulator to investigate seal formation in two Israeli soils as affected by soil-surface application of a high-molecular-weight, low-charge-density anionic PAM. They applied 10, 20, and 40 kg ha^{-1} of PAM and found that the beneficial effect of rates above 20 kg ha^{-1} were insignificant in maintaining high IR (Fig. 10). They stated that the comparable efficiency of 20 and 40 kg ha^{-1} could have resulted from technical problems related to PAM application. Because of difficulties experienced in dissolving PAM to form high-concentration solutions, increasing the amount of PAM applied is done by increasing the volume of the solutions used rather than increasing the solution concentration. Consequently, upon increasing the amount of PAM added, a difficulty arises in keeping the PAM at the upper layer of the soil (Shainberg et al., 1990). In comparison, when an anionic polymer with a medium molecular weight was used (Shaviv et al., 1986), 80 kg ha^{-1} of the polymer was the most effective treatment. These results suggest that higher-molecular-weight polymers are more effective in controlling seal formation and

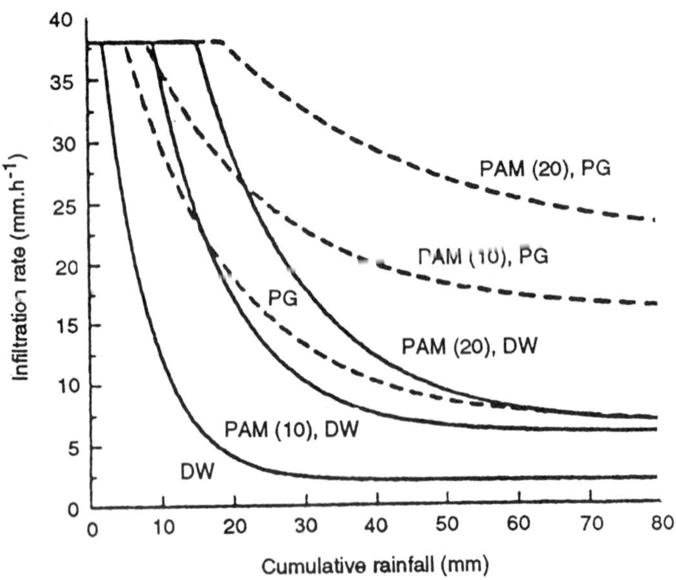

Fig. 10 Infiltration rate of loess as a function of cumulative rainfall, the level of PAM application (10, 20, or 40 kg ha^{-1}) under distilled water (DW) rainfall and phosphogypsum (PG) (5 t ha^{-1}) treatments. (From Shainberg et al., 1990.)

a smaller application is required. These two studies also found that the effect of the polymer was dramatically enhanced when the polymer was added in combination with PG. For instance, the infiltration rate in the PAM treatment was two to three times that of a nontreated soil. But, when PAM application was supplemented with 5 Mg ha^{-1} of PG, the final IR was about 10 times higher than the control (Fig. 10). PG spread at the soil surface dissolves and increases the electrolyte concentration in the soil solution above the flocculation value of the soil clays (Oster et al., 1980). Flocculation of the soil clay is apparently a precondition for the cementing and stabilization of aggregates at the soil surface by anionic polymers. This conclusion was supported by the results of Smith et al. (1990). These investigators found that ''raining'' a PAM-treated soil with water containing electrolytes resulted in comparable IR values to those obtained when the PAM-treated soil was amended with PG.

The efficacy of polymers in controlling seal formation depends on rain kinetic energy. Levin et al. (1991) studied the effect of combined application of PAM and PG on three different soil types under three rain energies (3.6, 8.0, and 12.4 J mm^{-1} m^{-2}) in the laboratory. The effect of PG and PAM + PG on increasing the cumulative infiltration from the entire rainstorm was generally similar at low-kinetic-energy rain. When high-kinetic-energy rain was used, the effect of PAM + PG was greater than that of PG alone (Table 3). Levin et al. (1991) hypothesized that at low-kinetic-energy rain clay dispersion was more important in the sealing process than aggregate disintegration. Hence, the presence of the polymer which increases aggregate strength was less important at low kinetic energy than PG, which prevented clay dispersion. Conversely, when high-kinetic-energy rain was used, prevention of aggregate breakdown by the beating impact of the raindrops was essential. Treating the soil with a soil stabilizing

Table 3 Ratio between Cumulative Infiltration (CIF) Data Obtained from Amended Soil Samples and Untreated Samples (Control)

Soil classification	Raindrop energy (J mm^{-1} m^{-2})	CIF (control) (mm)	CIF (amended soil)/ CIF (control)	
			PG	PAM + PG
Typic Rhodoxeralf	3.6	23.8	2.4	2.8
	12.4	9.6	2.5	7.1
Calcic Haploxeralf	3.6	24.6	1.6	2.7
	12.4	16.4	1.5	4.1
Typic Chromoxerert	3.6	31.2	1.8	2.1
	12.4	19.8	2.4	3.7

Source: Levin et al. (1991).

agent (PAM) in addition to a source of electrolytes (PG) is imperative for a successful control of seal formation under high-energy rainfall (Levin et al., 1991).

The effectiveness of polymers in improving IR was also observed in larger field-scale experiments. A study was conducted in Israel on a cultivated montmorillonitic clay loam Grumusol irrigated with a center-pivot sprinkler irrigation system (Levy, Ben-Hur, and Agassi, 1991). Percent runoff from the PAM-treated plots was significantly lower than that from control plots (Fig. 11). Field studies in South Africa (Stern et al., 1991b, 1992) showed that a combined treatment of PAM and PG is very effective in controlling runoff from kaolinitic and illitic soils. Annual runoff percentage from PAM + PG–treated plots was significantly lower than that from the nontreated (control) and PG-treated plots (Table 4). Similar efficiency of PAM was reported by Fox and Bryan (1992) working with two soils from Kenya. The results of these field experiments clearly demonstrate that spreading a small amount of PAM at the soil surface has a long-term effect on stabilizing the aggregates at the soil surface and reducing runoff, which lasts over the entire rain/irrigation season.

Additionally, runoff from PAM-treated Israeli smectitic soil and South African kaolinitic and illitic soils were similar (~20% runoff). Similar to PG (see previous discussion), the effect of PAM on runoff does not depend on clay mineralogy. This observation supports the notion that seal formation is quite a general phenomenon and that runoff prevention greatly depends on aggregate breakdown and clay dispersion.

2. *Polymer Addition to the Irrigation Water*

Adding small amounts of polymers to the irrigation water to form dilute concentration of the polymer (<50 g m^{-3}) under conditions of overhead sprinkler irrigation has been considered a logical alternative to surface application of the polymers. Studies on the effect of dilute concentration of polymer in the irrigation water on seal formation and runoff have indicated that the efficacy of the polymers depended mainly on polymer type, polymer charge density, and water quality. Ben-Hur and Letey (1989) studied the effect of type of charge and charge density of PSDs at very low concentrations (0–10 g m^{-3}) in the irrigation water on water infiltration of a sandy loam soil. They observed that addition of nonionic and anionic PSDs had no beneficial effect on the IR (Fig. 12). Conversely, 10 g m^{-3} of cationic PSD had a significant beneficial effect on the IR compared with the control. However, the higher-charge cationic PSD was more effective than the lower-charge one in maintaining high final IR (Fig. 12). Levy et al. (1992) compared the effects of low concentrations of an anionic PAM and a cationic PSD on two Israeli soils. They found that a lower concentration of PAM (10 g m^{-3}) was needed for an optimal effect on infiltration and runoff than

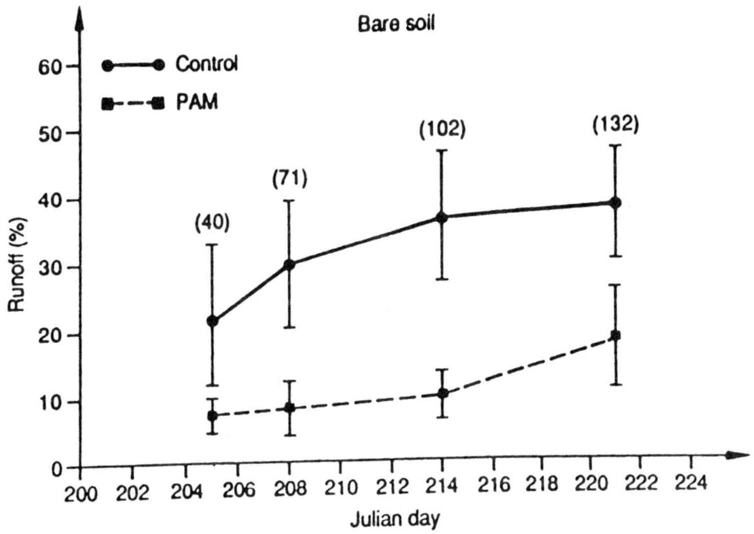

Fig. 11 Percent runoff out of the total amount of water applied at each irrigation event to the bare plots in the Negba vertisol. Vertical lines represent two standard deviations. Numbers in parentheses indicate cumulative depth of irrigation (mm). (From Levy et al., 1991.)

that required if PSD were used (20 g m^{-3}). For optimal treatments the final IR and runoff were generally higher in the PAM than in the PSD treatments (Levy et al., 1992).

The electrolyte concentration in the applied water greatly affects the efficacy of the polymers added to the water. For PSDs it was found that increasing the electrolyte concentration in the applied water increases the efficiency of the

Table 4 Effect of Combined Application of PAM (20 kg h^{-1}) and PG (5 Mg ha^{-1}) on Percent Runoff from Natural Rainstorms

Soil classification	Clay mineralogy	Annual rain (mm)	Runoff out of annual rain (%)	
			Control	PAM + PG
Paleudalf	K, I	214	40.2	20.7
Rhodustalf	K, I	107	72.5	27.9
Haplustalf	I, K	122	40.9	15.3

Source: Stern et al. (1991b).

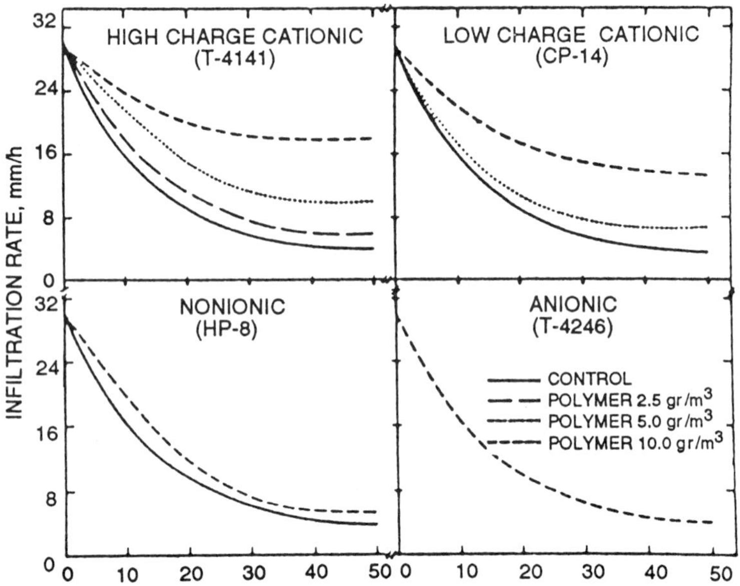

Fig. 12 Infiltration rate as a function of cumulative (cum.) water application for four polymer types and their different concentrations in the water applied (for the anionic polymer the control and the polymer treatments have the same line). (From Ben-Hur and Letey, 1989.)

polymer added to the water in maintaining high IR and controlling seal formation (Helalia and Letey, 1988). Ben-Hur et al. (1989) reported that the effectiveness of polymers added to distilled water in controlling seal formation was in the following decreasing order: high-charge cationic PSD>low-charge cationic PSD>very-low-charge anionic PAM. This order was comparable to the order of the polymers' adsorption on soil surfaces (Malik and Letey, 1991). Ben-Hur et al. (1989) concluded that adsorption of the polymer enhances its ability to flocculate the particles at the soil surface, which is important in preventing seal formation. The observed negligible effect of PAM on preventing seal formation when added to electrolyte-free irrigation waters is in agreement with results for surface application of PAM exposed to distilled-water rain (Shainberg et al., 1990). These researchers concluded that flocculation of the clay is a prerequisite for aggregate stabilization at the soil surface by anionic polymers.

The effect of polymers added to the irrigation water depends also on the sodicity level of the applied water (sodium adsorption ratio, SAR) and soil ESP. Studies indicate that the relative beneficial effect of PSD tends to decrease with increasing SAR (El-Morsy et al., 1991). Concerning soil ESP, Ben-Hur et al. (1992a) observed that for ESPs of 8.5 and 30.6, addition of 10 or 50 g m^{-3} of PSD to the irrigation water had no significant effect on the final IR.

The residual effect of cationic PSD on controlling seal formation and maintaining high IR values in subsequent simulated rainstorms was found to be negligible (Helalia and Letey, 1988; Ben-Hur et al., 1989; Levy et al., 1992). Conversely, the latter investigators reported that an anionic PAM had some residual effect on the IR in subsequent irrigations. Levy et al. (1992) postulated that the difference in the residual effects of the two polymers stems from the difference in their adsorption to the soil. The PSD, being a cationic polymer, is highly adsorbed to the soil clay and hence is adsorbed on a relatively small volume of soil. During consecutive irrigations, soil loss exceeds the volume of soil treated with PSD, and hence no polymer is left to stabilize the surface aggregates. PAM, on the other hand, is adsorbed to a lesser extent on soil clays than PSD is and thus is adsorbed on a larger volume of soil. The amount of soil removed by erosion in the subsequent irrigation is smaller than that treated by PAM. Thus, PAM-treated aggregates are left to stabilize the soil surface at subsequent irrigations (Levy et al., 1992). If soil material is considered to have a weakly charged surface (Malik and Letey, 1991), then the lower residual effect of PSD compared with PAM could be explained by partial degradation of the polymer in the course of the wetting and drying cycles.

E. Effect of Polymers on Soil Erosion

Soil loss from sandy loam samples treated with 20 kg ha^{-1} of PAM in combination with a source of electrolyte (either using irrigation-water rain, which contains electrolytes, or distilled-water rain together with PG addition to the soil surface), placed at a 15% slope and exposed to simulated rainfall, was studied by Smith et al. (1990). The aforementioned PAM treatments resulted in soil losses of <10% of the soil losses obtained in the untreated samples (Fig. 13). Similar results were reported by Levin et al. (1991), and Levy et al. (1992). It was argued that the beneficial effect of the combined treatment of PAM and a source of electrolytes was not only because of reduced runoff (see previous discussion) but also because soil particles at the soil surface were larger than those in the untreated soil and, hence, these particles were more difficult to detach, and a quicker sedimentation of the suspended particles occurred (Levin et al., 1991).

The beneficial effect of PAM on decreasing soil loss was also highly significant at steep slopes. Wallace and Wallace (1986) added 16 and 33 kg ha^{-1} of PAM to the surface of soil samples placed at a 30% slope and exposed to simulated irrigation-water "rain." They reported that soil loss decreased by 77.7% and 98.3%, respectively, in comparison with the control. Agassi and Ben-Hur (1992) compared the effect of 20 kg ha^{-1} of PAM and 70 kg ha^{-1} PSD

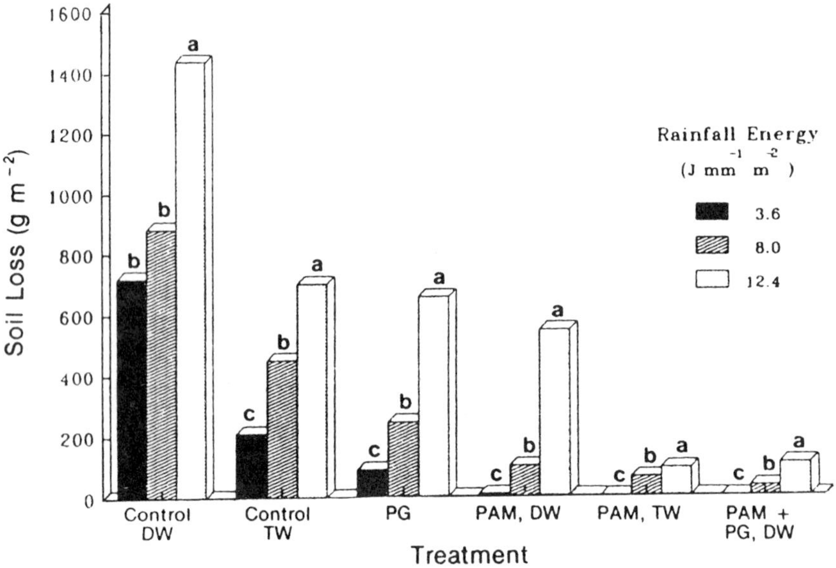

Fig. 13 Total soil loss after an 80-mm storm for three kinetic energy levels of rain. Within treatments, bars labeled with the same letter do not differ significantly at the 0.05 level. Treatments: PAM = anionic polyacrylamide; PG = phosphogypsum; DW = distilled water; TW = tap water. (From Smith et al., 1990.)

(both supplemented by 10 Mg ha^{-1} of PG) on soil losses from steep slopes (33–60%) 15–25 m long under natural rain conditions. Soil losses from the PAM and the PSD treatments were similar and decreased 6- to 10-fold in comparison with the control (Agassi and Ben-Hur, 1992). The latter results suggest that PAM efficiency is greater, as the amount of PSD used was 3.5 times that of the PAM. The greater efficacy of PAM in controlling soil loss compared with PSD has also been reported by Ben-Hur et al. (1992a) and Levy et al. (1992). Moreover, in a few laboratory studies it was found that the presence of cationic PSD resulted in an increase in soil loss compared with the control (Ben-Hur et al., 1990; Ben-Hur et al., 1992a).

The inability of PSD to reduce soil erosion was hypothesized to be related to the failure of PSD to form particles at the soil surface of a size and strength that would make them less susceptible to detachment (Levy et al., 1992). Another explanation related the effect of PSD on aggregate size distribution to changing it toward a size range favorable for transport by runoff water (Ben-Hur et al., 1990). The inefficiency of PSD to form big and strong enough aggregates that could withstand being eroded stems from its relatively short molecules (resulting from its medium molecular weight) and high adsorption. PAM, on the other hand, because of its long molecule (resulting from its high molecular weight) and limited adsorption, is efficient in cementing aggregates together and increasing their resistance to erosion.

The effect of polymers on preventing erosion that results from concentrated flow (e.g., rill, gallies, and furrow erosion) has received little attention compared with interrill erosion. Recently a field experiment was conducted where small amounts of anionic PAM were added to the irrigation water during the first 1–2 h of furrow irrigation (Lentz et al., 1992). The dilute PAM solutions (5–20 g m^{-3}) reduced sediment loss in the furrows by 45% to 98% in the initial irrigation. Cumulative reduction during the initial (treated) and two subsequent untreated irrigations was 42% to 58%. The variation in PAM efficacy in controlling soil loss resulted from differences in polymer concentration, duration of furrow exposure and water flow rate (Lentz et al., 1992). Although more studies are needed to verify these results, it has been established that with proper management, effective erosion control under furrow irrigation conditions of concentrated flow can be achieved with the addition of small amounts of PAM to the irrigation water.

IV. SUMMARY

Adding soil amendments, namely PG and polymers to the soil surface, can serve as an alternative to the traditional strategies for erosion control. However, further investigations are needed to fully establish whether their use for agricultural purposes is economically feasible.

Surface application of PG releases electrolytes to the soil-surface solution, interferes mechanically with the structure and continuity of the seal and serves as a mulch and partially protects the soil surface from the beating action of the raindrops, all of which lead to reduced runoff and decreased soil loss. Additionally, PG application reduces soil susceptibility to erosion by (1) decreasing runoff water velocity by maintaining rough soil surface, (2) enhancing flocculation and resultant deposition of the suspended clay-size particles in the runoff water, and (3) increasing surface aggregates' stability; thus fewer particles are detached by raindrops or overland flow.

Small amounts of polymers (10–20 kg ha^{-1}) sprayed directly onto the soil surface or added to the applied water lead to stabilization and cementation of aggregates at the soil surface and, hence, increase their resistance to seal formation and soil erosion. Compared with PG, the polymers generally have a more favorable effect on reducing runoff and soil erosion. Polymer efficiency depends on charge type and density and on the molecular weight of the polymer, as well as the electrolyte composition and concentration of the applied water. When anionic polymers are used, their efficacy is enhanced when they are applied in combination with a source of electrolyte (either in the "rain" water or PG addition). Of the polymers currently available and under study, anionic PAM has been found to be the most effective in controlling soil erosion and having the longest residual effect. This is probably because of PAM's long molecules, which promote the formation of relatively large and strong aggregates at the soil surface which are not easily detached and eroded under conditions of rain/irrigation or concentrated flow.

PG and PAM efficiency in reducing soil erosion extends over a wide range of slopes and to soils of varying mineralogy. Hence, these soil amendments may be considered as universal stabilizing agents for cultivated soils and earth structures worldwide.

REFERENCES

Agassi, M., and Ben-Hur, M. (1991). Effect of slope length, aspect and phosphogypsum on runoff and erosion from steep slopes. *Aust. J. Soil Res. 29*: 197–207.

Agassi, M., and Ben-Hur, M. (1992). Stabilizing steep slopes with soil conditioners and plants. *Soil Technol. 5*: 249–256.

Agassi, M., Morin, J., and Shainberg, I. (1982). Infiltration and runoff control in the semi-arid region of Israel. *Geoderma 28*: 345–355.

Agassi, M., Morin, J., and Shainberg, I. (1985a). Effect of raindrop impact energy and water salinity on infiltration rates of sodic soils. *Soil Sci. Soc. Am. J. 49*: 186–190.

Agassi, M., Morin, J., and Shainberg, I. (1985b). Infiltration and runoff in wheat fields in the semi-arid region of Israel. *Geoderma 36*: 263–276.

Agassi, M., Shainberg, I., and Morin, J. (1981). Effect of electrolyte concentration and soil sodicity on infiltration rate and crust formation. *Soil Sci. Soc. Am. J. 45*: 848–851.

Agassi, M., Shainberg, I., and Morin, J. (1986). Effect of powdered phosphogypsum on the infiltration rate of sodic soils. *Irrig. Sci. 7*: 53–61.

Agassi, M., Shainberg, I., and Morin, J. (1990). Slope, aspect and phosphogypsum effects on runoff and erosion. *Soil Sci. Soc. Am. J. 54*: 1102–1106.

Aly, S. M., and Letey, J. (1988). Polymer and water quality effects on flocculation of montmorillonite. *Soil Sci. Soc. Am. J. 53*: 1453–1458.

Azzam, R. A. I. (1980). Agricultural polymer, polyacrylamide preparation, application, and prospects. *Comm. Soil Sci. Plant Anal. 11*: 767–834.

Baver, L. D., Gardner, W. H., and Gardner, W. R. (1972). *Soil Physics*, 4th ed. Wiley, New York.

Ben-Hur, M., Clark, P., and Letey, J. (1992a). Exchangeable Na, polymer and water quality effects on water infiltration and soil loss. *Arid Soil Res. Rehab. 6*: 311–317.

Ben-Hur, M., Faris, J., Malik, M., and Letey, J. (1989). Polymer as soil conditioner under consecutive irrigation and rainfall. *Soil Sci. Soc. Am. J. 53*: 1173–1178.

Ben-Hur, M., and Letey, J. (1989). Effect of polysaccharides, clay dispersion and impact energy on water infiltration. *Soil Sci. Soc. Am. J. 53*: 233–238.

Ben-Hur, M., Letey, J., and Shainberg, I. (1990). Polymer effects on erosion under laboratory rainfall simulator conditions. *Soil Sci. Soc. Am. J. 54*: 1092–1095.

Ben-Hur, M., Malik, M., Letey, J., and Mingelgrin, U. (1992b). Adsorption of polymers on clays as affected by clay charge and structure, polymer properties, and water quality. *Soil Sci. 153*: 349–356.

Bradford, J. M., Ferris, J. E., and Remley, P. A. (1987). Interrill soil erosion processes. Effect of surface sealing on infiltration, runoff and soil splash detachment. *Soil Sci. Soc. Am. J. 51*: 1566–1571.

Chepil, W. S. (1954). The effect of synthetic conditioners on some phases of soil structure and erodibility by wind. *Soil. Sci. Soc. Am. Proc. 18*: 386–390.

El-Morsy, E. A., Malik, M., and Letey, J. (1991). Interactions between water quality and polymer treatment on infiltration rate and clay migration. *Soil Technol. 4*: 221–231.

Fox, D., and Bryan, R. B. (1992). Influence of a polyacrylamide soil conditioner on runoff generation and soil erosion: Field test in Baringo district Kenya. *Soil Technol. 5*: 101–119.

Gabriels, D. M., Moldenhauer, W. C., and Kirkham, D. (1973). Infiltration, hydraulic conductivity and resistance to water-drop impact of clod beds as affected by chemical treatment. *Soil Sci. Soc. Am. Proc. 37*: 634–637.

Gu, B., and Doner, H. E. (1992). The interaction of polysaccharides with Silver Hill illite. *Clays Clay Miner. 40*: 151–156.

Helalia, A. M., and Letey, J. (1988). Cationic polymer effects on infiltration rates with a rainfall simulator. *Soil Sci. Soc. Am. J. 52*: 247–250.

Hepper, C. M. (1975). Extracellular polysaccharides of soil bacteria. In *Soil Microbiology* (N. Walker, ed.). Butterworths, London, pp. 93–110.

Hesselnik, F. Th. (1977). On the theory of polyelectrolyte adsorption. The effect on adsorption behavior of the electrostatic contribution to the adsorption free energy. *J. Colloid Interface Sci. 60*: 448–466.

Hilgard, E. W. (1906). *Soils—Their Formation, Properties Composition and Relation to Climate and Plant Growth in the Humid and Arid Regions.* Macmillan, London.

Hillel, D. (1980). *Fundamentals of Soil Physics.* Academic Press, New York

Hurlbut, C. S., and Klein, C. (1971). *Manual of Mineralogy*, 19th ed. Wiley, New York.

Kazman, Z., Shainberg, I., and Gal, M. (1983). Effect of low levels of exchangeable Na (and phosphogypsum) on the infiltration rate of various soils. *Soil Sci. 135*: 184–192.

Keren, R., and Shainberg, I. (1981). Effect of dissolution rate on the efficiency of industrial and mined gypsum in improving infiltration in a sodic soil. *Soil Sci. Soc. Am. J. 47*: 1001–1004.

Keren, R., Shainberg, I., Frenkel, H., and Kalo, Y. (1983). The effect of exchangeable sodium and gypsum on surface runoff from loess soil. *Soil Sci. Soc. Am. J. 47*: 1001–1004.

Lane, L. J., Foster, G. R., and Nicks, A. D. (1987). Use of fundamental erosion mechanics in erosion prediction. *ASAE* Paper No. 87-2540, St. Joseph, MI.

Lentz, R. D., Shainberg, I., Sojka, R. E., and Carter, D. L. (1992). Preventing irrigation furrow erosion with small application of polymers. *Soil Sci. Soc. Am. J. 56*: 1926–1932.

Levin, J., Ben-Hur, M., Gal, M., and Levy, G.J. (1991). Rain energy and soil amendments effects on infiltration and erosion of three different soil types. *Aust. J. Soil Res. 29*: 455–465.

Levy, G. J., Ben-Hur, M., and Agassi, M. (1991). The effect of polyacrylamide on runoff, erosion, and cotton yield from fields irrigated with moving sprinkler systems. *Irrig. Sci. 12*: 55–60.

Levy, G. J., Levin, J., Gal, M., Ben-Hur, M., and Shainberg, I. (1992). Polymer's effect on infiltration and soil erosion during consecutive simulated sprinkler irrigations. *Soil Sci. Soc. Am. J. 56*: 902–907.

Malik, M., and Letey, J. (1991). Adsorption of polyacrylamide and polysaccharide on soil material. *Soil Sci. Soc. Am. J. 55*: 380–383.

McIntyre, D. S. (1958). Permeability measurements of soil crusts formed by raindrop impact. *Soil Sci. 85*:185–189.

Miller, W. P. (1987). Infiltration and soil loss of three gypsum-amended ultisols under simulated rainfall. *Soil Sci. Soc. Am. J. 51*:1314–1320.

Miller, W. P., and Scifres, J. (1988). Effect of sodium nitrate and gypsum on infiltration and erosion of a highly weathered soil. *Soil Sci. 148*: 304–309.

Morin, J., and Benyamini, Y. (1977). Rainfall infiltration into bare soils. *Water Resour. Res. 13*: 813–817.

Mortensen, J. L. (1959). Adsorption of hydrolyzed polysaccharide on kaolinite. II: Effect of solution electrolytes. *Soil Sci. Soc. Am. Proc. 23*: 199–202.

Mortensen, J. L. (1962). Adsorption of hydrolyzed polyacrylonitrile on kaolinite. In *Clays Clay Miner, Proc. 9th Nat. Conf.*, West Lafayette, IN (Ada Swineford, ed.), Pergamon Press, New York, pp. 530–545.

Nadler, A., and Letey, J. (1989). Adsorption isotherms of polyanions on soils using tritum labeled compounds. *Soil Sci. Soc. Am. J. 53*: 1375–1378.

Norton, L. D., Shainberg, I., and King, K. W. (1993). Utilization of gypsiferous amandments to reduce surface sealing in some humid soils of the eastern USA. In Catena Supplement no. 24, *Soil Sealing and Crusting* (J. W. E. Poesen and M. A. Nearing, eds.), Catena Verlag, Cremlingen, Germany.

Oster, J. D., Shainberg, I., and Wood, J. D. (1980). Flocculation value and gel structure of NA/Ca montmorillonite and illite suspensions. *Soil Sci. Soc. Am. J. 44*: 955–959.

Palmer, R. S. (1964). The influence of a thin water layer on water drop impact forces. *Int. Assoc. Sci. Hydrol. Publ. 65*: 141–148.

Poesen, J. W. A. (1987). The role of slope angle in surface seal formation. In *International Geography*, Part II (V. Gardiner, ed.), Wiley, New York.

Quirk, J. P., and Schofield, R. K. (1955). The effect of electrolyte concentration on soil permeability. *J. Soil Sci. 6*: 163–178.

Rose, C. W. (1985). Development in soil erosion and deposition models. In *Advances in Soil Science*, Vol. 2, Springer-Verlag, New York, pp. 1–63.

Ruehrwein, R. A., and Ward, D. W. (1952). Mechanism of clay aggregation by polyelectrolytes. *Soil Sci. 73*: 485–492.

Schamp, N., and Huylebroeck, J. (1973). Adsorption of polymers on clays. *J. Polym. Sci. Symp. 42*: 553–562.

Schamp, N., Huylebroeck, J., and Sadones, M. (1975). Adhesion and adsorption phenomena in soil conditioning. In *Soil Conditioners*, Soil Sci. Soc. Am. Special Pub. No. 7, Am. Soc. Agron., Madison, WI, pp. 13–23.

Schofield, R. K., and Samson, H. R. (1954). Flocculation of kaolinite due to the attraction of oppositely charged crystal faces. *Disc. Faraday Soc. 18*: 138–145.

Shainberg, I., Sumner, M. E., Miller, W. P., Farina, M. P. W., Pavan, M. A., and Fey, M. V. (1989). Use of gypsum on soils: A review. In *Advances in Soil Science*, Vol. 9, Springer-Verlag, New York, pp. 2–101.

Shainberg, I., Warrington, D. N., and Rengasamy, P. (1990). Water quality and PAM interactions in reducing surface sealing. *Soil Sci. 149*: 301–307.

Shanmuganathan, R. T., and Oades, J. M. (1983). Influence of anions on dispersion and physical properties of the A horizon of a red brown earth. *Geoderma 29*: 257–277.

Shaviv, A., Ravina, I., and Zaslavski, D. (1986). Surface application of anionic surface conditioners to reduce crust formation. In *Assessment of Soil Surface Sealing and Crusting* (Proc. Symp.) (F. Callebaut, D. Gabriels, and M. De Boodyt, eds.), Ghent, Belgium, pp. 286–293.

Smith, H. J. C., Levy, G. J., and Shainberg, I. (1990). Waterdroplet energy and soil amendments: Effect on infiltration and erosion. *Soil Sci. Soc. Am. J. 54*: 1084–1087.

Stern, R. (1990). Effects of soil properties and chemical ameliorants on seal formation, runoff and erosion. D. Sc. thesis, University of Pretoria, Pretoria, South Africa.

Stern, R., Ben-Hur, M., and Shainberg, I. (1991a). Clay mineralogy effect on rain infiltration, seal formation and soil loess. *Soil Sci. 152*: 455–462.

Stern, R., Laker, M. C., and van der Merwe, A. J. (1991b). Field studies on the effect of soil conditioners and mulch on runoff from kaolinitic and illitic soils. *Aust. J. Soil Res. 29*: 249–261.

Stern, R., van der Merwe, A. J., Laker, M. C., and Shainberg, I. (1992). Effect of soil surface treatments on runoff and wheat under irrigation. *Agron. J. 84*: 114–119.

Stutzmann, Th., and Siffert, B. (1977). Contribution to the adsorption mechanism of acetamide and polyacrylamide onto clays. *Clays Clay Miner. 25*: 392–406.

Theng, B. K. G. (1979). *Formation and Properties of Clay-Polymer Complexes*. Elsevier, Amsterdam.

Theng, B. K. G. (1982). Clay-polymer interactions: Summary and perspectives. *Clays Clay Miner. 30*: 1–10.

Walker, P. H., Hutka, J., Moss, A. J., and Kinnell, P. I. A. (1977). Use of a versatile experimental system for soil erosion studies. *Soil Sci. Soc. Am. J. 41*: 610–612.

Wallace, G. A., and Wallace, A. (1986). Control of soil erosion by polymer soil conditioners. *Soil Sci. 141*: 363–367.

Warrington, D., Shainberg, I., Agassi, M., and Morin, J. (1989). Slope and phosphogypsum effects on runoff and erosion. *Soil Sci. Soc. Am. J. 53*: 1201–1205.

Young, R. A., and Wiersma, J. L. (1973). The role of rainfall impact in soil detachment and transport. *Water Resour. Res. 9*: 1629–1636.

13

Reclamation of Gullies and Channel Erosion

Earl H. Grissinger
National Sedimentation Laboratory, Agricultural Research Service, U.S. Department of Agriculture, Oxford, Mississippi (Retired)

I. INTRODUCTION

Rehabilitation may imply different degrees of recovery to different individuals. According to the *World Book Dictionary*, rehabilitation may be defined as (a) to make over into a new form (b) to restore to a good condition, or (c) to bring back to the initial condition. Obviously, economics and food production pressures become important criteria governing the degree of rehabilitation that is feasible for a given site-specific problem. This site specificity could or would make detailed discussion excessively cumbersome and unwieldy. Therefore, food production pressures and economics will not be considered nor will various degrees of rehabilitation. Such decisions should be part of the planning procedure for any specific rehabilitation effort. This discussion will concentrate on ways and means of stopping further degradation due to gullying and stream channel erosion. Subsequent to abatement of specific problems, various soil conservation practices could be added as necessary to achieve planning goals. Specific within-field conservation practices are discussed elsewhere in this book and will only be considered here as necessary for discussions of specific gully and/or stream channel erosion abatement practices.

For ease of discussion, rehabilitation practices for ephemeral gullies will be considered separately from those for upland and valley-floor gullies and stream channels. Off-site effects of rehabilitation activities will be discussed, usually briefly, along with benefits of integrating such activities into a watershed systems approach using primarily gridded Geographic Information Systems (GIS) technology.

II. EPHEMERAL GULLIES

Ephemeral gullies are produced by concentrated flow erosion in swales or other topographically controlled locations. This type of erosion represents a bridge between rill erosion and the development of classical gullies. Ephemeral gullies are obliterated by annual tillage operations but reform in the same locations year after year within a given drainage area. With time, the topsoil depth over the entire swale is reduced by the repeated cycles of ephemeral gully growth due to concentrated flow erosion followed by gully filling from adjacent areas due to annual cultivation. These gullies may be either a continuous extension of the drainage network or they may be discontinuous. Both types of gullying impair on-site source areas but off-site water-quality and sedimentation problems are more critically impacted where the downstream drainage is continuous.

Objectives associated with the various degrees of rehabilitation must be precisely defined prior to development of specific conservation practices. Questions that must be answered range from (a) should soil loss be reduced to tolerable levels or to levels suitable for sustainable agriculture, and (b) should soil loss be reduced by trapping sediments after erosion has occurred or by reducing erosion in the first place. Implicit in the first alternative is the practical problem that, in many cases, ephemeral gully erosion is initially chronic in magnitude but becomes acute with time as the gully formation recurs in the same location. For short-term observation periods, sustainability cannot be evaluated solely on soil loss values but must include assessments of concentrated flow erosion within the study area. For example, in the loess area of northern Mississippi, the average annual soil loss from gaged no-till watersheds was 0.5 t/A, whereas that from a comparable but conventionally-tilled watershed was 9.6 t/A [1]. Although soil loss was tolerable in the no-till watersheds, an ephemeral gully had evolved in just four years in one of these watersheds to a size that inhibited routine cultivation. Clearly agricultural sustainability was being degraded. This example also illustrates the potential benefits of controlling erosion and thus maintaining productivity. The alternative mode of sediment abatement is trapping off-site resulting in impaired productivity on-site. Additionally the easily suspended fraction of the sediment load is usually lost through the trap structure.

Ideally, for sustainable agriculture, conservation tillage, residue management and related methods for minimizing rill and interrill erosion would be coupled with methods to minimize ephemeral gully development. This minimization could be based on managing runoff rates and magnitudes, by increasing the erosion resistance of the boundary material in the concentrated flow path, and/or by protecting the boundary material from a direct attack by the flow. Usual methods employed to accomplish one or more of these minimizations include the aforementioned conservation tillage and residue management in combination with contouring, terracing, grassed waterways, subsurface drainage systems, stripcropping, and sedimentation basins to control water and sediment. These control measures have been thoroughly documented in the literature, for example see Laflen et al. [2 Chapter 23], and will not be further discussed. The use of geotextiles has not been thoroughly examined and should also be considered for some applications to protect critical areas from concentrated flow [3] or as cellular inserts into the soil surface to increase erosion resistance. Cellular geotextiles have a meaningful advantage over the more standard cellular blocks in that the geotextiles do not create an obstacle to the use of no-till equipment. Other measures include cross-slope ditches and vegetal trapping of sediment. Cross-slope ditches are effective in noncultivated areas and may have application to no-till farming. In this application, grass waterways are imposed across slopes to deliver runoff to pipe or concrete-lined drainage systems. In addition to sedimentation basins, grass strips [4–6] and stiff grass hedges [7,8] have potential to trap sediments. Development of grass hedge applications is relatively new and this usage will probably expand in the future.

Development of procedures for the application of GIS-based runoff models to conservation planning for field-sized watershed is a rapidly evolving area of research. One application for this technology is the site-specific minimization of furrow slope by manipulation of furrow flow alignment. Results for a gaged field-size watershed in the Goodwin Creek Research Watershed, northern Mississippi, are presented in Tables 1 and 2. In this application, topography and furrow alignment are input as separate GIS layers, generally comparable to the procedure used by Zevenbergen [9]. Furrow slopes are then calculated for a series of alternate furrow alignments relative to the ground surface topography. Obviously, due to the GIS orthogonal grid pattern, the furrow alignment cannot be directly changed. Rather, the ground surface topography is reformatted to the desired rotation, creating the same relative positioning as if the furrows were rotated over a stationary ground surface.

In this example, the topographic slope ranged from 5.31 to 5.44 for the various rotations (0°, 15°, 30°, and 45°). These slopes were not significantly different based on the Smirnov T criteria. Furrows aligned either north to south (NS) or east to west (EW) had slopes that were significantly less than the topographic slopes (Table 1). Based on the Smirnov T criteria (Table 2), the 0° and 15° NS and EW alignments were superior to the other data sets for simple rotations. However, best results were obtained by performing separate optimi-

Table 1 Average Topographic and Furrow Slopes for a Field-Sized Watershed

	Average slope	
Rotation/furrow alignment	Topographic (%)	Furrow (%)
0°	5.43	
0°/NS		3.10
0°/EW		3.60
15°	5.44	
15°/NS		3.11
15°/EW		3.58
30°	5.32	
30°/NS		3.52
30°/EW		3.57
45°	5.31	
45°/NS		3.71
45°/EW		3.48
0°/C		1.56
15°/C		1.68

zations in the two halves of the watershed, that is the halves separated by the concentrated flow path in the swale. These minimizations are noted as 0°/C and 15°/C (Table 1). Again, using the Smirnov T, these dual minimizations are significantly better than the 0° and 15° NS and EW alignments (the T/T critical at $\alpha = 0.05 > 2$).

This minimization of the furrow slope inherently establishes rill flow routing in the watershed, and when coupled with the appropriated runoff simulation

Table 2 Values of Smirnov T Divided by T Critical at $\alpha = 0.05$ for Furrow Slopes[a]

Rotation/furrow alignment	0°/NS	15°/NS	0°/EW	15°/EW	30°/EW	30°/NS	45°/EW	45°/NS
0°/NS				1.08	1.41	1.45	1.74	1.88
15°/NS					1.31	1.23	1.66	1.76
0°/EW							1.18	1.26
15°/EW							1.18	1.28
30°/EW								
30°/NS								
45°/EW								
45°/NS								

[a]Nonsignificant values (less than 1) are not shown.

model, could be used to calculate concentrated flow in the ephemeral gully channel. For a design storm, soil and agronomic management system, the concentrated flow magnitudes thus calculated could be used to evaluate the stability of the concentrated flow channel. In essence, procedures such as described could be used to develop design criteria to evaluate the suitability of using a grassed waterway as specified in Agricultural Handbook No. 667 [10]. Alternatively, this type of calculation could be used to estimate the amount of slope reduction necessary to achieve stability, and/or the increase in material resistance to erosion necessary to achieve stability. Slope reduction could be achieved by incorporating relatively small low-drop structures into the flow path. Channel boundary material resistance to erosion could be increased vegetatively or by the use of cellular geotextiles.

In the author's opinion, efficient implementation of this type of procedure is conditional upon advances in two areas. One of these is the prediction of critical shear stress or comparable stability parameter for consolidated cohesive materials. As discussed briefly in Chapter 8, the surface roughness of this type of material adds a degree of additional freedom to those involved in comparable studies of disturbed, disaggregated, or unconsolidated materials. In situ test equipment comparable to that developed by Hanson [11] may have utility but additional testing is needed. This testing must address the effects of sample morphology or structure on material erodibility and of surface roughness on interface flow conditions. A more general concern presented by Kuti and Yen [12] is the lack of accord between equilibrium conditions and rates of erosion prior to attainment of the equilibrium. Their results beg the question of applicability of stability evaluations using small test samples as is the norm for many current procedures. These and other aspects of stability of consolidated cohesive materials have been discussed previously by Grissinger [13].

The second area in need of further refinement involves evaluation of the flow resistance term and particularly the partitioning of flow resistance between the vegetal roughness elements and the surface roughness of cohesive materials at various degrees of consolidation. Obviously, this is needed for the calculation of applied shear such as used in WEPP [14] as briefly discussed in Chapter 8. Further evaluation of the flow resistance term is also needed for flow routing calculations for gridded field-sized watersheds so that partial source areas [15] can be adequately expressed in the simulation. The extreme influence of ground cover on runoff rates and magnitudes is illustrated in Fig. 1. Runoff from the well-canopied pasture was subdued relative to that from an adjacent clean-tilled soybean field. The total runoff volume was about 20% lower for the pasture and, more importantly, the peak runoff rate was only about one-third that of the soybean field. On-site and off-site consequences of peak flow reductions of this magnitude are obvious.

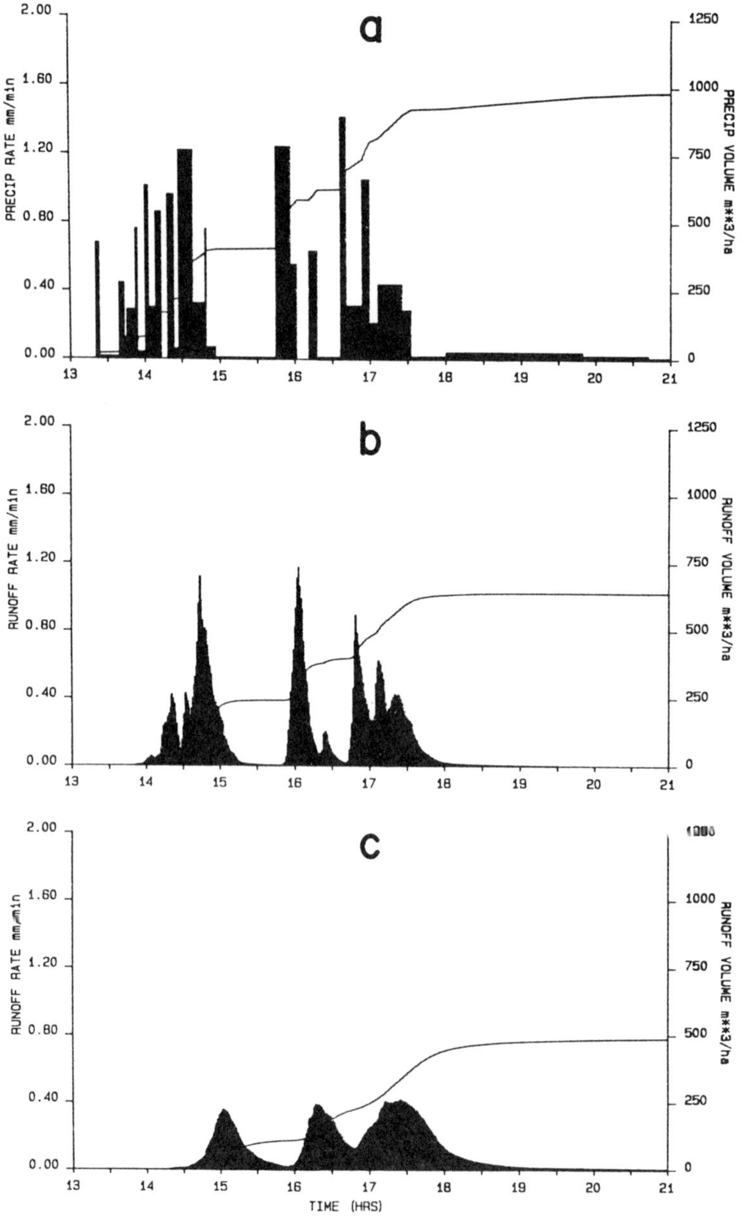

Fig. 1 Precipitation (a) and runoff from a clean tilled soybean field (b) and an adjacent pasture (c). Both areas are about 5 acres.

III. GULLIES AND CHANNELS

Many of the rehabilitation practices and procedures in common usage today have evolved by trial and error in reaction to local conditions. For example upland gullying problems in western Tennessee in the early 1900s and the recommended reclamation methods were summarized by Purdue [16–18] and Maddox [19,20]. Rehabilitation practices were generally comparable to many present-day activities, but details were obviously different. They discussed the use of terraces, brush and larger dams to trap sediment, and revegetation in order to both cure then-current problems and to prevent future problems. Techniques were labor intensive and at least some landowners had their own nurseries for producing transplanting stock of bermuda grass, of black locust, walnut, yellow poplar, sweetgum, and sycamore seedlings and/or of kudzu, honeysuckle, and Himalaya vines. Off-site consequences of upland gullying were also discussed by Maddox [20], but the enormity of such off-site problems was not fully documented until Happ, Rittenhouse, and Dobson's [21] classic article on accelerated valley sedimentation, about 25 years after Purdue's and Maddox's work.

Both Purdue and Maddox stressed the importance of adapting remedial procedures to best meet the needs of each situation. Obviously, this is just as important today as it was about 80 years ago. Both failure processes and the site-specific controls of these processes must be considered in adapting rehabilitation procedures to individual cases. Although the rehabilitation procedures differ markedly for upland gullying, valley-floor gullying, and stream channel failure problems, the processes are not that dissimilar. Rather than simply review the various rehabilitation practices usually employed to mitigate channel and gullying problems, this discussion will center on process similarities and control variable dissimilarities for selected problem sites.

In general, bank failure is a three-stage process. These stages are a prefailure conditioning stage, the actual failure, and lastly the removal of slough by concentrated flow. The conditioning stage is largely driven by seasonal weather conditions related to the magnitude of seep flow, freeze-thaw, wetting-drying, and the bank profile water content. On the other hand slough removal varies with the magnitude of individual storm events. Failures are primarily gravity driven and can be expressed by any of the usual slope stability equations. The applicability of slope stability analyses to bank failure problems has been established by the studies of Bradford and Piest [22], Bradford et al. [23,24], Thorne [25], Piest et al. [26], Little et al. [27], and Osman and Thorne [28]. Lithologic process controls by units of the valley-fill depositional sequence have been documented by Grissinger and co-workers [29–32]. Specific variables affecting slope stability of cohesive materials include bulk unit weight, critical bank height, length of the tension crack, bank slope angle and material cohesion and friction angle. For noncohesive materials, the angle of repose is critical. For ease of presentation, the stability of noncohesive banks will not be considered further.

Of the six variables affecting cohesive material bank stability, the bank height and slope angle are site properties whereas the other four (bulk unit weight, length of tension crack, cohesion, and friction angle) are material properties. The numerical value of individual material variables is not constant, however, but varies depending upon the prefailure conditions, particularly the seasonal precipitation characteristics. In essence, the pecularities of the material conditioning impose a failure probability aspect to the analysis, that is the probability that failure will occur is established by the probability of occurrence of specific prefailure conditions that reduce the factor of safety to a value of less than 1. The site properties are also variable. Bank heights vary with channel entrenchment or aggradation and bank angles vary with erosion of bank toe material. In the generalized case of channel/gully erosion, the initial base level lowering (entrenchment) precedes repeated cycles of mass failure followed by toe erosion to an oversteepened bank angle. Immediately after failure the bank is stable, but renewed toe erosion initiates a repeat cycle. This cyclicity continues until the channel/gully widens to such a degree that the flow is no longer able to remove the slough at an accelerated rate.

The various remediation procedures address one or more of the process controls. In general, these procedures are designed to protect the bank toe from erosion so as to maintain a stable bank angle, to increase material strength, to reduce or limit bank height, and/or to manage or train concentrated flows. For upland gullying, remediation usually involves a combination of flow diversion, grassed waterways, terraces, and plugs, ponds and larger structures [33,34]. For relatively large gully heads, drop inlets, drop spillways, or concrete chutes will be necessary to reduce the channel grade to achieve stability. In all cases, revegetation with grass and/or pines is an important practice and may be the simplest and cheapest way to control relatively small gullies with comparable-sized contributing area. Revegetation not only protects the soil surface from concentrated flow shear but also improves soil strength characteristics. If livestock are abundant, fencing may be necessary but should be kept to a minimum due to costs. Similarly, reshaping is effective but costly, and should be used only where necessary to promote revegetation. Subsurface drainage to relieve seepage and related problems associated with perched water bodies should be used with discretion as warranted by near-surface lithologic conditions. Select case histories of gully controls in southwestern Wisconsin have been documented by Woolhiser and Miller [35] to evaluate postremediation downstream channel changes. They reported that "... a strict technical prediction of degradation on the basis of materials present, stable slope, and flow regime is not always a complete answer to the problem" and that procedures must be developed that will incorporate the site-specific "environmental factors influencing the channel degradation process."

For downstream valley-floor gully and stream channel erosion problems, runoff rates and magnitudes are larger, requiring a different mix of conservation practices. Grade control structures including both low and high drops, can be used to both reduce the channel grade and to reduce or limit bank heights to design values. Material strengths can be increased vegetatively where growth is not limited by submergence or other growth inhibiting conditions. Alternately, strength can be increased by the use of soil cements, grouts, asphalts, and other soil stabilizers. However, the most common stabilizing practices involve procedures to protect the bank toe from erosion and complementary procedures to train flows so that they do not directly impinge onto the bank. Usual toe protection methods include riprap, rubble, gabions, mattresses, sacks, blocks, used tires, etc. Fences, jacks, bulkheads, dikes, groins, etc., are frequently used to train flows away from the banks, and riser pipes are used to transfer runoff from floodplain positions to the channel without causing excessively erosive conditions. Details of these remediation measures have been documented in various reports by the U.S. Army Corps of Engineers [36–39].

The application of any of these remediation measures to a specific valley-floor gully or stream channel erosion problem varies with details of the site condition, particularly with bank material lithology. For example, valley-floor gullies in western Iowa typically are incised to glacial till [24,26]. Failures are either popout or rotational but are primarily related to the magnitude of groundwater perched above the glacial till [22,23]. Clearly, seepage control must be a key element in the remediation package. In northern Mississippi however, the valley-floor gullies typically bottom in late-Quaternary weakly consolidated materials that have minimum cohesive strength [30,31]. Slab failures accentuated by tension crack development are the most common mode of instability [27,40]. For this situation, toe protection must be the key element, either directly through the use of riprap or comparable materials, or indirectly through flow training with groins, dikes, jacks, etc.

The one area that has received insufficient attention for the control of valley-floor gully and stream channel erosion problems is management of runoff volumes and rates in critical parts of the watershed. The question remains as to whether or not runoff can be abated by structural and vegetative means to a degree sufficient to inhibit removal of slough material, thus promoting bank stability. Probably, a better question would be to what degree this type of runoff management could be used as part of the remediation package, and under what conditions would it be feasible. Field studies have been implemented in the Goodwin Creek Research Watershed [41] to evaluate the influences of upstream watershed conditions on channel bed and bank stability. Clearly, however, parallel and complementary numerical simulations are necessary to fully evaluate the influence of long-term watershed changes and the hydrologic response to such changes. Ideally, the simulation would be interfaced with a GIS for spatial control of on-site effects and for at least qualitative evaluation of off-site effects.

Even though remediation is a site-specific abatement process, system properties related to sediment movement and temporary storage [42,43] must be considered in an intensive watershed rehabilitation program.

IV. SUMMARY

In this discussion of gully and stream channel rehabilitation, processes and process controls have been emphasized more than individual remediation practices. Site-specific material and environmental conditions must be considered in designing individual abatement activities. In general, these activities are designed to train and/or manage concentrated flows, to protect bank and or bed materials from such flows, to increase the material resistance to erosion, and to reduce or limit bank height.

Several areas have been identified for further study. These include

1. The influence of surface roughness of consolidated cohesive materials on interface flow conditions, particularly on partitioning of flow roughness elements
2. Equipment and procedures to evaluate erosion resistance of consolidated cohesive materials
3. Application of GIS procedures to erosion abatement activities on field-scale and watershed-scale levels
4. The possible application of watershed hydrologic controls for management of sediment transport and stability of valley-floor gullies and stream channels

REFERENCES

1. S. M. Dabney, C. E. Murphree, G. B. Triplett, E. H. Grissinger, L. D. Meyer, L. R. Reinschmiedt, and F. E. Rhoton, Conservation production systems for silty uplands, Proceedings of the 1993 Southern Conservation Tillage Conference for Sustainable Agriculture, Monroe, Louisiana, pp. 43–48 (1993).
2. J. M. Laflen, R. E. Highfill, M. Amemiya, and C. K. Mutchler, Structures and methods for controlling water erosion. In *Soil Erosion and Crop Productivity* (R. F. Follett and B. A. Stewart, eds.), Soil Science Society of America, Madison, Wisconsin, pp. 432–442 (1985).
3. J. Poesen and G. Govers, Gully erosion in the loam belt of Belgium: Topology and control measures. In *Soil Erosion on Agricultural Land* (J. Boardman, I. D. L. Foster, and J. A. Dearing, eds.), John Wiley and Sons, New York, pp. 513–530 (1990).

4. C. A. Ohlander, Defining the sediment trapping characteristics of a vegetative buffer. Special case: Road erosion, Proceedings of the 3rd Federal Interagency Sedimentation Conference, Denver, Colorado, pp. 2-77–2-82 (1976).
5. L. G. Wilson, *Trans. ASAE, 10*:35–37 (1967).
6. D. E. Line, Sediment trapping effectiveness of grass strips, in Proceedings of the 5th Federal Interagency Sedimentation Conference, Las Vegas, Nevada, pp. PS-56–PS-63 (1991).
7. D. Kemper, S. M. Dabney, L. Kramer, D. Dominick, and T. Keep, *J. Soil Water Conservation, 47(4)*:284–288 (1992).
8. S. M. Dabney, K. C. McGregor, L. D. Meyer, E. H. Grissinger, and G. R. Foster, Vegetative barriers for runoff and sediment control, Proceedings of the International Symposium on Integrated Resource Management, Chicago, Illinois, pp. 60–70 (1993).
9. L. W. Zevenbergen, *Modelling Erosion Using Terrain Analysis*, Ph.D. thesis, Department of Geography and Earth Science, Queen Mary College, University of London, London, p. 345 (1989).
10. *Stability Design of Grass-Lined Open Channels*, Agricultural Handbook 667, U.S. Department of Agriculture, Agricultural Research Service, Superintendent of Documents, Washington, DC, p. 167 (1987).
11. G. J. Hanson, ASCE Paper No. 89-2151, p. 15 (1989).
12. E. O. Kuti and C. Yen, *J. Hydraulic Research, 14(3)*:195–206 (1976).
13. E. H. Grissinger, Bank erosion of cohesive materials. In *Gravel-Bed Rivers* (R. D. Hey, J. C. Bathurst, and C. R. Thorne, eds.), John Wiley and Sons, New York, pp. 273–287 (1982).
14. *USDA-Water Erosion Prediction Project: Hillslope Profile Version*, NSERL Report No. 2, USDA-ARS National Soil Erosion Research Laboratory, West Lafayette, Indiana, p. 263 (1989).
15. I. A. Campbell, The partial area concept and its application to the problem of sediment source areas. In *Soil Erosion and Conservation* (S. A. El-Swaify, W. C. Molderhauer, and A. Lo, eds.), Soil Conservation Society of America, Ankeny, Iowa, pp. 128–138 (1985).
16. A. H. Purdue, The waste from hillside wash. In *The Resources of Tennessee II(6)*, The Tennessee State Geological Survey, Nashville, pp. 250–254 (1912).
17. A. H. Purdue, The gullied lands of west Tennessee. In *The Resources of Tennessee III(3)*, The Tennessee State Geological Survey, Nashville, pp. 119–136 (1913).
18. A. H. Purdue, A double waste from hillside wash. In *The Resources of Tennessee IV(1)*, The Tennessee State Geological Survey, Nashville, pp. 36–37 (1914).
19. R. S. Maddox, Progress in reclaiming waste lands in west Tennessee. In *The Resources of Tennessee*, The Tennessee State Geological Survey, Nashville, pp. 217–224 (1916).
20. R. S. Maddox, Forests, gullies, and reconstruction. In *The Resources of Tennessee IX(1)*, The Tennessee State Geological Survey, Nashville, pp. 23–31 (1919).
21. S. C. Happ, G. Rittenhouse, and G. C. Dobson, *Some Principles of Accelerated Stream and Valley Sedimentation*, Technical Bulletin 695, U.S. Department of Agriculture, Superintendent of Documents, Washington, DC, p. 134 (1940).

22. J. M. Bradford and R. F. Piest, *Soil Science Society of America Proceedings, 41(1)*: 115–122 (1977).
23. J. M. Bradford, D. A. Farrell, and W. E. Larson, *Soil Science Society of America Proceedings, 37(1)*:103–107 (1973).
24. J. M. Bradford, R. F. Piest, and R. G. Spomer, *Soil Science Society of America Proceedings, 42(2)*:323–328 (1978).
25. C. R. Thorne, *Proceedings of Bank Erosion in River Channels*, Ph.D. thesis, School of Environmental Science, University of East Anglia, Norwich, United Kingdom, p. 447 (1978).
26. R. F. Piest, J. M. Bradford, and G. M. Wyatt, *J. ASCE, 101(HY 1)*:65–80 (1975).
27. W. C. Little, C. R. Thorne, and J. B. Murphey, *Trans. ASAE, 25(5)*:1321–1328 (1982).
28. M. A. Osman and C. R. Thorne, *J. ASCE, 114(HY 2)*:134–150 (1988).
29. E. H. Grissinger, Bank erosion of cohesive materials. In *Gravel-Bed Rivers* (R. D. Hey, J. C. Bathurst, and C. R. Thorne, eds.), John Wiley and Sons, New York, pp. 273–287 (1982).
30. E. H. Grissinger and J. B. Murphey, Present channel stability and late-Quaternary valley deposits in northern Mississippi. In *Modern and Ancient Fluvial Systems* (J. D. Collinson and J. Lewin, eds.), Special Publication No. 6, International Association of Sedimentologists, pp. 241–250 (1983).
31. E. H. Grissinger, J. B. Murphey, and R. L. Frederking, Geomorphology of upper Peters Creek Catchment, Panola County, Mississippi. Part II: Within-channel characteristics. In *Modeling Components of the Hydrologic Cycle* (V. P. Singh, ed.), Water Resources Publications, Littleton, Colorado, pp. 267–282 (1982).
32. E. H. Grissinger, J. B. Murphey, and W. C. Little, *Southeastern Geology, 23(3)*: 147–162 (1982).
33. H. G. Jepson, *Prevention and Control of Gullies*, U.S. Department of Agriculture Farmers' Bulletin No. 1813, Superintendent of Documents, Washington, DC, p. 59 (1939).
34. C. J. Francis (revised in 1973 by R. C. Barnes), *How to Control a Gully*, U.S. Department of Agriculture Farmers' Bulletin No. 2171, Superintendent of Documents, Washington, DC, p. 16 (1961).
35. D. A. Woolhiser and C. R. Miller, Case histories of gully control structures in southwestern Wisconsin, ARS Publication No. 41-60, U.S. Department of Agriculture, Agricultural Research Service, Washington, DC, p. 28 (1963).
36. *Streambank Protection Guidelines*, U.S. Army Corps of Engineers, Waterways Experiment Station, Vicksburg, Mississippi, p. 60 (1983).
37. *Interim Report to Congress: Section 32 Program, Streambank Erosion Control Evaluation and Demonstration Act of 1974*, U.S. Army Corps of Engineers, Washington, DC, p. 12 with 9 Appendixes (1978).
38. *Final Report to Congress: The Streambank Erosion Control Evaluation and Demonstration Act of 1974*, U.S. Army Corps of Engineers, Washington, DC, p. 180 (1981).
39. M. P. Keown, N. R. Oswalt, E. B. Perry, and E. A. Dardeau, Jr. *Literature Survey and Preliminary Evaluation of Streambank Protection Methods*, Technical Report

H-77-9, U.S. Army Corps of Engineers, Waterways Experiment Station, Vicksburg, Mississippi, p. 262 (1977).
40. E. H. Grissinger and W. C. Little, Similarity of bank problems on dissimilar streams, Proceedings of the 4th Federal Interagency Sedimentation Conference, Las Vegas, Nevada, pp. 5-51–5-60 (1986).
41. *Stream Channel Stability*, A Comprehensive Report With 14 Appendixes Prepared for the U.S. Army Corps of Engineers, Vicksburg District, by the National Sedimentation Laboratory, NTIS Nos. AD A101 385 through AD A101 399, National Technical Information Center, Springfield, Virginia, p. 1630 (1981).
42. R. F. Piest and A. J. Bowie, Gully and streambank erosion, Proceedings of the 29th Annual Meeting of the Soil Conservation Society of America, Ankeny, Iowa, pp. 188–196 (1974).
43. S. A. Schumm, M. D. Harvey, and C. C. Watson, *Incised Channels; Morphology, Dynamics, and Control*, Water Resources Publications, Littleton, Colorado, p. 200 (1984).

14

Reclamation of Salt-Affected Soil

J. D. Oster
University of California, Riverside, California

I. Shainberg
Institute of Soils and Water, The Volcani Center, Agricultural Research Organization, Bet-Dagan, Israel

I. P. Abrol
Indian Council of Agricultural Research, New Delhi, India

I. INTRODUCTION

Reclamation of salt-affected soils through tillage, water, crop, and amendment practices is an increasingly important tool for improving crop productivity in many areas of the world. Traditionally, reclamation has been driven by the need to turn marginally arable lands to agricultural use by reducing the levels of salinity, exchangeable sodium, and boron in the soil, thereby increasing both crop yields and the number of crop species that can be grown in a specific area. Adverse levels of these substances can, however, be caused by other factors as well, including underirrigation (not providing enough water for leaching), inadequate drainage, and the use of moderately saline-sodic waters for irrigation. Use of such poor-quality waters is increasing due to growing municipal demands for the available supplies of good-quality water and the need to dispose of municipal wastewaters and agricultural drainage waters. As a result, a solid understanding of soil reclamation will become an increasingly important component of water and soil management to ensure the long-term sustainability of irrigated agriculture.

Water movement into and through soils is the key factor in reclamation and management of salt-affected soils. Infiltration rates (IR) and hydraulic conductivities (K) decrease with decreasing soil salinity and with increasing exchangeable sodium. Infiltration rates are more strongly affected by low soil salinity and exchangeable sodium levels than are hydraulic conductivities because of the mechanical impact and stirring action of the applied water and the freedom for soil particle movement at the soil surface. Farmers must maintain adequate IR and *K* through various combinations of crop, soil amendment, and tillage practices. Furthermore, drainage (either natural or artificial) must be adequate so that salts, exchangeable sodium, and boron are removed from the rootzone.

Our emphasis in this chapter is on the application of the basic soil and plant sciences to reclamation practices used by farmers in developed and developing countries. Physical, chemical, and mineralogical properties of soil are discussed in Sec. II. Specific reclamation examples and methods related to soil, water, and cropping practices are discussed in Secs. III and IV. The final section describes farming practices used in California, Israel, and India for reclamation. We have drawn heavily on recent reviews (Gupta and Abrol, 1990; Keren, 1990; Rhoades and Loveday, 1990; and Sumner, 1993).

The American Society of Agronomy granted permission to republish this chapter, which was previously written and accepted for publication in the ASA Drainage Monograph. Minor revisions were made to coordinate with Chapter 8 of this book, particularly in regard to the effects of magnesium on IR (Keren, 1991). This chapter differs from Chapter 8 in three major aspects: (1) an emphasis on the linkage between research-based knowledge and farming practices related to reclamation, (2) the role of cropping as a major component of reclamation, and (3) documentation of reclamation practices used by farmers in developed and developing countries.

II. SALINITY AND SODICITY EFFECTS ON SOIL PHYSICAL PROPERTIES

Reclamation research and practice has a long history throughout many areas of the world. Most of this historical information remains valuable today. One of the problems with making use of this wealth of information, however, is a history of confusing and sometimes contradictory terminology that has been used to describe the different types of salt-affected soils. Therefore, we begin with a brief overview of some basic terms of soil classification, both in order to provide a historical perspective on reclamation research and to describe how such terms will be used in this chapter.

Historically, the physical behavior of salt-affected soils has been described in terms of the combined effects of soil salinity as measured by the electrical conductivity of a saturation extract (ECe) and exchangeable sodium percent (ESP) on flocculation and soil dispersion. The U.S. Salinity Laboratory Staff (1954) described the physical properties of a saline soil (ECe > 4 dS/m; ESP < 15) as follows: "Owing to the presence of excess salts and the absence of significant amounts of exchangeable sodium, saline soils generally are flocculated; and, as a consequence, the permeability is equal to or higher than that of similar nonsaline soils.'' A saline-alkali soil (ECe > 4 dS/m; ESP > 15) was described as similar to a saline soil ''as long as excess salts were present.'' However, ''upon leaching, the soil may become strongly alkaline (pH readings above 8.5), the particles disperse, and the soil becomes unfavorable for entry and movement of water and for tillage.''

By 1979, the term ''alkali'' was listed as obsolete by the Soil Science Society of America (though it is still used by farm advisors and others), and the word ''sodic'' was listed in its place (Anonymous, 1979), with the definition, ''a soil having an ESP > 15.'' Outside the United States, the word ''alkali'' is generally used in a narrower context referring only to soils (1) with both high sodicity and high pH (ESP > 15, pH > 8.3) and (2) containing soluble bicarbonate and carbonate (Na/[Cl + SO4] > 1) (Gupta and Abrol, 1990). In this context the term ''alkali'' has merit because swelling and dispersion increase as both ESP and pH increase (Gupta et al., 1984; Suarez et al., 1984), and soil solutions where sodium and bicarbonate plus carbonate are the predominant ions tend to have low salinities and high pH values. The pH of a sodic soil, on the other hand, can be either greater or less than 7 (Kelley, 1951, pp. 78–81) and such soils can be either saline or nonsaline. Use of the term ''alkali'' thus allows practical distinctions between saline, sodic, and alkali soils in terms of soil management (Bhumbla and Abrol, 1979; Gupta and Abrol, 1990). In this chapter, ''saline'' and ''sodic'' are used as defined by the Soil Science Society of America (Anonymous, 1979), while ''alkali'' is used as defined in this paragraph.

Despite the usefulness of such simple numerical criteria, it is important to recognize their limitations. For example, research conducted since 1954 has documented many instances in which the tendency for swelling, aggregate failure, and dispersion increases as salinity decreases even if the ESP is less than 3—that is, nonsaline soil can behave like a sodic soil (Shainberg and Letey, 1984; Sumner, 1993). These tendencies increase as ESP increases, requiring increasingly higher salinities to stabilize the soil. The boundary between stable and unstable conditions varies from one soil to the next. In addition, the stability boundary for water entry *into* a soil (infiltration) is different than that for water movement *through* the soil (unsaturated and saturated hydraulic conductivity): soil surfaces are more sensitive to low salinity, magnesium (Keren, Chapter 8), and exchangeable sodium than is the soil underneath. In short, the numerical criteria used to differentiate between saline, saline-alkali, and nonsaline-alkali soils (U.S. Salinity Laboratory staff, 1954) give only one point on what is actually a salinity-sodicity continuum. It is important to understand this continuum and its impact on management guidelines for reclamation and subsequent water and soil management.

A. Clay Swelling and Dispersion

Clay swelling and dispersion are the two mechanisms which account for changes in hydraulic properties and soil structure (Shainberg and Letey, 1984; Quirk, 1986). Swelling that occurs within a fixed soil volume reduces pore radii, thereby reducing both saturated hydraulic conductivity (*K*) (Quirk and Schofield, 1955; McNeal et al., 1966; Rengasamy et al., 1984) and unsaturated *K* (Russo and Bresler, 1977; Xiao et al., 1992). Swelling results in aggregate breakdown or slaking (Cass and Sumner, 1982; Abu-Sharar et al., 1987), and clay particle movement, which in turn leads to blockage of conducting pores (Quirk and Schofield, 1955; Rowell et al., 1967; Rhoades and Ingvalson, 1969; Felhendler et al., 1974).

Soil clay content is important in influencing the stability of soil structure and hydraulic properties because of the large surface area of clay particles, their thin platy shape, and their negative lattice charge, which is balanced by exchangeable cations. The type of clay is also important. A dominant clay mineral in semiarid and arid regions is montmorillonite. Kaolinite is common in more humid regions, while illite is common to both regions. The latter minerals, in their pure state, swell and disperse much less than montmorillonite. However, kaolinitic and illitic soils that contain low percentages of montmorillonite tend to be dispersive (Schofield and Samson, 1954; Frenkel et al., 1978).

The negative lattice charge of the clay and the exchangeable cations which reside in the soil solution immediately adjacent to the clay surface form two layers, each with an opposite charge. This situation is commonly referred to as a diffuse double layer. The thickness of the diffuse double layer depends on the exchangeable cation composition and the salt concentration of the soil solution. Exchangeable cations are subject to two opposing tendencies: (1) electrostatic attraction to negatively charged clay surfaces, and (2) diffusion from a high concentration at the clay surface to a lower concentration in the soil solution. The result is an exponentially decreasing exchangeable ion concentration with distance from the clay surface. Since divalent ions are attracted to the surface with a force considerably greater than monovalent ions, the thickness of the diffuse double layer is more compressed for divalent ions. Increasing the salt concentration also compresses the double layer, because it reduces the tendency of exchangeable ions to diffuse away from the surface.

When two clay platelets approach each other, their diffuse double layers overlap and work must be done to overcome the electrical repulsion forces between the two positively charged exchangeable ion atmospheres. This repulsive force is also called the "swelling pressure." The greater the compression of the exchangeable ions toward the clay surface, the smaller the overlap of ionic atmospheres and the smaller the swelling pressure. Consequently, both clay swelling and the swelling pressure between particles decrease with increasing salt concentration of the bulk solution and increasing valence of the exchangeable cations. Sodium montmorillonite swells freely in dilute salt solutions, and single clay platelets tend to persist in these solutions. However, when divalent cations are adsorbed on montmorillonite surfaces, individual platelets aggregate into packets, or quasicrystals (Quirk and Aylmore, 1971), four to nine layers thick (Shainberg and Letey, 1984; Sposito, 1984). When this happens, only the external surfaces of calcium and magnesium quasicrystals participate in swelling, and swelling is much less than for sodium montmorillonite.

The distribution of Na/Ca in mixed cation systems deserves special consideration. Swelling of calcium montmorillonite quasicrystals is small and increases only slightly with small increases in ESP, whereas their electrophoretic mobility (Bar-On et al., 1970) and the salt concentration required for flocculation increase greatly (Oster et al., 1980). These observations are explained by the "demixing" of the exchangeable ions: the initial increments of sodium adsorb on the external surfaces of calcium quasicrystals, whereas adsorbed calcium remains in the interlayers between individual clay platelets. Consequently, the size of the quasicrystals does not increase with the initial increments of sodium adsorption, though their electrophoretic mobility increases greatly. The sodium ions on the external surfaces of the quasicrystals impart to them a mobility similar to that of sodium montmorillonite and cause a disproportionate increase in the salt concentration required for flocculation.

B. Dispersion of Illites and Kaolinites

Under similar ESP and salt concentrations, illite suspensions were more dispersed than montmorillonite suspensions (Oster et al., 1980). Illite particles, which consist of platelets stacked to a thickness of about 10 nm, have rough edges which mismatch upon close approach, resulting in smaller attraction forces. Thus, higher salt concentrations are required for flocculation of illite than for montmorillonite.

Schofield and Samson (1954) found that a pure sodium kaolinite flocculated at pH < 7 under conditions where dispersed illite and montmorillonite remained dispersed. At this pH, planar faces of kaolinite crystals were negatively charged and the edge surfaces were positively charged, with resulting attraction between positive and negative charged surfaces causing flocculation. Deflocculation of the salt-free suspension occurred with the addition of NaOH. At pH > 8, the edges became negatively charged, resulting in deflocculation and dispersion. Suarez et al. (1984) reported similar results for two out of three soils studied. At pH 9, the saturated *K*-values of a montmorillonitic soil and a kaolinitic soil were lower than at pH 6, whereas pH did not affect *K* for a vermiculitic soil. In addition to increased clay dispersion with increasing pH, Gupta et al. (1984) reported that additions of farmyard manure to the same soil also increased clay dispersion, a likely result of the adsorption of organic polyanions on positively charged surfaces of clay minerals (Durgin and Chaney, 1984; Shanmuganathan and Oades, 1983).

Positive edge charges of soil clays can also be decreased by adsorption of sodium montmorillonite (Schofield and Samson, 1954). Frenkel et al. (1978) found that the *K* of a kaolinitic soil was not affected by an ESP of 20, but when the soil was mixed with 2% montmorillonite, the soil mixture was very susceptible to dispersion. Stern et al. (1991) and Ben-Hur et al. (1992) arrived at similar conclusions for South African soils which were predominantly kaolinitic and illitic but contained small amounts of montmorillonite.

Aluminum and iron hydroxides and oxides, common constituents of most soils, occur as coatings on clay minerals. It is generally accepted that the various forms of aluminum and iron present in soils promote clay flocculation and reduce clay swelling and dispersion under sodic conditions (Deshpande et al., 1968; McNeal et al., 1968; Goldberg and Glaubig, 1987).

C. Effects of Salinity and Sodicity on Soil Hydraulic Properties

Soil *K* depends on both the ESP of the soil, or the sodium adsorption ratio, SAR* of the soil solution, and the salinity of the soil solution (Fireman and Bodman, 1939; Quirk and Schofield, 1955; McNeal and Coleman, 1966). Here it is important to note that the SAR of the soil solution approximately equals the ESP of the soil in the range from 0 to 40. The higher the SAR and the lower the salinity, the larger is the reduction in *K*. Typical effects of salinity and SAR on the *K* of soils from the western United States are shown in Fig. 1. Each soil responds differently to the same combination of salinity and SAR because of differences in clay content, clay mineralogy, iron and aluminum oxide content, and organic matter content.

Quirk and Schofield (1955) introduced the concept of "threshold concentration," or salt concentration at which a 10–15% decrease in *K* may occur for a given SAR. A plot of threshold concentration against SAR for a British soil (threshold concentration, Quirk and Schofield, 1955, Fig. 1) resulted in an approximately linear line for $0 < \text{SAR} < 60$. Salt concentrations, or soil salinities, to the right of the line for the British soil were greater than the threshold for a given SAR, and the *K* was stable. Hydraulic conductivities were not stable at concentrations to the left of the line, because salinity levels were inadequate to prevent swelling and/or dispersion.

---

*SAR has units of $(\text{mmol/L})^{1/2}$ because $\text{SAR} = C_{\text{Na}}/\sqrt{C_{\text{Ca}} + C_{\text{mg}}}$, where ion concentrations C are in mmol/L. Following conventional usage, SAR values in the text will not include these units.

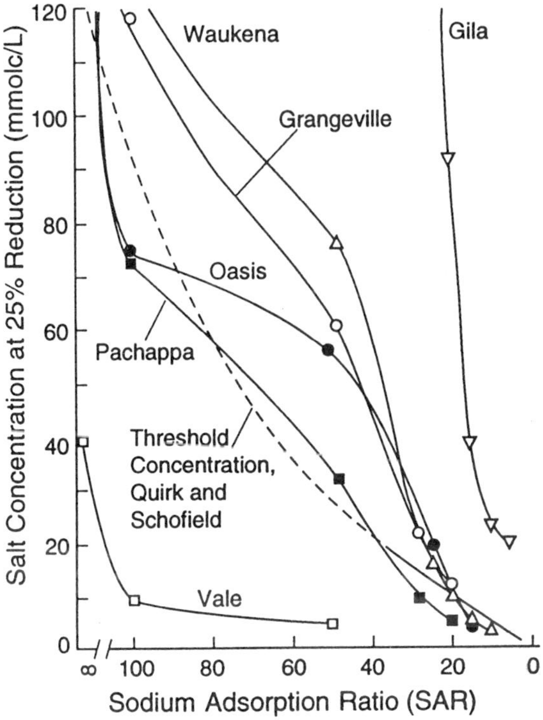

Fig. 1 Combinations of salt concentration and sodium adsorption ratio (SAR) required to produce a 25% reduction in hydraulic conductivity for selected soils from the western U.S. (Adapted from McNeal and Coleman, 1966.)

With the exceptions of the Vale and Aiken soils (McNeal and Coleman, 1966; Aiken not shown on graph), 25% reductions in *K* were associated with higher salinities than the "threshold concentration" associated with 10–15% reductions for the British soil (Quirk and Schofield, 1955). In other words, the Oasis, Grangeville, Pachappa, and Waukena soils were somewhat more sensitive to salinity than the British soil and Gila was much more sensitive. In summary, swelling and dispersion increase with increasing SAR and decreasing salinity, thereby influencing the physical properties of each soil in a unique manner (Pratt and Suarez, 1990). Significant reductions (10–25%) in saturated *K* for soils with SAR (or ESP) values of 15 can be expected if soil salinity is less than 5 to 50 mmolc/L (0.5 to 5 dS/m). Based on research conducted since 1966, similar reductions can be expected for soils with SAR (or ESP) values as low as 3 if soil salinity is less than 0.2 to 1 dS/m.

D. Effect of Salinity and Sodicity on Infiltration Rate

When water is applied to the soil surface, whether by rainfall or irrigation, some penetrates the surface and flows into the soil, while the remainder fails to penetrate and instead either accumulates at the surface or runs off. Generally, the rate of water entry into the soil, or IR, is high during the initial stages of infiltration but decreases exponentially with time to approach a constant rate. Two main factors cause this drop in IR: (1) a decrease in the matric potential gradient which occurs as infiltration proceeds, and (2) the formation of a seal or crust at the soil surface. In cultivated soils from semiarid regions, the organic matter content is low, soil structure is unstable, and sealing is a major determinant affecting the steady-state IR (Duley, 1939; Morin and Benyamini, 1977). Seal formation at the soil surface is in turn due to two processes: (1) physical disintegration of soil aggregates and soil compaction caused by the impact of water, especially waterdrops; (2) chemical dispersion and movement of clay particles and the resultant plugging of conducting pores. Both of these processes act simultaneously, with the first enhancing the second (Agassi et al., 1981).

Infiltration rates are especially affected by the sodicity (SAR) and electrical conductivity (EC) of irrigation water, because of the mechanical and stirring action of falling water drops, overland water flow, and the relative freedom of particle movement at the soil surface (Rengasamy et al., 1984). In studies in which waters of different qualities were applied to cropped columns of a loam soil, Oster and Schroer (1979) obtained a considerably better correlation of final IR to the SAR and EC of the applied water than to the SAR and EC of the soil solution averaged over the entire column length.

In arid regions where irrigation is essential for maintaining agricultural production, occasional rain may lower soil solution EC below the flocculation value, resulting in soil dispersion and severe reductions in IR. Similar conditions occur where nonsaline irrigation waters (~0.10 dS/m) are used for irrigation, such as along the east side of the Central Valley of California. The effect of soil ESP on the IR of a sandy loam subjected to rainfall is presented in Figs. 2a and 2b. Increasing the ESP of the sandy loam soil (Fig. 2A) from 1.0 to 2.2 dropped the final IR from 7.5 to 2.3 mm/h; increasing the ESP to 4.6 dropped the final IR to 0.6 mm/h. Similar results were obtained with the loessial silt loam (Fig. 2B). For both soils, spreading phosphogypsum at the soil surface was effective in reducing seal formation and the associated drop in IR (Fig. 2). Phosphogypsum dissolves and increases the EC and calcium content of the soil solution, reducing clay dispersion and seal formation.

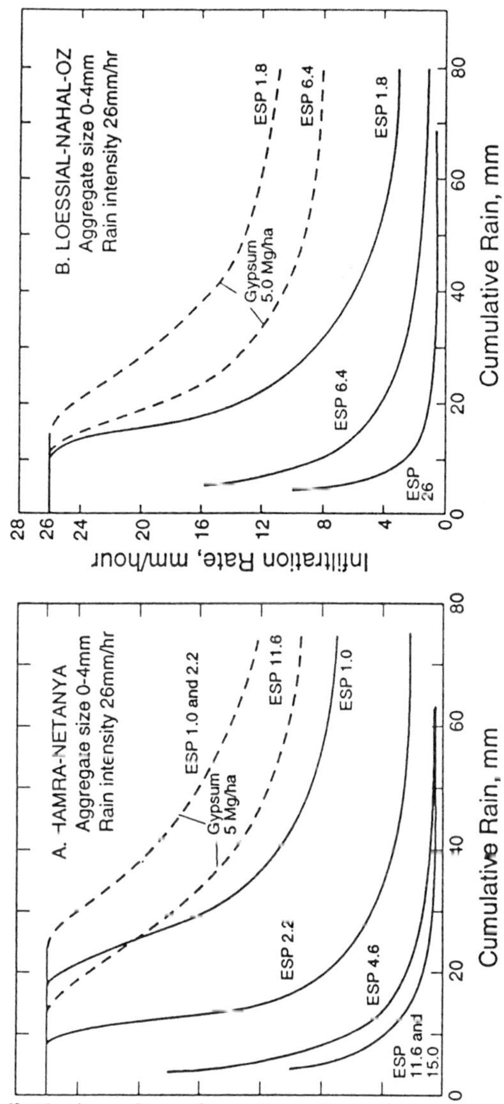

Fig. 2 Effects of ESP and phosphogypsum on the infiltration rate of a Hamra-Netanya sandy loam (a) and a loessial-Nahal-Oz soil (b). (From Kazman et al., 1983.)

E. Effects of Soil Mineral Equilibria on Soil EC, SAR, and pH

What constraints do soil minerals impose on the chemical composition of the water and exchanger phases of a soil ungoing reclamation? Upon leaching with rainfall or with nonsaline irrigation water, the EC of the soil solution will not decrease below a level which depends, to a considerable degree, on soil mineral dissolution. Dissolution, in turn, depends on the soluble minerals present in the soil and on associated chemical equilibria that involve the compositions of the solid, exchanger, liquid, and gaseous phases of the soil (Oster and Halvorson, 1978; Abrol et al., 1979). Arid land soils release 3 to 5 mmolc/L of calcium and magnesium to the soil solution as a result of the dissolution of plagioclase feldspars, amphiboles, pyroxenes, and other minerals (Rhoades et al., 1968). Dissolution of calcite and gypsum can maintain calcium, bicarbonate, and sulfate concentrations at even higher levels, depending on the exchangeable-ion composition and the partial pressure of carbon dioxide, P_{CO_2} (Oster and Rhoades, 1975; Oster, 1982). For soils that contain calcite, gypsum, or both, salinity increases linearly with SAR of the soil solution for a fixed Mg:Ca ratio and P_{CO_2} (Fig. 3) after leaching has removed the soluble salts.

The line labeled Mg = 0; P_{CO_2} = 0.032 kPa in Fig. 3 represents the calculated linkage between EC and SAR of the soil solution for a calcareous Na/Ca soil at partial pressure of 0.032 kPa (0.00032 atm) that does not contain any chloride or sulfate salts. For a similar soil, the next lower solid line represents the EC/SAR linkage where concentrations of calcium and magnesium are equal and the P_{CO_2} is 10 kPa (0.1 atm). These two lines bracket likely minimum soil salinities which can occur during reclamation of calcareous sodic soils, particularly with rainfall. During reclamation with irrigation water, chloride and sulfate salts in the irrigation water would increase the soil salinity of a calcareous sodic soil.

Minimum salinities for soil solutions saturated with gypsum (Fig. 3) are greater than for solutions saturated with calcite, because gypsum is considerably more soluble. Magnesium also increases the minimum salinity, but P_{CO_2} has no effect. Because of common-ion effects between calcium, bicarbonate, and sulfate, the minimum salinities for gypsiferous soils are independent of P_{CO_2} and are nearly the same as those for soils that contain both gypsum and calcite (Oster, 1982). During reclamation with irrigation water, only the chloride salts in the water would further increase the minimum soil salinity of a gypsiferous sodic soil.

Like minimum salinities, pH also depends on equilibrium constraints imposed by the presence of calcite (Gupta et al., 1981) or gypsum. The corresponding pH for the uppermost line in Fig. 3 ranges from 8.4 for a SAR of zero to 9.6 for a SAR of 100. For the next lower solid line, the pH ranges from 6.9 for a SAR of zero to 7.6 for a SAR of 100. The difference in pH between the two

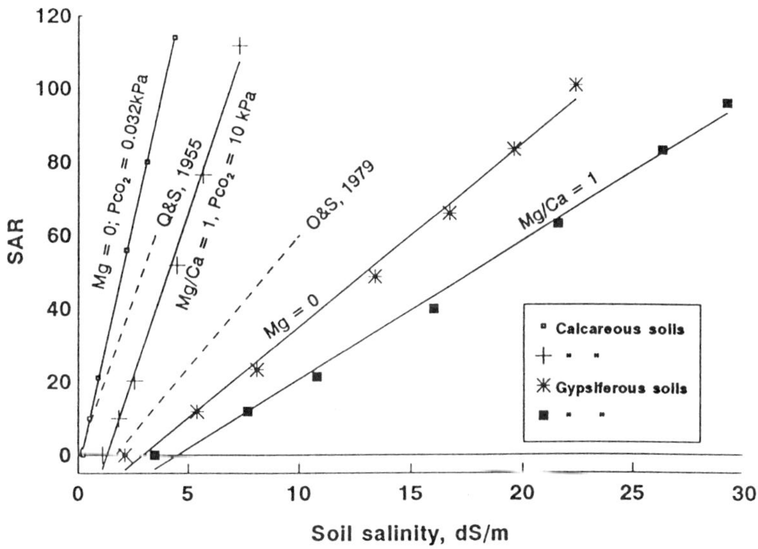

Fig. 3 Impact of SAR, P_{CO_2}, and Mg:Ca ratio on minimum soil solution salinities for calcareous and gypsiferous soils. The broken line labeled Q&S represents the combinations of EC and SAR which resulted in a 10–15% reduction in K of a British soil (Quirk and Schofield, 1955); the dotted line labeled O&S represents the combinations of EC and SAR which resulted in a 25% reduction in steady-state IR of a North Dakota loam soil (Oster and Schroer, 1979).

lines results from the higher partial pressures of carbon dioxide for the lower line (10 kPa vs. 0.032 kPa), as well as the higher concentration of magnesium. The P_{CO_2} in the root zone is a dynamic parameter which depends on microbial and root respiration and on soil water content. Values may range from 0.32 kPa near the soil surface to about 10 kPa in the lower portions of a rapidly respiring root system (Buyanovsky and Wagner, 1983). Increases in P_{CO_2} benefit the K of sodic soils for two reasons: (1) the minimum salinity is higher, and (2) the pH is lower.

F. The Physics and Inorganic Chemistry of *K* and IR: A Synthesis

The lines in Fig. 3 labeled "Q&S, 1955" and "O&S, 1979" provide quick comparisons of threshold salinities for K and IR with minimum salinities for calcareous and gypsiferous soils. EC/SAR combinations to the right of the "Q&S, 1955" line are likely to result in less than 10–30% reductions in K. Similarly, EC/SAR combinations to the right of the "O&S, 1979" line are likely to result in less than 20–30% reductions in steady-state IR. These comparisons can be summarized as follows:

1. Minimum salinities for calcareous soils during reclamation with rain or low-salinity irrigation waters may not be adequate to meet threshold salinity requirements for K, unless P_{CO_2} is enhanced by cropping (Robbins, 1986) or the soil contains significant levels of magnesium (Alperovitch et al., 1981). Further, minimum salinities for calcareous soils will usually not meet threshold IR for soils where the SAR of the near-surface soil solution exceeds 5 unless the irrigation water contains sufficient chloride and sulfate salts to meet the threshold requirements.

2. For soils that contain gypsum, minimum salinities can be expected to meet the threshold salinity requirements for K. This may not be true for IR, however. Salinity at the surface will depend on soil gypsum content, quantity of gypsum applied to the surface, dissolution kinetics of gypsum and the soils infiltration rate (Oster, 1982). It will also depend on the presence of any chloride salts in the irrigation water.

III. RECLAMATION OF SALINE SOILS

A. General Considerations

In ideal soils under piston-flow conditions, the EC of soil water and applied water would be equal after the passage of one pore volume of rainfall or irrigation water. However, soils are not ideal for two reasons: (1) Water flows faster through soil cracks and large pores between soil aggregates than through smaller pores within aggregates. This difference in localized water flow velocities is commonly referred to as bypass flow, or preferential flow. (2) Salt diffusion coefficients and chemical reactions are not sufficiently rapid to allow salt concentrations to be the same among all pores of differing sizes, particularly during periods of rapid water movement.

Salt transport models have provided considerable insight into these processes (Dutt et al., 1972; Tanji et al., 1972; Robbins et al., 1980; Bresler et al., 1982). However, suitable mathematical methods are not available to describe the multiple concurrent processes of water flow, including spatially variable IR and K, and all the chemical reactions—exchange, salt dissolution and precipitation, CO_2 liquid/gas equilibria—involved in salt transport on a field scale. Consequently, reclamation guidelines are largely based on experimental relations obtained from field and lysimeter reclamation experiments.

Method of water application and soil type are the primary variables affecting the amount of water required to reclaim saline soils. Hoffman (1986) summarized leaching results of several field reclamation studies. For saline soils, 60% or more of the salts initially present will be leached (or removed) by a depth of water equal to the depth of soil (Fig. 4) under continuously ponded conditions. Hoffman proposed the following relationship between the fraction of initial salt concentration remaining in the profile, c/c_0, and the depth of water infiltrated, D_w, through a given depth of soil, D_s:

$$\left(\frac{c}{c_0}\right)\left(\frac{D_w}{D_s}\right) = K' \tag{1}$$

where K' differs with soil type. For reclamation with continuous ponding, the K'-values for peat, clay loam, and sandy loam soils are 0.45, 0.3, and 0.1, respectively. The data in Fig. 4 for sandy loams are from two experiments in India (Leffelaar and Sharma, 1977; Khosla et al., 1979) and one in Iraq (Hulsbos and Boumans, 1960). The data for clay loams are from three tests in Utah on a clay loam, a silty clay loam, and a clay (Reeve et al., 1948) and two tests in California, on a clay loam (Reeve et al., 1955) and a silty clay (Oster et al., 1972). Data on peat soils are from Turkey (Beyce, 1972), with a field experiment conducted in California (Prichard et al., 1985) providing additional data that support the peat soil relationship. As defined, D_w in Eq. (1) does not include

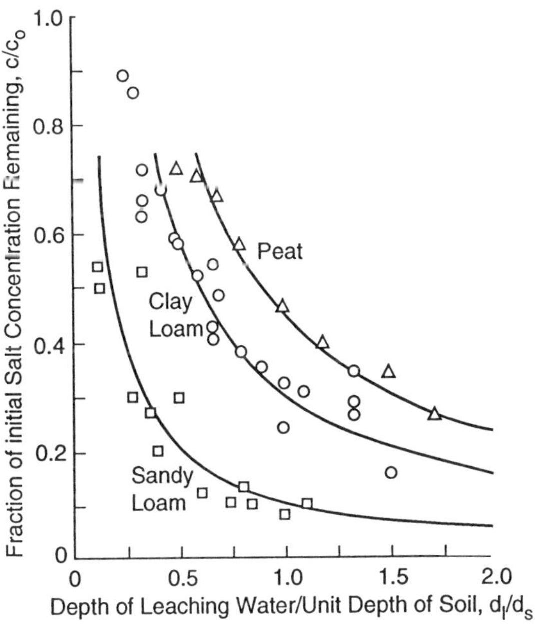

Fig. 4 Depth of leaching water per unit depth of soil required to reclaim a saline soil by continuous ponding. (Adapted from Hoffman, 1986.)

evaporation losses. Consequently, where evaporation is greater than 0.1 D_w, D_w should be corrected for evaporation (Minhas and Khosla, 1986).

Equation (1) is valid only when $D_w/D_s > K'$. Differences in K' reflect differences in saturated volumetric water content and leaching efficiency among soils. Sandy loam soils with low-saturated-water contents have higher leaching efficiencies than finer-textured soils. Soil pores in sandy soils are more uniform in diameter than in clay loam or clay soils. These finer-textured soils can have large cracks with large pores between aggregates and along crack surfaces when dry, and fine pores within aggregates when wet. Such bypass channels reduce leaching efficiency.

B. Intermittent Application: Ponding and Sprinkler Irrigation

The water requirement for leaching can be reduced by intermittent applications of ponded water, particularly for fine-textured soils. This reduces the constant, K' in Eq. (1) to 0.1 for silty clay, loam, and sandy loam soils (Fig. 5). Agreement among experiments is excellent considering that soil texture ranged from silty clay (Oster et al., 1972) to loam and sand (Talsma, 1967) and that the depth of water applied each cycle varied from 50 to 150 mm, with corresponding ponding

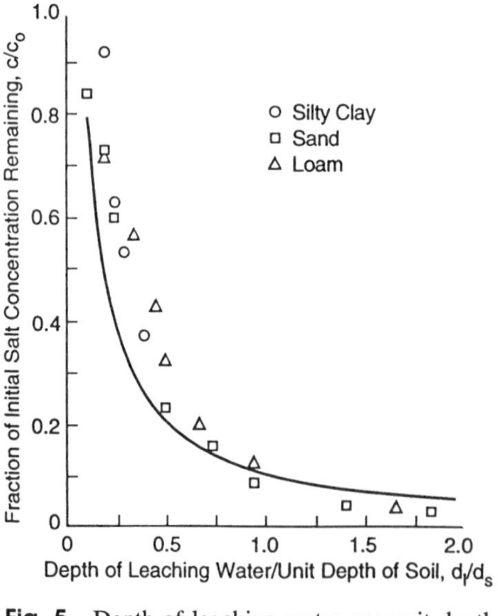

Fig. 5 Depth of leaching water per unit depth of soil required to reclaim a saline soil by ponding water intermittently. (Adapted from Hoffman, 1986.)

intervals varying from weekly to monthly. For the fine-textured soil, the water requirement for intermittent ponding is only about one-third of that required by continuous ponding to remove 70% of the soluble salts. However, intermittent ponding techniques may be slower than continuous ponding. Although 50% more water was required for reclamation with continuous ponding, the time required for leaching a clay loam soil was much longer for intermittent ponding (Miller et al., 1965). For a silty clay soil, however, the same reclamation was achieved in three months using continuous ponding, intermittent ponding, and sprinkler irrigation (Oster et al., 1972).

With sprinkler irrigation, reclamation can occur under continuously unsaturated conditions if water application rates are controlled so ponding does not occur. This circumvents inefficiencies caused by bypass flow in large cracks. Leaching occurs only within the depth of soil wet by sprinkler irrigation. If sprinkler irrigation occurs more frequently than intermittent ponding, the time-averaged water content within the depth wet by sprinkler irrigation can be higher than under intermittent ponding. The counteracting effects of reduced bypass flow and increased water content are consistent with the observed order of leaching efficiency (intermittent ponding > sprinkler > continuous ponding) in a field experiment conducted on a silty clay (Oster et al., 1972).

C. Reclamation of Saline Soils under Cropped Conditions

Leaching is also necessary for reclamation under cropped conditions, because salt removal by agronomic crops is insignificant. The average amount of salt contained in five mature forage crops (alfalfa, barley, corn silage, sudangrass, and sweetclover) is about 0.4 Mg/ha (Longenecker and Lyerly, 1974). Another example involves the salt content of *Diphlachne fusca* (L.) Beauv. or *Leptochloa fusca* (L.) Kunth (Booth, 1983), called Karnal grass in India and Kallar grass in Pakistan, a crop grown during reclamation of saline-sodic soils. Its salt content ranges from 4% to 8% when grown in soils with ECe values of about 20 dS/m (Malik et al., 1986). Assuming a yield of 10 Mg/ha, the salt content of forage removed from such a field would range from 0.4 to 0.8 Mg/ha. This amount is small when compared to the salt content of saline soils, or of the irrigation water used to grow the crop. The salt content of a root zone 1 m deep that has an electrical conductivity of 20 dS/m is about 40 Mg/ha; the salt content of 10 ML (1 ha-m) of irrigation water with a salinity of 1 dS/m is about 6 Mg. Plants that are very efficient in removing salt from saline soils, such as sea-blithe (*Suaeda fruiticosa*), remove less than 3 Mg/ha per harvest (Chaudri et al., 1964).

Although cropping during reclamation is a common practice, only a few crops can tolerate high-salinity levels, particularly during early stages of crop growth. Consequently, leaching before planting any crop is advisable when soil salinity is high (ECe > 10–15 dS/m) in the upper 1 m of soil. High-salinity levels in the seedbed delay seed germination and reduce plant vigor during seedling establishment. It is also difficult to maintain adequate soil water contents and prevent crusting if rainfall or additional irrigation is required during a prolonged period of seed germination and plant establishment. Thus, to ensure good crop establishment, salinity levels in the seed zone need to be considerably below threshold salinity levels for the crop (Maas and Grattan, 1995), which are generally based on crop response to salinity after plant growth is well established. In the San Joaquin Valley of California, for example, salinity levels (ECe) greater than 2 dS/m in the seed zone of cotton (threshold salinity of 7.7 dS/m) add five to seven days to the normal 10 days required for germination and seedling establishment (Fulton, private communication). Crops such as rice and Karnal grass that can be germinated or transplanted under ponded conditions circumvent problems with soil salinity during seedling establishment. For example, prewetted rice seed, which in California is ''flown on'' (distributed onto ponded fields from airplanes), is primarily exposed to the low salinities of the ponded water, as is the extensive surface root system that develops after seedling establishment. Similarly, transplanted rice plants can be established in nurseries with low soil salinities and transplanted into ponded saline soils at a more tolerant growth stage.

Provided that appropriate salinity conditions for the seed zone can be met, the choice of salt-tolerant crops depends in part on yield reductions that are acceptable and on the existing average rootzone salinity (Table 1). Soil salinities for 0%, 20%, and 50% yield reductions were calculated from data reported by Maas and Grattan (1995). The crops included in Table 1 are a select list of moderately sensitive to tolerant food, fiber, forage, and vegetable crops that are commonly grown and for which accurate salt tolerance data are available. Vegetable crops are included because they grow primarily during the cooler parts of the year and thus have relatively low water needs. In addition, vegetable crops are shallow rooted and are not generally affected by high salinity or boron levels below the rootzone.

It should be noted that many moderately sensitive to tolerant crops included in the tables reported by Maas and Grattan (1995) are not listed in Table 1. Thus, the genetic diversity of crop salt tolerances provides many options for growing a range of crops during soil reclamation.

D. Reclamation of Soils with High Levels of Boron

Boron is toxic to most crops at soil solution concentrations exceeding a few milligrams per liter (Maas and Grattan, 1995). Boron is more difficult to leach than salts because it is adsorbed on clay minerals, hydroxy oxides of Al, Fe, and Mg, and organic matter (Keren and Bingham, 1985). Further, in soils with

Table 1 Yield Potentials as a Function of Average Root Zone Salinity

Crop	Average root zone salinities (dS/m) at specific yield potentials		
	50%	80%	100%
A. Grain/forage/fiber			
Triticale (grain)	26	14	6
"Probred" wheat (grain)	25	15	9
Wheat (forage)	24	12	4
Durum wheat (forage)	22	10	2
Karnal grass[a]	20	8	3
Durum wheat (grain)	19	11	6
Barley (grain)	18	12	8
Cotton	17	12	8
Rye (grain)	16	13	11
Sugarbeet	16	10	7
Bermuda grass	15	10	7
Sudan grass	14	8	3
Wheat (grain)	13	9	6
Barley (forage)	13	9	6
Berseem clover	10	5	2
Narrow leaf birdsfoot trefoil	10	8	7
Sorghum	10	8	7
Alfalfa	9	5	2
Rice (paddy)	7	5	3
Corn (forage)	9	5	2
Corn (grain)	6	3	2
B. Vegetables			
Asparagus	29	14	4
Zucchini squash	10	7	5
Celery	10	5	2
Red beet	10	6	4
Spinach	9	5	2
Eggplant	8	4	1
Broccoli	8	5	3
Tomato	8	4	2

[a]Malik et al. (1986).
Source: From Maas and Grattan (1995).

high levels of native boron, reduced boron concentrations immediately following leaching may be only temporary. Concentrations can increase with time, or regenerate, due to slow release of previously adsorbed boron (Rhoades et al., 1970; Bingham et al., 1972), especially during the early stages of soil reclamation. Figure 6 shows reclamation results obtained under field conditions for two California soils high in native boron (Reeve et al., 1955; Bingham et al., 1972). Using Eq. (1) to fit the data results in a K'-value of 0.6 (Hoffman, 1986). The water required to remove a given fraction of boron by intermittent ponding or sprinkling is thus about two times greater than that required to remove soluble salts by the least effective method, continuous ponding. In addition, further periodic leaching may be required if boron regeneration occurs.

Crops that are among the most tolerant of boron include (from least to most tolerant) barley, sugar beet, tomato, sorghum, garlic, cotton, celery, onion, and asparagus. Yield potentials of 80% occur at soil-solution concentrations of 8 mg/L for barley, 12 mg/L for garlic, 16 mg/L for celery, and 20 mg/L for onion (Maas and Grattan, 1995).

IV. RECLAMATION OF SODIC SOILS

This section describes specific cases in which exchangeable sodium affects the reclamation of salt-affected soils. The focus is on maintaining adequate soil salinity to counteract the adverse effects of exchangeable sodium on soil hydraulic properties. Salinity is the key to rapid reclamation as was best demonstrated by Reeve and Doering (1966) in the reclamation of clay loam soil using the high-saltwater dilution method. The subsections are ordered in terms of decreasing soil salinity: reclamation using the high-saltwater dilution method, reclamation of gypsiferous soils, reclamation of sodic soils using gypsum amendments, and reclamation of calcareous sodic soils using cropping techniques.

A. Reclamation with Saline Water (High-Saltwater Dilution)

Hypersaline irrigation water can cause large increases in soil K by itself without the need of a soil for tillage, or cropping, or both. Equilibrations with successive dilutions of a hypersaline saline water will reduce both ESP and ECe. For a sandy loam soil with an ESP of 37, irrigation with Salton Sea water (560 mmolc/L) resulted in an K of 5 mm/h (Reeve and Bower, 1960) as compared to 0.2 mm/h for irrigation with Colorado River water (11 mmolc/L). In a field experiment conducted on a native, untilled clay loam ($70 < \text{ESP} < 80$, and $16 < \text{ECe} < 21$ in the upper 900 mm of soil), hydraulic conductivities of 0.05, 0.07, 4.17, and 10.95 mm/h were obtained for Colorado River water (11 mmolc/L), Colo-

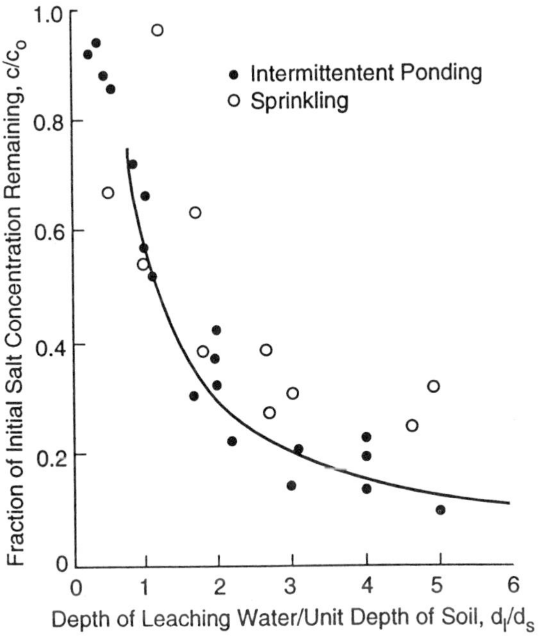

Fig. 6 Depth of leaching water per unit depth of soil required to reclaim a soil high in boron. (Adapted from Hoffman, 1986.)

rado River water saturated with gypsum (38 mmolc/L), seawater (611 mmolc/L), and calcium chloride solution (655 mmolc/L), respectively (Reeve and Doering, 1966). After continuous ponding for 238 days the ESP of the upper 150 mm of soil was reduced from 70 to 57 by Colorado River water and from 74 to 32 by gypsum-saturated Colorado River water, but in both cases the ESP at greater depths remained the same or even increased. In contrast, the ESP of the entire upper 900 mm of soil was reduced from 70 to 22 in 167 days by continuous ponding with successive dilutions of seawater (final SAR = 11 and EC = 2 dS/m), and from 70 to 18 in three days with successive dilutions of calcium chloride solutions (SAR $\geq$ 0, final EC = 5 dS/m). Following reclamation with calcium chloride, the *K* for Colorado River water was the same as for the last calcium chloride solution. In other words, after the ESP was reduced, the electrolyte concentration of Colorado River water was sufficient to maintain soil *K*.

Muhammed et al. (1969) investigated the possibility of reducing the water requirement for reclamation of the same clay loam by saturating the successive dilutions of seawater with gypsum and applying only enough water to achieve one-half of full equilibration for each successive dilution step. Application of their results (Oster, 1993) to the data obtained by Reeve and Doering (1966) results in a water requirement of 1500 mm water/900 mm of soil, or 1.7 mm water/mm soil. This scaled water requirement is similar to that for an 82% reduction in salinity under ponded conditions for clay loam soils (Eq. (1)).

Use of the high-saltwater dilution method should be considered wherever saline waters are available. However, the EC should exceed 20 dS/m, and the ratio of divalent cation concentration to total cation concentration should exceed 0.3 (with concentrations expressed in mmolc/L). If necessary, this ratio can be increased by the addition of gypsum, using, for example, the technique proposed by Keisling et al. (1978). Potential sources of highly saline waters include underlying groundwaters, nearby saline inland lakes, and ocean estuaries. The major problems with this technique are the facilities required to collect, convey, and treat the saline water, and the need to collect and dispose of highly saline drainage water in order to avoid contamination of surface and ground waters.

B. Reclamation of Gypsiferous Sodic Soils

Because of the high salinities of sodic soils which contain gypsum (Fig. 3), these soils are usually reclaimed successfully by leaching without additional amendments. For example, infiltration rates after 48 h ranged from 7 to 9 mm/h during reclamation of two clay loam soils ($30 < ESP < 40$) under ponded conditions using an irrigation water with an EC of 2.5 dS/m and an SAR of 6 (Reeve et al., 1948). The gypsum content in the upper 600 mm of soil was sufficient to replace most of the exchangeable sodium. In the same study, surface application of gypsum increased the infiltration rate from 8 to 14 mm/h for a clay soil which did not contain sufficient gypsum.

For a lysimeter experiment, Jury et al. (1979) reported similar success with sandy loam and clay loam sodic soils which contained gypsum, both from continuous applications of water using surface ponding and from daily applications at an unsaturated rate of about 0.4 mm/h, with irrigation water salinities of about 0, 0.5, and 1.3 dS/m. Infiltration rates were unaffected by irrigation-water salinity even in the most extreme case, a clay loam soil reclaimed with distilled water: it sustained an infiltration rate of 4.5 mm/h for 3.6 pore volumes of drainage. The K' value for Eq. (1) with this data is 0.29 (Oster, 1993), which agrees with the 0.30 constant obtained by Hoffman (1986) for saline soil reclamation during ponding for five soils. Three of the five reclamation studies in Hoffman's database were those conducted on gypsiferous soda soils by Reeve et al. (1948), suggesting that gypsum in sodic soils does not affect the water requirement for reclamation. This K'-value for gypsiferous soils applies across a spectrum of soil textures ranging from sandy loam to clay. The data of Jury et al. (1979) also suggest that bypass water flow, hydrodynamic dispersion, and gypsum-dissolution kinetics do not affect the water requirement under conditions where water application rates range from 0.04 to 8 mm/h.

C. Reclamation of Sodic Soils with Underlying Gypsiferous Layers

In locations where sodium-affected surface soils are underlain by soil containing significant quantities of gypsum, deep plowing has been effective in breaking up and mixing the layers while supplying soluble calcium to aid reclamation. The depth of plowing required may vary from 0.5 to more than 1.0 m, depending on the concentration and depths of the sodic- and calcium-rich layers. A procedure is available to predict the optimum depth of plowing to maintain adequate permeability during the reclamation process (Rasmussen and McNeal, 1973).

D. Reclamation of Sodic Soils with Gypsum and Other Amendments

Gypsum, either incorporated into the soil or left on the surface, is the calcium source most commonly used to reclaim sodic soils and to improve soil water infiltration that has been decreased by low salinities. Sources include mineral deposits and phosphogypsum, a by-product of the phosphate fertilizer industry. When gypsum is incorporated into the soil, the reduction in ESP upon irrigation and leaching is primarily limited to the soil depth interval where it is present (Fig. 7A). This is a consequence of the greater selectivity of exchange sites for calcium than for sodium. The exchange phase of the soil, in the presence of gypsum, is an effective sink for calcium, which replaces exchangeable sodium.

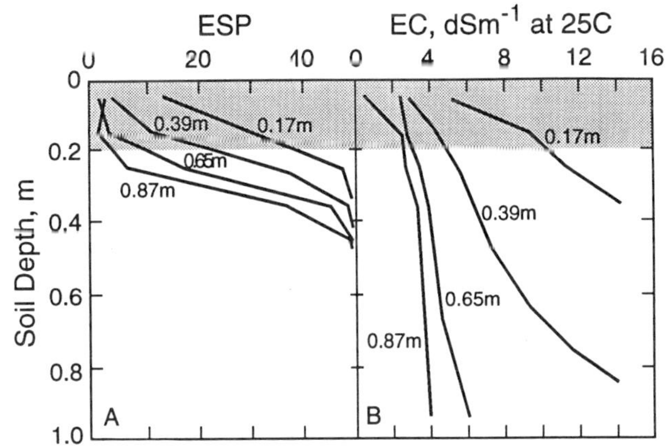

Fig. 7 Computer-modeling results (ESP and EC) for reclamation with gypsum of a soil with a cation-exchange capacity of 200 mmolc/kg using an irrigation water with an EC of zero. The initial ESP of the soil was 50 at all depths; the numbers next to each line are the depths of applied water in meters. (Adapted from Oster and Frenkel, 1980.)

The salinity of the soil solution (Fig. 7B) within and below the gypsum-amended layer decreases as the ESP within the amended layer decreases (Khosla et al., 1979; Oster and Frenkel, 1980; Frenkel et al., 1989). Only as reclamation approaches completion within the amended soil layer does the gypsum that dissolves begin to replace exchangeable sodium at greater depths. During this second phase of reclamation, the required threshold salinity levels in lower soil layers may not be maintained, resulting in decreased K. A decrease in K during partial soil reclamation is consistent with K data reported by Robbins (1986) obtained from 1-m-long lysimeters obtained during a reclamation study of a calcareous sodic silt loam soil: (1) greatest reclamation occurred within the upper 0.2 m of the uncropped gypsum-treated lysimeters, the layer in which adequate gypsum was incorporated to reclaim the upper 0.5 m of soil (Fig. 8A); and (2) the water flow rate declined to near zero in the gypsum-treated lysimeters after reclamation in the upper 0.2 m was complete and one pore volume of drainage had been collected.

The gypsum required (GR) for reclamation, in Mg/ha, can be calculated using Eq. 2:

$$GR = 0.0086\, F D_s \rho_b (\mathrm{CEC})(\mathrm{ESP}_i - \mathrm{ESP}_f) \tag{2}$$

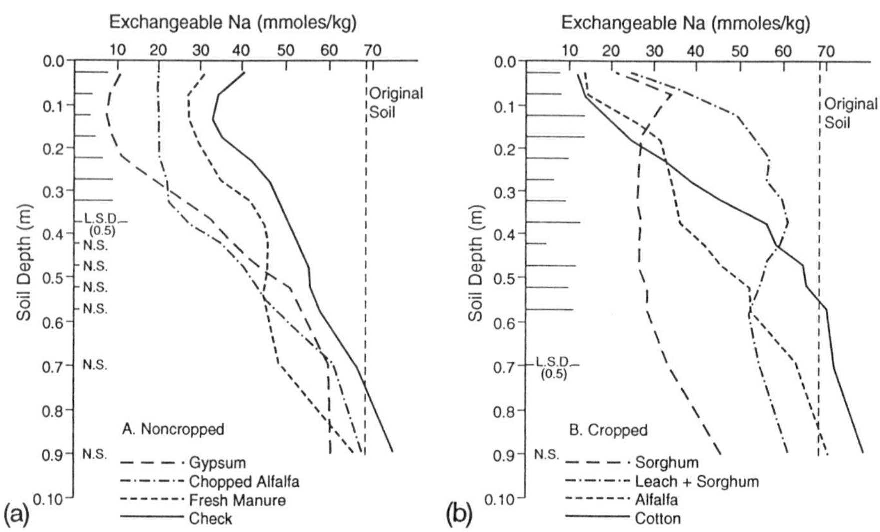

Fig. 8 Final distribution of exchangeable Na levels with soil depth resulting from reclamation treatments for noncropped (a) and cropped (b) treatments. (From Robbins, 1986.)

where

F = Ca-Na exchange efficiency factor (unitless)
D_s = soil depth (m)
ρ_b = soil bulk density (Mg/m^3)
CEC = cation-exchange capacity (mmolc/kg)
ESP_i, ESP_f = initial and final exchangeable sodium percentages.

The efficiency factor ranges from 1.1 for an ESP_f of 15 to 1.3 for an ESP_f of 5 (Oster and Frenkel, 1980). SAR can be substituted for ESP in the range $0 < SAR < 50$.

A laboratory method can also be used to determine GR, but the values tend to be high. The Schoonover procedure (U.S. Salinity Laboratory Staff, 1954, procedure 22d) determines the amount of calcium required to replace all the exchangeable sodium plus any that is precipitated by soluble bicarbonates and carbonates in the soil. Since these soluble ions are leached during sodic-soil reclamation, Abrol et al. (1975) proposed a modified procedure. It includes leaching 5 g of soil with 20 mL of 60% ethanol to remove soluble bicarbonate and carbonate, followed by equilibration with 100 mL of a calcium sulfate solution (30 mmolc/L). The difference between the calcium concentration in the calcium sulfate solution and the calcium plus magnesium concentration (mmolc/L) in a clear filtrate of the soil suspension, times 20, equals CEC (ESP_i-ESP_f) in Eq. (2), or CEC(ESP_i) since ESP_f is zero after equilibration with the calcium sulfate solution.

Other amendments used for calcareous sodic soils include sulfuric acid and acid formers such as sulfur, lime sulfur, pyrite, and iron and aluminum sulfates. These amendments react with calcite, thereby providing a soluble source of calcium within the soil. Equivalent quantities of chemically pure amendments relative to a ton of gypsum or sulfur are given in Table 2.

Table 2 Amounts of Amendments Equivalent to 1 Ton of Either Gypsum or Sulfur

		Gypsum	Sulfur
Gypsum	$CaSO_4 \cdot 2H_2O$	1.00	5.38
Calcium chloride	$CaCl_2 \cdot 2H_2O$	0.85	4.59
Sulfur	S_8	0.19	1.00
Iron sulfate	$FeSO_4 \cdot 7H_2O$	1.61	8.69
	$Fe_2(SO_4)_3 \cdot 9H_2O$	1.09	5.85
Aluminum sulfate	$Al_2(SO_4)_3 \cdot 18H_2O$	1.29	6.94
Iron pyrite	FeS_2, 30%S	0.35	1.87

The following chemical reactions illustrate how various amendments react with calcareous sodic soils. In these reactions X represents the soil exchange complex.

1. Inorganic Reactions
   Gypsum ($CaSO_4 \cdot 2H_2O$)
   $$2NaX + CaSO_4 \cdot 2H_2O \Leftrightarrow CaX_2 + Na_2SO_4 + 2H_2O$$
   Sulfuric acid (H_2SO_4)
   $$2NaX + H_2SO_4 + CaCO_3 \Leftrightarrow CaX_2 + Na_2SO_4 + CO_2 + H_2O$$
   Iron sulfate ($FeSO_4 \cdot 7H_2O$)
   $$4NaX + 2FeSO_4 + 2CaCO_3 + 1/2\ O_2 \Leftrightarrow 2CaX_2 + 2Na_2SO_4 + Fe_2O_3 + 2CO_2$$
2. Microbiologically Mediated Reactions
   Sulfur (S_8)
   $$4NaX + 2S + 2CaCO_3 + 3O_2 \Leftrightarrow 2CaX_2 + 2Na_2SO_4 + 2CO_2$$
   Iron pyrite (FeS_2)
   $$8\ NaX + 2FeS_2 + \frac{15}{2}O_2 + 4CaCO_3 \Leftrightarrow 4\ CaX_2 + 4Na_2SO_4 + Fe_2O_3 + 4CO_2$$

Cost, availability, and time required for reaction with soil determine which chemical amendment will be best for specific circumstances. Calcium chloride, likely the most costly amendment, is the quickest to dissolve and react with the soil. Iron and aluminum sulfates and gypsum also react quickly when mixed with the soil. The finer the particle size of the solid amendments and the greater the uniformity of soil incorporation, the faster and more efficient the reclamation reactions.

Degree of crystallinity is another factor: phosphogypsum is a more porous crystal than mined gypsum and therefore dissolves faster. This factor may also operate in the case of sulfuric acid, which reacts quickly with calcite and has been observed to be more effective than equivalent amounts of gypsum (Overstreet et al., 1951; Miyamoto et al., 1975). After the initial acid dissolution and exchange reactions are complete, additional calcite dissolution by sulfuric acid can result in gypsum precipitation. This gypsum is relatively noncrystalline, and its uniformity of incorporation exceeds any that could be achieved by mechanical incorporation of finely ground mined gypsum.

Sulfur and iron pyrite must be oxidized by soil microorganisms and are therefore both classified as slow-acting amendments. Since microbial activity increases with decreasing pH, it is important not to dilute these amendments by incorporating them too deeply into the soil. In other words, shallow tillage may be preferable to deep tillage. Since these reactions increase with increasing soil temperature (Overstreet et al., 1955), application is best during the spring and summer months.

E. Reclamation of Nonsaline Sodic (Alkali) Soils by Growing Crops

Reclamation practices often involve growing crops together with the application and infiltration of excess water during the cropping season. Benefits from cropping include the following: (1) Increased pCO_2 due to plant root respiration and decomposition of organic matter (Kelly, 1951; Chhabra and Abrol, 1977; Robbins, 1986; Gupta et al., 1989), which in turn increases soil salinity and decreases pH in calcareous soils (see Sec. II.E.). Robbins (1986) demonstrated that "satisfactory" K could be maintained with cropping treatments, whereas this was not the case for gypsum-treated soils which were not cropped. (2) In situ production of polysaccharides, which, in conjunction with differential dewatering at the root/soil interface, promotes aggregate stability (Boyle et al., 1989). (3) The physical effects of root action (Chhabra and Abrol, 1977), including removal of entrapped air from the larger conducting pores (McNeal et al., 1966), generation of alternate wetting and drying cycles, and creation of macropores and release of CO_2 upon decomposition. (4) Financial or other benefits from crops grown during reclamation that can help support farming operations.

In a lysimeter study, Robbins (1986) measured the partial pressure of CO_2 in the root zone during reclamation of a nonsaline (ECe = 2.4), sodic (ESP = 33), calcareous silt loam (CEC = 210 mmolc/kg) under cropped conditions. He demonstrated conclusively the benefits of increased soil atmosphere CO_2 concentrations for *K*, with concurrent reduction of exchangeable sodium. In the case of lysimeters cropped with a hybrid sorghum, the amount of sodium removed equaled that of noncropped lysimeters amended with gypsum. Reclamation for the gypsum-amended lysimeters (Fig. 8A), however, occurred primarily in the zone where gypsum was incorporated, whereas it occurred throughout the root zone (Fig. 8B) for lysimeters cropped with sorghum and alfalfa during the entire reclamation process. Leaching the soil before cropping (Fig. 8B; leach + sorghum) resulted in low *K* and subsequent poor plant growth.

F. Crop Selection

Crops grown during sodic soil reclamation must tolerate both poor soil physical properties (Gupta and Abrol, 1990) and sodium-induced calcium deficiency characterized by necrotic curled leaf tips and heavily serrated leaves (Carter et al., 1979; Maas and Grieve, 1987; Grieve and Maas, 1988). High sodium concentrations reduce the amount of calcium that is available for plant uptake (Cramer and Lauchli, 1986). Calcium concentrations that are adequate under nonsodic conditions (1–2 mmolc/L) can thus become inadequate when the Na/Ca ratio of the soil solution is high (Bernstein, 1975). Seedling growth stages of cereals are particularly susceptible to calcium deficiency, though considerable variability occurs among crops (Grieve and Maas, 1988). Variability also occurs among genotypes of sorghum (Grieve and Maas, 1988), rice (Grieve and Fujiyama, 1987), triticale (Norlyn and Epstein, 1984), and wheat (Kingsbury and Epstein, 1986). The data for rice are of interest because it is often recommended for growing during reclamation of sodic soils in India (Gupta and Abrol, 1990). For the rice cultivars "M9" and "M-201," Grieve and Fujiyama (1987) reported severe Ca deficiencies at an osmotic potential of −0.4 MPa (about 10 dS/m) and at an Na/Ca molar ratio of 78 (SAR = 87), whereas Yeo and Flowers (1985) reported no Ca deficiencies at Na/Ca molar ratios as high as 500 (SAR = 160) for cultivar "IR2153." Based on limited data for seedlings of different crops, the general order of increasing susceptibility to Ca deficiency is rice cultivar "IR2153," cotton, barley, wheat, rye, sorghum, cowpea (Grieve, private communication, 1993). The wide diversity among crops and crop genotypes increases the likelihood that field trials conducted under nonsaline sodic conditions will identify local crops that are adaptable to sodic soil conditions.

Extensive areas of the Indo-Gangetic plains of northern India have calcareous sodic soils. Field plot studies at the Central Soil Salinity Research Institute near Karnal conducted since 1970 provide the crop tolerance results shown in Table 3 (adapted from Gupta and Abrol, 1990). Differential, nonsaline sodic conditions were obtained by applying different amounts of gypsum to the soil and leaching by ponding for about 20 days (Abrol and Bhumbla, 1979). The ESP values ranged from about 10 to 70, with a corresponding range in SAR of about 7 to 160. Crops were grown under both rainfall and irrigated conditions, with recommended agronomic, fertilization, and plant protection practices used for all treatments. Rankings in Table 3 are based on 50% yield reductions, with rankings of rice, barley, and wheat similar to those reported by Grieve. However,

Table 3 Relative Tolerance of Crops to Sodicity

ESP range	Crops[a]
10–15	Safflower, mash, peas, lentil, pigeon-pea, curd bean
16–20	Bengal gram, soybean
20–25	Groundnut, cowpea, onion, pearl millet
25–30	Linseed, garlic, guar
30–50	Indian mustard, wheat, sunflower, berseem, hybrid napier, guinea grass
50–60	Barley, Sesbania, saftal, Penicums
60–70	Rice, Para grass
70+	Karnal, Rhodes, and Bermuda grasses

[a]Yields are about 50% of the potential yields in the respective sodicity ranges.
Source: Gupta and Abrol (1990).

because of crop genotype variations and varying experimental conditions, this similarity could be fortuitous.

In addition to the sodicity-tolerant crops listed in Table 3, grasses can also be grown on sodic soils. *Diplachne fusca* (Karnal or Kallar grass), *Chloris gayana* (Rhodes grass), and *Brachiaria mutica* (Para grass) have been reported as highly tolerant to sodicity (Kumar and Abrol, 1986). In addition, Karnal and Para grasses grow well under ponded conditions.

V. FIELD RECLAMATION PRACTICES

A. California

1. Reclamation of Saline and Sodic Soils

One interesting example of reclamation that combines tillage, ponding, and cropping is represented by practices in the Imperial Valley (ETo, 1900 mm; rainfall, 0–50 mm; T_{max}, 45°C, T_{min}, −2°C) through the early 1960s. Reclamation practices for calcareous and gypsiferous, saline-sodic soils (clay loam to silty clay) with Colorado River water (ECe = 1.2 dS/m, SAR = 1.5) included tillage, land leveling, leaching by continuous ponding, and cropping. Tillage of dry fallow soils included either ripping or slip plowing (to depths of 1 or 2 m, respectively), followed by disking and contour diking. Ripping loosens plow pans, clay layers, and hardpans, whereas slip plowing mixes layers of different textures. Unusually large amounts of water will infiltrate during the first irrigation following either ripping or slip plowing of dry soils. Water was ponded between large border dikes during the fall and early winter months until salinities at the soil surface were less than 12 dS/m (R. S. Ayers, personal communication, 1992). The field was then prepared for furrow irrigation after the upper 30 cm was dry. Cotton, grown between April and November, was planted on the side of the furrow slightly above the water level maintained during the germination irrigation (p. 44f in Ayers and Westcot, 1985). This provided additional leaching of the seed zone and prevented problems with seedling emergence due to soil crusting. Cotton was grown in rotation with barley, which was grown between December and June, until soil salinities were sufficiently low (based on spatial uniformity of crop growth and yields) to include alfalfa in rotation.

In the Central Valley of California the climate is cooler than in the Imperial Valley (ETo 1400 mm; T_{max} 43°C; T_{min} −7°C), and rainfall (150–460 mm) occurs primarily during the winter months. Virgin soils were calcareous, containing a wide range of salinity, sodicity, gypsum, and boron, particularly along the west side of the valley. Some of the virgin soils were nonsaline-sodic (Kelly, 1951; Overstreet et al., 1955). Cropping during reclamation was a common practice in this area. Chemical amendments were used selectively, with reclamation being possible without them on many soils (Overstreet et al., 1955). Amendments such as gypsum and sulfuric acid (on calcareous soils) increased the rate of reclamation, with Overstreet et al. (1951, 1955) reporting field data that indicate sulfuric acid is superior to gypsum. However, recent laboratory studies indicate that gypsum, sulfates of aluminum and iron, and sulfuric acid are equally effective when applied to moderately sodic calcareous, silty clay soils (Miyamota and Enriquez, 1990).

Generally speaking, California farmers tend to practice reclamation over the long term. For example, annual or semiannual gypsum applications of 2 to 4 Mg/ha to an entire field are continued as long as crop growth is uneven and yields are low. Amendment applications to small areas within a field with continually poor plant growth are made by some farmers, though this is not a typical practice.

Barley, a winter crop, was usually the first crop grown on new ground during reclamation in the 1950s and 1960s, with border irrigation used to supplement annual rainfall. After one or more barley crops, cotton was often added to the rotation. Cotton fields were ripped before planting, amended with gypsum if necessary, listed to create furrows, and preirrigated. Large amounts of water (250–350 mm) infiltrated during preirrigation, resulting in considerable leaching and reclamation. The following quote summarizes recommendations provided by the University of California at the time:

> Field crops, particularly barley, wheat, sorghum, cotton, and sugar beets, are real tools for use in reclamation, or as transition crops to 'get acquainted' with soils and to get from 'where we are' to 'where we want to be' from the standpoint of soil salinity, sodium, and boron. By utilizing *more water* on these crops than is actually needed, salts, sodium, and boron can be leached beyond reach of roots, and the soils can be prepared for later plantings of more sensitive high-income crops. (Dean's Committee, University of California, 1968)

Reclamation using such methods required considerable time, since excess water necessary to reclaim the upper 5 ft of soil ranged from five feet for salinity to 15 ft for boron (Bingham et al., 1972). Additional water is still generally required to compensate for spatial variability of soils and associated infiltration rates (Jaynes and Hundsaker, 1989; Wichelns and Oster, 1990), but with continuing efforts success eventually occurred. In the San Joaquin Valley of California, vineyards planted in the 1970s now exist in fields where reclamation of saline-sodic soils began in the 1950s.

Reclamation by one or two large, annual irrigations which usually occur before planting or during the first crop irrigation is a form of intermittent reclamation that has also been successful when practiced over the long term. Farmers along the west side of the San Joaquin Valley of California use ripping tools to till soils to depths of 0.5–0.75 m. Following disking, land planing, and listing to prepare deep (0.3–0.4 m) furrows, 0.25–0.35 m of water infiltrates during the subsequent irrigation. Gypsum may be applied at rates of 2 to 4 Mg/ha either before or after tillage, depending on farmer experience with gypsum on a particular field or on soil analysis that indicates it would be beneficial. For cotton, the heaviest irrigation occurs one to three months before planting; then, several days before planting, the tops of the beds are removed (''decapped'') and the cotton is planted into the exposed wet soil. In saline fields, the depth of decapitation is increased to ensure that seeds are planted at a depth where the soil is less saline and wetter than if the decapitation were shallower.

2. *Use of Gypsum to Maintain Infiltration Rates*

Addition of gypsum to the soil surface (1 to 2 Mg/ha) or to irrigation water (3–5 mmolc/L) for the purpose of maintaining infiltration rates is also a common practice in California. Doneen (1948) reported that 270,000 Mg of gypsum were applied in the San Joaquin Valley to improve infiltration in 1945. The addition of gypsum to Friant-Kern irrigation water (EC < 0.2 dS/m) or to the nonsodic soils irrigated with it was a common practice in the 1950s on the east side of the valley between Fresno and Bakersfield (R. S. Ayers, personal communication, 1980), and such practices are still common today. In recent years, machines to inject gypsum slurries into the irrigation water have also become available, and their use is increasing.

Gypsum particle size is an important consideration for both surface applications to soil and injection through the irrigation water. For land application, a typical particle size of commercially available gypsum (92% pure) in California is 87% finer than 2.4 mm, 52% finer than 0.3 mm, and 25% finer than 0.07 mm. Finer material is needed for injection into irrigation water, with a typical particle-size distribution of 99–100% finer than 0.15 mm, 93–97% finer than 0.07 mm, and 3–78% finer than 0.04 mm.

B. Israel

1. *Sustained Irrigation with Saline-Sodic Groundwater*

In the western Negev region of Israel there is a saline aquifer with EC values ranging from 2.5 to 8.5 dS/m and SAR values of 15 to 26. The dominant soils are silty loams and the climate is Mediterranean, with winter rainfall ranging between 250 and 400 mm. Cotton is the dominant crop.

The effect of 16 years of irrigation with water from a well at Kibbutz Nahal-Oz (EC of 4.6 dS/m and SAR of 26) provides a typical example of ongoing reclamation practices (Keren et al., 1990). Irrigation during the summer (450 mm) results in ESP values in the upper 600 mm of soil of 20–26. There is no deterioration in soil hydraulic properties during the summer due to the high EC of the irrigation water. However, deterioration does occur during the rainy season due to the low salt concentrations of the rainwater. To offset this, phosphogypsum is spread annually on the soil surface, following tillage in the fall, at a rate of 5 Mg/ha. This prevents seal formation and maintains high infiltration rates which, in turn, provide sufficient infiltration of rainfall to leach salts from the root zone. Fall application of phosphogypsum and leaching during the rainy season, coupled with adequate irrigation with the saline-sodic water to meet crop needs during the summer months, has resulted in seed cotton yields averaging 5 Mg/ha between 1979 and 1988. These yields were similar to those obtained when only nonsaline-sodic water was used for irrigation.

C. India

1. Reclamation of Nonsaline-Sodic or Alkali Soils

In the Indo-Gangetic plains of India, most farmers begin reclamation in the monsoon season (July–September; 600–900-mm rainfall) by growing rice. Exchangeable sodium levels of the soil can be as high as 100%. Excess exchangeable sodium, high pH, lack of adequate zinc and calcium, and the resulting poor soil physical conditions and nutritional properties are the chief causes for poor productivity.

Recommended practices for this area (Mehta, 1985; Singh, 1985; Gupta and Abrol, 1990) include the following: (1) dividing fields into subplots of 0.4 ha each, bordered by 0.4-m-high dikes (bunds); (2) leveling, locating high and low spots in the subplots using a shallow irrigation, and releveling with a slope of 0.1% toward the drainage channel; (3) avoiding tillage while the soils are wet because of the poor physical properties of sodic soils; (4) applying finely powdered gypsum (100% finer than 2 mm, 75% finer than 1 mm, and 35% finer than 0.125 mm) at rates of about 10–15 Mg/ha, depending on soil sodicity and texture to reduce the ESP of the surface 150 mm of soil to 15–20; (5) incorporating the gypsum in the top 60–80 mm of soil by shallow tillage; (6) leaching for 15–20 days prior to transplanting rice, with recommended varieties including P2-21, IR 8, PR 106, and Basmati 370; (7) installing tubewells and utilizing the pumped water for irrigation, particularly where water tables are high; and (8) planting and transplanting three-leaf rice seedlings (35–45 days old) grown on reclaimed soil (rice seedlings are sensitive to sodicity at the 1–2 leaf stage).

If properly fertilized with nitrogen and zinc (Chhabra and Abrol, 1977; Singh, 1985; Gupta and Abrol, 1990), yields of dwarf rice varieties during reclamation can approach levels achieved in fully reclaimed soils when ESP values for the surface 150 mm are 60 or less (soil solution SAR values < 100). The intention of the initial amendment application is to reduce exchangeable sodium to less than 25–30% so that excellent yields of rice and moderate yields of wheat can be obtained even during the first year. Farmers often do not apply gypsum, but instead resort to prolonged leaching and accompanying application of farmyard manures. When this is done, rice yields are reduced, and rice may have to be grown for three to five years before wheat can be grown with even moderate yields. Reclamation is not considered to be complete until the upper 600 mm of soil are fully reclaimed, i.e., soil ESP or soil solution SAR levels less than 5. At this juncture, wheat yields approach maximum potentials, though the yields of more sensitive crops such as peas may still be reduced because of moderate sodicity levels below soil depths of 600 mm.

Diphlachne fusca or *Leptochloa fusca* (Karnal or Kallar grass) grows well under ponded conditions in saline-sodic soils (Kumar, 1985; Malik et al., 1986). Stem cuttings or root stolons are transplanted into flooded fields. This is a perennial plant and can be planted at any time of the year, but the best planting time is March, since growth is maximum during the summer. If irrigation is continued during the winter, one cutting can be obtained during winter months (as compared to three cuttings between spring and fall). Experiments conducted in Pakistan (Malik et al., 1986) indicate that nitrogen and phosphorus fertilizers have little effect on growth. Growing this grass for three to four years results in sufficient reclamation to grow rice and wheat, without any need for further reclamation practices; however, market demand for the grass is small, limiting the usefulness of this reclamation method.

VI. CONCLUDING COMMENTS

The major thrust of this chapter has been on the technical side of the reclamation of saline-sodic soils—the scientific basis for reclamation, the relative effectiveness of different reclamation practices, and examples of successful reclamation practices from different areas of the world. However, there are two additional major factors that lie outside of agronomy as traditionally practiced in the United States which cannot be ignored by agricultural scientists, farm advisors, consultants, and farmers.

Optimum reclamation practices depend on the soil problems requiring correction, the crops grown in the region, and the equipment available for soil tillage. In developed countries, mechanical power is available for deep tillage and other land preparation techniques (spreading and incorporation of chemical amendments, land leveling, preparation of high earthen dikes, etc.) required to prepare large areas of land (60–250 ha) so that large amounts of irrigation water can be applied and allowed to infiltrate. Cropping in conjunction with repeated tillage and heavy irrigation of saline soils or soils high in boron is a common practice, as is cropping of sodic soils where tillage is combined with application of amendments. Farmers in developing countries, who have only limited access to mechanical power, must rely more on crops like rice which can be grown in small fields under ponded conditions. Cropping in conjunction with reclamation is a popular method in both types of countries because it provides concurrent income; i.e., it is a "pay-as-you-go" option.

Reclamation also has significant environmental consequences: making the soil better in one place degrades soil and water resources somewhere else. Reclamation requires irrigation, drainage, and a place for salt disposal. Whether artificial tile drainage is installed or not, salts will be displaced downward into soil strata and groundwaters beneath the irrigated land. Extensive reclamation often results in shallow saline water tables which may require installation of tile drainage. The resulting saline drainage water generally increases the salinity of receiving surface waters. These negative environmental impacts of reclamation require purposeful and long-term planning and education. Is the eventual environmental degradation acceptable? Will the increased productivity compensate for the degradation? Reclamation and subsequent irrigation without short- and long-term resolutions of drainage-water-disposal issues becomes, in the long run, the bane of irrigated agriculture. Agriculturists working in such areas must know how to reclaim soils with minimum environmental impacts; how to provide the necessary information to local farmers and help them learn how to use it; how to tell others what the trade-offs are between food production and environmental consequences; and how to tell future generations what the economic and environmental tradeoffs are and will continue to be.

ACKNOWLEDGMENT

We thank Drs. Amrhein, Jurinak, and McNeal, who provided many excellent suggestions, and Jonathan Langford, as well, for editing and helping prepare the final manuscript.

REFERENCES

Abrol, I. P., and Bhumbla, D. R. (1979). Crop response to differential gypsum applications in a highly sodic soil and the tolerance of several crops to exchangeable sodium under field conditions. *Soil Sci. 127*: 79–85.

Abrol, I. P., Dahiya, I. S., and Bhumbla, D. R. (1975). On the method of determining gypsum requirement of soils. *Soil Sci. 120*: 30–36.

Abrol, I. P., Gupta, R. K., and Singh, S. B. (1979). Note on the solubility of gypsum and sodic soil reclamation. *J. Indian Soc. Soil Sci. 27*: 482–483.

Abu-Sharar, T. M., Bingham, F. T., and Rhoades, J. D. (1987). Stability of soil aggregates as affected by electrolyte concentration and composition. *Soil Sci. Soc. Am. J. 51*: 309–314.

Agassi, M., Morin, J., and Shainberg, I. (1981). Effect of electrolyte concentration and soil sodicity on infiltration rate and crust formation. *Soil Sci. Soc. Am. J. 45*: 848–851.

Alperovitch, N., Shainberg, I., and Keren, R. (1981). Specific effect of magnesium on the hydraulic conductivity of soils. *J. Soil Sci. 32*: 543–554.

Anonymous (1979). *Glossary of Soil Science Terms*. Soil Science Society of America, Madison, WI.

Ayers, R. S., and Westcot, D. W. (1985). Water quality for agriculture. FAO Irrigation and Drainage Paper 29 Rev. 1, Rome, Italy.

Bar-On, P., Shainberg, I., and Michaeli, I. (1970). The electrophoretic mobility of Na/Ca montmorillonite particles. *J. Colloid Interface Sci. 33*: 471–472.

Ben-Hur, M., Stern, R., van der Merwe, A. J., and Shainberg, I. (1992). Slope and gypsum effects on infiltration and erodibility of dispersive and nondispersive soils. *Soil Sci. Soc. Am. J. 56*: 1571–1576.

Bernstein, L. 1975. Effects of salinity and sodicity on plant growth. *Ann. Rev. Phytopath. 13*: 295–312.

Beyce, O. (1972). Experiences in the reclamation of saline and alkali soils and irrigation water qualities in Turkey. FAO Irrigation and Drainage Paper No. 16, pp. 63–84.

Bhumbla, D. R., and Abrol, I. P. (1979). Saline and sodic soils. In *Soils and Rice Symp.*, IRRI, Los Banos, Laguna, Philippines, pp. 719–738.

Bingham, F. T., Marsh, A. W., Branson, R., Mahler, R., and Ferry, G. (1972). Reclamation of salt-affected high-boron soils in western Kern County. *Hilgardia 41(8)*: 195–211.

Booth, F. E. M. (1983). Survey of economic plants for arid and semi-arid tropics (SEPASAT). Royal Botanical Gardens. Kew Richmond, Survey TW 9,3AB, England. SEPASAT DRAFT DOSSIER on *Leptochloa fusca*.

Boyle, M., Frankenberger, Jr., W. T., and Stolzy, L. H. (1989). The influence of organic matter on aggregation and water infiltration. *J. Prod. Agric. 2*: 290–299.

Bresler, E., McNeal, B. L., and Carter, D. L. (1982). *Saline and Sodic Soils: Principles-Dynamics-Modelling*. Springer-Verlag, Berlin, Heidelberg, New York.

Buyanovsky, G. A., and Wagner, G. H. (1983). Annual cycles of carbon dioxide level in soil air. *Soil Sci. Soc. Am. J. 47*: 1139–1145.

Carter, M. R., Webster, G. R., and Cairns, R. R. (1979). Calcium deficiency in some solonetzic soils of Alberta. *J. Soil Sci. 30*: 161–174.

Cass, A., and Sumner, M. E. (1982). Soil pore structural stability and irrigation water quality. II. Sodium stability data. *Soil Sci. Soc. Am. J. 46*: 507–512.

Chaudri, I. I., Shah, B. H., Naqvi, N., and Mallick, I. A. (1964). Investigation on the role of *Suaeda fruticosa forsk* in the reclamation of saline and alkaline soils in West Pakistan Plains. *Plant Soil 21*: 1–7.

Chhabra, R., and Abrol, I. P. (1977). Reclaiming effect of rice grown in sodic soils. *Soil Sci. 124*: 49–55.

Cramer, G. R., and Lauchli, A. (1986). Ion activities in solution in relation to Na^+-Ca^{2+} interactions at the plasmalemma. *J. Exp. Bot. 37*: 320–330.

Dean's Committee, University of California. (1968). Report 1. Agricultural development of new lands on the west side of the San Joaquin Valley: Land, crops, and economics. College of Agr. and Env. Sci. Davis, CA.

Deshpande, T. L., Greenland, D. J., and Quirk, J. P. (1968). Changes in soil properties associated with the removal of iron and aluminum oxides. *J. Soil Sci. 19*: 108–122.

Doneen, L. D. (1948). The quality of irrigation water and soil permeability. *Soil Sci. Soc. Am. Proc. 13*: 523–526.

Duley, F. L. (1939). Surface factors affecting the rate of intake of water by soils. *Soil Sci. Soc. Am. Proc. 4*: 60–64.

Durgin, P. B., and Chaney, J. F. G. (1984). Dispersion of kaolinite by dissolved organic matter from Douglas-fir roots. *Can. J. Soil Sci. 64*: 445–455.
Dutt, G. R., Terkeltoub, R. W., and Rauschkolb, R. S. (1972). Prediction of gypsum and leaching requirements for sodium-affected soils. *Soil Sci. 114*: 93–103.
Felhendler, R., Shainberg, I., and Frenkel, H. (1974). Dispersion and the hydraulic conductivity of soils in mixed solutions. Trans. 10th Int. Cong. *Soil Sci. 1*: 103–112.
Fireman, M., and Bodman, G. B. (1939). The effect of saline irrigation water upon the permeability and base status of soils. *Soil Sci. Soc. Am. Proc. 4*: 71–77.
Frenkel, H., Gerstl, Z., and Alperovitch, N. (1989). Exchange-induced dissolution of gypsum and the reclamation of sodic soils. *J. Soil Sci. 40*: 599–611.
Frenkel, H., Goertzen, J. O., and Rhoades, J. D. (1978). Effects of clay type and content, exchangeable sodium percentage, and electrolyte concentration on clay dispersion and soil hydraulic conductivity. *Soil Sci. Soc. Am. J. 42*: 32–39.
Goldberg, S., and Glaubig, R. A. (1987). Effect of saturating cation, pH, and aluminum and iron oxides on the flocculation of kaolinite and montmorillonite. *Clays Clay Miner. 35*: 220–227.
Grieve, C. M., and Fujiyama, H. (1987). The response of two rice cultivars to external Na/Ca ratio. *Plant Soil 103*: 245–250.
Grieve, C. M., and Maas, E. V. (1988). Differential effects of sodium/calcium ratio on sorghum genotypes. *Crop Sci. 28*: 659–665.
Gupta, R. K., and Abrol, I. P. (1990). Salt affected soils: Their reclamation and management for crop production. *Adv. Soil Sci. 11*: 223–288.
Gupta, R. K., Bhumbla, D. K., and Abrol, I. P. (1984). Effect of sodicity, pH, organic matter, and calcium carbonate on the dispersion behavior of soils. *Soil Sci. 137*: 245–251.
Gupta, R. K., Chhabra, R., and Abrol, I. P. (1981). The relationship between pH and exchangeable sodium in a sodic soil. *Soil Sci. 131*: 215–219.
Gupta, R. K., Singh, R. R., and Abrol, I. P. (1989). Influence of simultaneous changes in sodicity and pH on the hydraulic conductivity of an alkali soil under rice culture. *Soil Sci. 147*: 28–33.
Hoffman, G. J. (1986). Guidelines for reclamation of salt-affected soils. *Appl. Agr. Res. 1*: 65–72.
Hulsbos, W. C., and Boumans, J. H. (1960). Leaching of saline soils in Iraq. *Netherlands J. Agri. Sci. 8(1)*: 1–10.
Jaynes, D., and Hundsaker, D. J. (1989). Spatial and temporal variability of water content and infiltration on a flood-irrigated field. *Trans. ASAE 26*: 1422–1429.
Jury, W. A., Jarrell, W. W., and Devitt, D. (1979). Reclamation of saline-sodic soils by leaching. *Soil Sci. Soc. Am. J. 43*: 1100–1106.
Kazman, Z., Shainberg, I., and Gal, M. (1983). Effect of low levels of exchangeable Na and applied phosphogypsum on the infiltration rate of various soils. *Soil Sci. 135*: 184–192.
Keisling, T. C., Rao, P. S. C., and Jessup, R. E. (1978). Gypsum dissolution and sodic soil reclamation as affected by water flow velocity. *Soil Sci. Soc. Am. J. 46*: 726–732.
Kelley, W. P. (1951). *Alkali Soils: Their Formation, Properties, and Reclamation.* Van Nostrand Reinhold, New York.

Keren, R. (1990). Reclamation of saline-, sodic-, and boron-affected soils. In *Agricultural Salinity Assessment and Management*, ASCE Manuals and Reports on Engineering Practices No. 71 (K. K. Tanji, ed.), ASCE, New York, pp. 410–431.

Keren, R. (1991). Specific effect of magnesium on soil erosion and water infiltration. *Soil Sci. Soc. Am. J. 55*: 783–787.

Keren, R., and Bingham, F. T. (1985). Boron in water, soils, and plants. In *Advances in Soil Science* (B. A. Stewart, ed.), Springer-Verlag, New York, Vol. 1, pp. 229–276.

Keren, R., Sadan, D., Frenkel, H., Shainberg, I., and Meiri, A. (1990). Irrigation with sodic and brackish water and its effect on the soil and on cotton yields: A summary of 15 years. *Hassadeh 70*: 1822–1829.

Khosla, B. K., Gupta, P. R. and Abrol, I. P. (1979). Salt leaching and the effect of gypsum application in a saline-sodic soil. *Agri. Water Management 2*: 193–202.

Kingsbury, R. W., and Epstein, E. (1986). Salt sensitivity in wheat. A case for specific ion toxicity. *Plant Physiol. 80*: 651–654.

Kumar, A. (1985). Karnal grass for reclaiming alkali soils. Tech. Bull. No. 5, Central Soil Salinity Research Inst., Karnal, India.

Kumar, A., and Abrol, I. P. (1986). Grasses in alkali soils. Tech. Bull. No. 11, Central Soil Salinity Research Inst., Karnal, India.

Leffelaar, P. A., and Sharma, R. P. (1977). Leaching of a highly saline-sodic soil. *J. Hydrol. 32*: 203–218.

Longenecker, D. E., and Lyerly, P. J. (1974). Control of soluble salts in farming and gardening. Texas Agr. Exp. Sta. Bull. 876.

Maas, E. V., and Grattan, S. R. (1995). Crop yields as affected by salinity. ASA Drainage Monograph (J. Van Schilfgaarde and W. Skaggs, eds.), in press.

Maas, E. V., and Grieve, C. M. (1987). Sodium-induced calcium deficiency in salt-stressed corn. *Plant Cell Environ. 10*. 559–564.

Malik, K. A., Aslam, Z., and Naqvi, M. (1986). Kallar grass: A plant for saline land. Nuclear Institute for Agriculture and Biology, Faisalabad, Pakistan.

McNeal, B. L., and Coleman, N. T. (1966). Effect of solution composition on soil hydraulic conductivity. *Soil Sci. Soc. Am. Proc. 30*: 308–312.

McNeal, B. L., Layfield, D. A., Norvell, W. A., and Rhoades, J. D. (1968). Factors influencing hydraulic conductivity of soils in the presence of mixed salt solutions. *Soil Sci. Soc. Am. Proc. 32*: 187–190.

McNeal, B. L., Norvell, W. A., and Coleman, N. T. (1966). Effect of solution composition on the swelling of extracted clay soils. *Soil Sci. Soc. Am. J. 30*: 313–317.

McNeal, B. L., Pearson, G. A., Hatcher, J. T., and Bower, C. A. (1966). Effect of rice culture on the reclamation of sodic soils. *Agron. J. 58*: 238–240.

Mehta, K. K. (1985). Alkali soils: Steps for reclamation. Better Farming in Salt Affected Soils, Brochure #2, Central Soil Salinity Research Institute, Karnal, India.

Miller, R. J., Nielson, D. R., and Biggar, J. W. (1965). Chloride displacement in Panoche clay loam in relation to water movement and distribution. *Water Resour. Res. 1*: 63–73.

Minhas, P. S., and Khosla, B. K. (1986). Solute displacement in a silt loam soil as affected by the method of water application under different evaporation rates. *Agric. Water Management 12*: 63–75.

Miyamota, S., and Enriquez, C. (1990). Comparative effects of chemical amendments on salt and Na leaching. *Irrig. Sci. 11*: 83–92.

Miyamoto, S., Prather, R. J., and Stroehlein, J. L. (1975). Sulfuric acid and leaching requirements for reclaiming sodium-affected calcareous soils. *Plant Soil 43*: 573.

Morin, J., and Benyamini, Y. (1977). Rainfall infiltration into bare soils. *Water Resour. Res. 13*: 813–817.

Muhammed, S., McNeal, B. L., Bower, C. A., and Pratt, P. F. (1969). Modification of the high-salt water method for reclaiming sodic soils. *Soil Sci. 108*: 249–256.

Norlyn, J. D., and Epstein, E. (1984). Variability in salt tolerance of four triticale lines at germination and emergence. *Crop Sci. 24*: 1090–1092.

Oster, J. D. (1993). Sodic soil reclamation. In *Towards the Rational Use of High Salinity Tolerant Plants*, Vol. 1 (H. Leith and A. Al Massom, eds.), Kluwer, pp. 485–490.

Oster, J. D. (1982). Gypsum usage in irrigated agriculture: A review. *Fertilizer Res. 3*: 73–89.

Oster, J. D., and Frenkel, H. (1980). The chemistry of the reclamation of sodic soils with gypsum and lime. *Soil Sci. Soc. Am. J. 44*: 41–45.

Oster, J. D., and Halvorson, A. D. (1978). Saline seep chemistry. In *Proc., Dryland-Saline-Seep Control, Subcommission on Salt Affected Soils*, 11th Int. Soil Sci. Cong., Edmonton, Canada.

Oster, J. D., and Rhoades, J. D. (1975). Calculated drainage water compositions and salt burdens resulting from irrigation with river waters in the western United States. *J. Environ. Qual. 4*: 73–79.

Oster, J. D., and Schroer, F. W. (1979). Infiltration as influenced by irrigation water quality. *Soil Sci. Soc. Am. J. 43*: 444–447.

Oster, J. D., Shainberg, F. W., and Wood, J. D. (1980). Flocculation value and gel structure of Na/Ca montmorillonite and illite suspensions. *Soil Sci. Soc. Am. J. 44*: 955–959.

Oster, J. D., Willardson, L. S., and Hoffman, G. J. (1972). Sprinkling and ponding techniques for reclaiming saline soils. *Trans. ASAE 15*: 115–117.

Overstreet, R., Martin, J. C., and King, H. M. (1951). Gypsum, sulfur, and sulfuric acid for reclaiming an alkali soil of the Fresno series. *Hilgardia 21*: 113–127.

Overstreet, R., Martin, J. C., Schulz, R. K., and McCutcheon, O. D. (1955). Reclamation of an alkali soil of the Hacienda series. *Hilgardia 24(3)*: 53–68.

Pratt, P. F., and Suarez, D. L. (1990). Irrigation water quality assessments. In *Agricultural Salinity Assessment and Management* (K. K. Tanji, ed.), ASCE Manuals and Reports on Engineering Practice No. 71, ASCE, New York, pp. 220–236.

Prichard, T. L., Hoffman, G. J., and Oster, J. D. (1985). Reclamation of saline organic soil. *Irrig. Sci. 6*: 211–220.

Quirk, J. P. (1968). Particle interaction and soil swelling. *Israel J. Chem. 6*: 213–234.

Quirk, J. P. (1986). Soil permeability in relation to sodicity and salinity. *Phil. Trans. R. Soc. Lond. A 316*: 297–317.

Quirk, J. P., and Aylmore, L. A. G. (1971). Domain and quasi-crystalline regions in clay systems. *Soil Sci. Soc. Am. J. 35*: 652–654.

Quirk, J. P., and Schofield, R. K. (1955). The effect of electrolyte concentration on soil permeability. *J. Soil Sci. 6*: 163–178.

Rasmussen, W. W., and McNeal, B. L. (1973). Predicting optimum depth of profile modification by deep plowing for improving saline-sodic soils. *Soil Sci. Soc. Am. Proc. 37*: 432–437.

Reeve, R. C., Allison, L. E., and Peterson, D. F. (1948). Reclamation of saline-alkali soils by leaching. Utah Agr. Exp. Sta. Bull. 335.

Reeve, R. C., and Bower, C. A. (1960). Use of high-salt waters as a flocculant and source of divalent cations for reclaiming sodic soils. *Soil Sci. 90*: 139–144.

Reeve, R. C., and Doering, E. J. (1966). The high-salt-water dilution method for reclaiming sodic soils. *Soil Sci. Soc. Am. Proc. 30*: 498–504.

Reeve, R. C., Pillsbury, A. F., and Wilcox, L. V. (1955). Reclamation of saline and high boron soil in the Coachella Valley of California. *Hilgardia 24*: 69–91.

Rengasamy, P., Greene, R. S. B., Ford, G. W., and Mehanni, A. H. (1984). Identification of dispersive behavior and the management of red brown earths. *Aust. J. Soil Res. 22*: 413–432.

Rhoades, J. D., and Ingvalson, R. D. (1969). Macroscopic swelling and hydraulic conductivity properties of four vermicullitic soils. *Soil Sci. Soc. Am. J. Proc. 33*: 364–367.

Rhoades, J. D., Ingvalson, R. D., and Hatcher, J. T. (1970). Laboratory determination of leachable soil boron. *Soil Sci. Soc. Am. Proc. 34*: 938–941.

Rhoades, J. D., Krueger, D. B., and Reed, J. J. (1968). The effect of soil-mineral weathering on the sodium hazard of irrigation waters. *Soil Sci. Soc. Am. Proc. 32*: 643–647.

Rhoades, J. D., and Loveday, J. (1990). Salinity in irrigated agriculture. In *Irrigation of Agricultural Crops* (B. A. Stewart and D. R. Nielsen, eds.), Agronomy 30-1089–1142.

Robbins, C. W. (1986). Sodic calcareous soil reclamation as affected by different amendments and crops. *Agron. J. 78*: 916–920.

Robbins, C. W., Wagenet, R. J., and Jurinak, J. J. (1980). A combined salt transport-chemical equilibrium model for calcareous and gypsiferous soils. *Soil Sci. Soc. Am. J. 44*: 1191–1194.

Rowell, D. L., Payne, D., and Ahmad, N. (1967). The effect of the concentration and movement of solutions on the swelling, dispersion and movement of clay in saline and alkali soils. *J. Soil Sci. 20*: 176–188.

Russo, D., and Bresler, E. (1977). Analysis of saturated and unsaturated hydraulic conductivity in mixed sodium and calcium soil systems. *Soil Sci. Soc. Am. J. 41*: 706–710.

Schofield, R. K., and Samson, H. R. (1954). Flocculation of kaolinite due to the attraction of oppositely charged crystal faces. *Faraday Disc. Chem. Soc. 18*: 135–145.

Shainberg, I., and Letey, J. (1984). Response of soils to sodic and saline conditions. *Hilgardia 52*: 1–57.

Shanmuganathan, R. T., and Oades, J. M. (1983). Influence of anions on dispersion and physical properties of the A horizon of a red-brown earth. *Geoderma 29*: 257–277.

Singh, K. N. (1985). Cultural practices for growing rice in alkali soils. *Better Farming in Salt Affected Soils*, Brochure #9, Central Soil Salinity Research Institute, Karnal, 132001 India.

Sposito, G. (1984). *The Surface Chemistry of Soils*. Oxford University Press, New York.

Stern, R., Ben-Hur, M., and Shainberg, I. (1991). Clay mineralogy effect on rain infiltration, seal formation and soil losses. *Soil Sci. 152*: 455–468.

Suarez, D. L., Rhoades, J. D., Savado, R., and Grieve, C. M. (1984). Effect of pH on saturated hydraulic conductivity and soil dispersion. *Soil Sci. Soc. Am. J. 48*: 50–55.

Sumner, M. E. (1993). Sodic soils: New perspectives. *Aust. J. Soil Res. 31*: 683–750.

Talsma, T. (1967). Leaching of tile-drained saline soils. *Aust. J. Soil Res. 5*: 37–46.

Tanji, K. K., Doneen, L. D., Ferry, G. V., and Ayers, R. S. (1972). Computer simulation analysis on reclamation of salt-affected soils in San Joaquin Valley, California. *Soil Sci. Soc. Am. Proc. 36*: 127–133.

U.S. Salinity Laboratory Staff. (1954). *Diagnosis and Improvement of Saline and Alkali Soils*. U.S. Government Printing Office, Washington, DC.

Wichelns, D., and Oster, J. D. (1990). Potential economic returns to improved irrigation infiltration uniformity. *Agric. Water Management 18*: 253–256.

Xiao, Z., Prendergast, B., and Rengasamy, P. (1992). Effect of irrigation water on soil hydraulic conductivity. *Pedosphere 2*: 237–244.

Yeo, A. R., and Flowers, T. J. (1985). The absence of an effect of the Na/Ca ratio on sodium chloride uptake by rice (*Oryza sativa* L.). *New Phytol. 99*: 81–90.

15

Reclamation of Sodic-Affected Soils

Rami Keren
Institute of Soils and Water, The Volcani Center, Agricultural Research Organization, Bet-Dagan, Israel

I. INTRODUCTION

Under arid or semiarid conditions and in regions of poor natural drainage, there is a real hazard of salts accumulation in soils. The processes by which soluble salts enter the soil solution and cause salinity and sodicity include the application of waters containing salts, weathering of primary and secondary minerals in soils, and organic matter decay. The importance of each source depends on the type of soil, the climate conditions, and the agricultural management.

Accumulation of dispersive cations such as Na in the soil solution and the exchange phase affect the soil physical properties such as structural stability, hydraulic conductivity, infiltration rate, and soil erosion. Sodic soils conditions increase the erodibility and reduce the productivity of soils. Reclamation of such soils could stabilize soils and free agriculture from natural hazards (surface crusting, soil erosion, low hydraulic conductivity, poor aeration, seedling emergence, and root development). For sodic soils, reclamation generally proceeds by increasing Ca ions on the absorbed phase at the expense of Na and Mg ions. The replaced Na and Mg are removed either below the root zone or out of the profile by leaching water. Thus, reclamation requires a certain flow of water through the soil profile, and, to be effective, an appropriate profile hydraulic conductivity (HC) must be achieved. The end result of reclamation must be a decrease in soil erodibility and a sufficient and stable soil profile porosity which provides a favorable physical environment for plants.

II. EVALUATING THE NEED FOR RECLAMATION

Many soils in arid to humid regions have unstable structures which make them difficult to manage, owing to their tendency to swell and disperse. Since soil permeability decreases with the square of the pore radius, a small reduction in size of the large pores due to swelling and clay movement has a large effect on soil permeability for water and gases.

The favorable effect of exchangeable Ca and the deleterious effect of exchangeable Na on soil swelling and dispersion is well known. Quirk and Schofield (1955) combined the exchangeable sodium percentage (ESP) and electrolyte concentration of the percolating solution parameters in their "threshold concentration" concept, which was defined as the electrolyte concentration which caused a certain decrease in soil permeability at a given ESP value. The dependence of soil HC on the total electrolyte concentration and sodium adsorption ratio (SAR) of the percolating solution was assessed by McNeal and Coleman (1966) and McNeal et al. (1968). Felhendler et al. (1974), Frenkel et al. (1978), Keren and Singer (1988), and Pupisky and Shainberg (1979) had also showed that soils responded differently to the same combination of electrolyte concentration and ESP. A combination of salt concentration and ESP at which a 25% reduction in hydraulic conductivity occurred is given in Fig. 1 for several soils. Thus, a unique threshold concentration exists for each soil.

The extent of swelling and dispersion of clays depends on the clay mineralogy, the composition of the adsorbed ions and the salt concentration in solution (Shainberg et al., 1971; Arora and Coleman, 1979; Low, 1980; Oster et al., 1980; Chiang et al., 1987; Goldberg and Glaubig, 1987; Keren and Singer, 1988) Yousaf et al. (1987) concluded that the electrolyte concentration–SAR threshold values for HC and clay dispersion of arid soils are closely related. However, whether clay will leave the system, or move and seal the soil depends on the extent of swelling (Keren and Singer, 1988). Clay swelling is a continuous process that increases gradually with decreases in solution electrolyte concentration, whereas clay dispersion is possible only at solution electrolyte concentration below the flocculation value (FV). The degree of clay swelling before replacing the solution with one that has electrolyte concentration below the FV of the clay determines whether clay will move in the conducting pores or whether clay particles will be trapped in the narrow pores (sieving effect) and the HC of the soil will further decrease.

Rengasamy et al. (1984) proposed a scheme for classifying soils in respect to sodicity. They divided the soils into three classes.

Class I dispersive soils are soils which disperse spontaneously. These soils will have severe problems associated with crusting, reduced porosity even under zero tillage. These soils are nonsaline but sodic (SAR >3.0).

Class 2 potentially dispersive soils are soils which disperse after mechanical shaking, e.g., by intensive cultivation or raindrop impact. The electrolyte con-

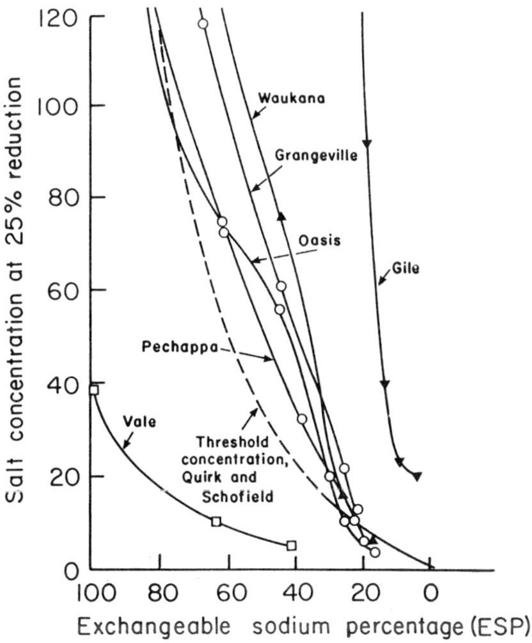

Fig. 1 Combination of salt concentrations and ESP at which a 25% reduction in hydraulic conductivity occurred. (From McNeal and Coleman, 1966.)

centration required to keep these soils stable varies with their SAR and mineralogy. Soils with SAR <3 and very low electrolyte concentration (below the flocculation value of the clay) and soils with SAR >3 but with electrolyte concentration exceeding the flocculation value of the clays belong to this class.

Class 3 flocculated soils are soils which do not disperse even after mechanical shaking.

The HC of a soil can be maintained even at high ESP levels provided that the electrolyte concentration of the percolating solution is above a threshold value. Conversely, when a solution with very low electrolyte concentration is used, even an ESP 5 caused a two-orders-of-magnitude decrease in the HC (McIntyre, 1979). Thus, in arid and semiarid regions where irrigation with saline water is practiced and soils are saline and sodic, the deleterious effect of the exchangeable Na is more evident during the irrigation with high-quality water (or rain) than that during the irrigation with saline water. This HC reduction with the decrease in electrolyte concentration was much smaller as the soil water content decreased (Russo and Bresler, 1977a,b).

At any particular ESP level, the influence of exchangeable Na is greater with exchangeable Mg as the complementary ion than it is with Ca (van der Merwe and Burger, 1969; Emerson and Bakker, 1973; Keren, 1990, 1991). In well-weathered soils, which do not contain $CaCO_3$, a specific effect of exchangeable Mg was observed, for both HC and clay dispersion, when Na/Mg soils are leached with distilled water.

An important soil property which determines the HC of soils leached with waters of very low salinity is the soils' potential for releasing electrolytes. Calcareous soils and soils containing primary minerals such as feldspar are examples of soils which release electrolytes into solution readily in the presence of exchangeable Na. The electrolyte concentration in soil solution is maintained above the FV in these soils, so clay does not disperse and the HC change is limited. In calcareous soils, therefore, exchangeable Mg does not have a direct effect on the HC and clay dispersion (Alperovitch et al., 1981). However, the presence of exchangeable Mg enhances the dissolution of $CaCO_3$, and the electrolytes prevent clay dispersion and HC reduction in the Na/Mg calcareous soils.

When soils are exposed to rainwater, susceptibility to low levels of ESP is enhanced (Oster and Schroer, 1979). Water infiltration is more sensitive to ESP than the HC of the soil profile for three reasons: (1) the mechanical impact of the water drops, (2) the absence of the soil matrix which slows clay movement, and (3) concentration of electrolytes in the surface soil solution is determined solely by the composition of the applied water, because dissolution of $CaCO_3$ and primary minerals is too slow to affect the surface solution concentration. Thus, when water with low electrolyte concentration is applied (rainwater or snowwater), salt concentration in the soil-surface solution remains low even for calcareous soils, and clay dispersion is possible only if the salt concentration is below the FV of the clay. On the other hand, soil solution concentration deeper in the soil profile is affected by $CaCO_3$ and primary mineral dissolution. A concentration above 3 molc m^{-3} might be enough to prevent clay dispersion at low ESP values.

As water wets the soil surface, cohesive forces between the soil particles are reduced and the waterdrop impact breaks the soil aggregates. The low electrolyte concentration in rainfall causes clay dispersion and increases aggregate breakdown at the soil surface. The soil loss by erosion is significantly affected by absorbed Na (Singer et al., 1982; Shainberg and Letey, 1984). The higher the ESP the higher is runoff and erosion (Fig. 2).

Studies imply that also Mg is a deleterious ion in some circumstances. Adsorbed Mg has a specific effect on soil erosion and infiltration for montmorillonitic noncalcareous and calcareous soils (Keren, 1989, 1990), and the erosion rate for Na/Mg soil is greater than the rate for Na/Ca soil (Fig. 3).

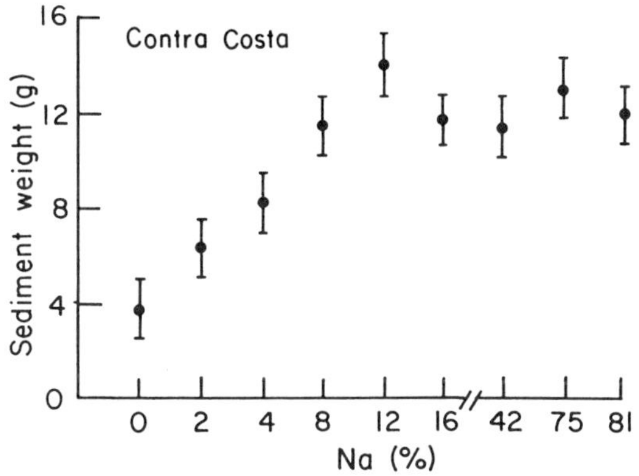

Fig. 2 Soil detachment as a function of ESP for the Contra Costa soil. (From Singer et al., 1982.)

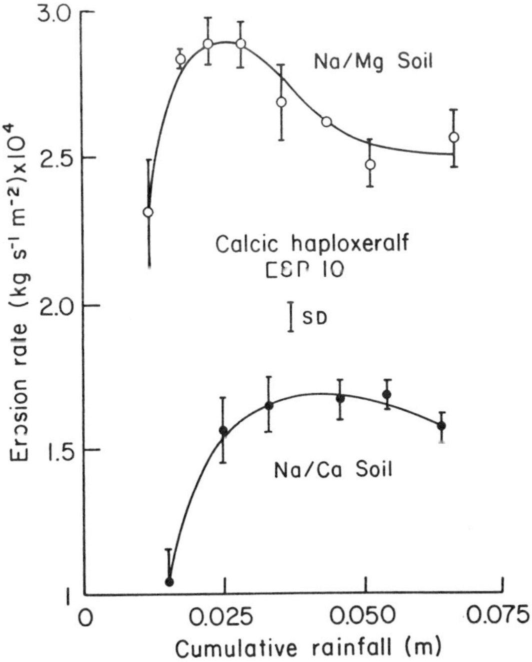

Fig. 3 Erosion rate of a Calcic Haploxeralf soil adsorbed with Na-Mg and Na-Ca ions at ESP 10, as a function of cumulative rainfall at a waterdrop kinetic energy of 12.5 kJ m^{-3}. (From Keren, 1991.)

It is obvious, therefore, that any attempt to assess sodicity hazard in soils, the soil texture, the clay mineralogy, the exchangeable sodium percentage and the total electrolyte concentration of percolating solution should be considered.

III. RECLAMATION OF SODIC SOILS

In order to reduce runoff and erosion in sodic soils, reclamation is a common practice. Reclamation of sodic soils involves replacement of exchangeable Na by Ca. The source of Ca for replacing adsorbed Na may be the soil itself, involving the dissolution of Ca-containing minerals, or external sources such as gypsum, calcium chloride, or irrigation water containing Ca ions.

Although chemical amendments may be effective in restoring reduced water penetration caused by either excessive sodium or too low salt concentration in the applied water, they will not be effective if low permeability is caused by soil texture, compaction, or water-restricting layers. Since reclamation requires a certain flow of water through the soil profile, an appropriate infiltration rate (IR) and hydraulic conductivity must be achieved.

In soils having stable surface structures, decreases in IR result from the inevitable decrease in the metric suction gradient which occurs as infiltration proceeds. Decreases in soil IR from an initially high rate can also result from gradual deterioration of soil structure and the formation of a surface crust. When a crust of very low HC is formed, its reduced permeability determines the IR of the soil, and the wetting front depth has only a slight effect on the infiltration rate (Morin et al., 1989). Preventing crust formation and reducing runoff could be achieved by improving soil structure and aggregate stability of the soil by using soil amendments such as gypsum and organic polymers.

Swelling and dispersion of soil surface may be prevented by spreading gypsum (or another readily available electrolyte source) at the soil surface (Keren and Shainberg, 1981). The main mechanism by which gypsum affects the IR of soils exposed to rainwater is by its dissolution and release of electrolyte into the soil solution. The dissolution rate of gypsum is important because of the short contact time between rainwater and gypsum particles at the soil surface.

Gypsum source, application rate, and fragment size all have their effect on crust formation and IR (Fig. 4). The IR without gypsum decreased sharply as the cumulative rainfall increased, reaching a steady-state value of 2 mm h^{-1}. Conversely, with 3.4 t ha^{-1} of powdered phosphpgypsum (PG, a by-product of the phosphorus industry) and mined gypsum on the soil surface, the steady-state IR values were 7.5 and 5.5 mm h^{-1}, respectively. When coarse fragments (4–5.7 mm) of mined gypsum were used, the steady-state IR value was about 2 mm h^{-1} (similar to control), independent of the amount of gypsum applied (3.4 or 6.8 t ha^{-1}). Conversely, coarse fragments of PG were effective in pre-

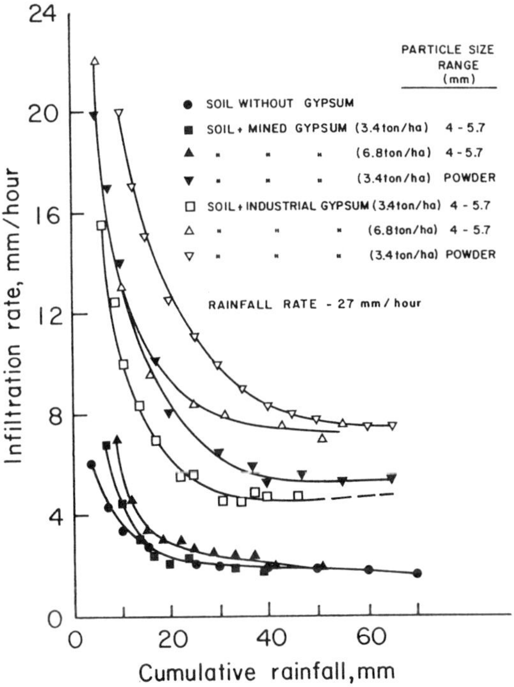

Fig. 4 The effect of industrial and mined gypsum (at amount and particle sizes indicated) on infiltration rate of loess soil as a function of the cumulative rainfall. (From Keren and Shainberg, 1981.)

venting the drop in IR of the soil, and their effectiveness increased with an increase in the amount of gypsum applied. The higher efficiency of PG in maintaining high IR is attributed to its high rate of dissolution (Keren and Shainberg, 1981).

Surface runoff from a deep-silt-loam loess soil (containing 20% of $CaCO_3$) at two ESP values (4.6 and 19.3) and the beneficial effect of PG were studied in field plots under natural rainfall conditions (Keren et al., 1983). Application of PG was found to be very effective in increasing IR (Table 1). Spreading PG over the soil was more effective than mixing it in the upper 10 cm of soil. Although the PG affects the ESP of the soil in the upper 0–15 cm layer very little (Table 2), a large effect was found in IR. The presence of gypsum on the soil surface increases soil stability by increasing electrolyte concentration in the percolated solution and by cation-exchange effect. The soil with the lower sodicity was more responsive to the gypsum treatment because low electrolyte concentration is enough to prevent aggregate disintegration and clay dispersion,

Table 1 Surface Runoff, During Rainstorms, from Loess Soil at Two Levels of ESP as Affected by Gypsum Treatments

			Surface runoff, % of rainfall				
			ESP 4.6		ESP 19.3		
Storm number	Amount of rainfall (mm)	Time between storms (d)	Control	Gypsum spread over[a]	Control	Gypsum mixed in[a]	Gypsum spread over[a]
1	16	—	0	0	23.1	2.6	0
2	20	40	6.5	0	13.3	3.0	0.7
3	19	20	1.3	0.5	27.3	10.5	3.8
4	59	6	21.5	3.8	41.0	21.2	17.2
5	44	9	13.4	2.1	45.0	22.5	12.4
6	12	1	12.5	2.5	40.0	25.0	12.5
7	12	15	10.0	2.5	31.7	13.8	6.7
Total (mm)	182		22.8	3.8	64.2	30.1	18.7
Percent of annual rainfall			12.5	2.1	35.3	16.5	10.3

[a]The rates of phosphogypsum application were 5 and 10 t/ha for the soils with ESP 4.6 and 19.3, respectively.
Source: Keren et al. (1983).

Table 2 Exchangeable Na Percentage[a] in the Profile of the Sodic Loess Soil as Affected by Gypsum Treatments

Treatment	Soil layer depth (cm)				
	0–15	15–30	30–45	45–60	60–90
1. Control[b]	19.3 ± 4.8	19.0 ± 4.9	16.4 ± 5.1	16.4 ± 4.8	14.2 ± 3.9
	16.9 ± 2.9	19.4 ± 4.0	19.0 ± 2.8	19.3 ± 3.4	19.6 ± 4.0
2. Gypsum	17.4 ± 5.0	20.1 ± 4.6	20.3 ± 6.0	20.8 ± 6.0	21.3 ± 6.7
mixed in	12.8 ± 5.0	17.7 ± 3.6	19.6 ± 3.3	17.4 ± 2.9	18.0 ± 2.3
3. Gypsum	16.9 ± 5.8	17.0 ± 5.8	19.3 ± 6.0	20.9 ± 4.5	21.1 ± 4.5
spread over	11.5 ± 4.1	16.3 ± 3.6	19.4 ± 3.5	19.3 ± 7.0	20.4 ± 4.3
CEC cmol (Na^+) kg^{-1} soil	18.2 ± 1.5	17.8 ± 1.3	18.5 ± 2.0	18.9 ± 2.2	18.1 ± 1.1

[a]Average values with standard deviation (12 and 43 repetitions for ESP and CEC, respectively).
[b]Top value is for Autumn 1981 (after irrigation with 400 mm of saline water, before gypsum application), and bottom number is for Spring 1982 (after the rainy season).
Source: Keren et al. (1983).

whereas soils with high sodicity may erode even in solutions with moderate concentration.

Small rates of organic polymers (20–40 kg ha^{-1}), mainly polyacrylamide (PAM) and polysaccharides, for improving the structure of soil surface can also be used (Helalia and Letey, 1988; Ben-Hur and Letey, 1989; Shainberg et al., 1990). In the case of anionic PAM, the presence of electrolytes which ensured the flocculation of the soil clay is essential for the effectiveness of PAM application (Shaviv et al., 1985; Shainberg et al., 1990).

A. Soil Amendments

Reclamation of sodic soils involves replacement of exchangeable Na by Ca. The typical source of Ca is an amendment that either contains soluble Ca or dissolves Ca upon reaction in the soil. Amendments that provide soluble Ca include gypsum, lime, and $CaCl_2$. Common amendments that produce Ca in calcareous soils by enhancing the conversion of $CaCO_3$ to the more soluble $CaSO_4$ include sulfuric acid, sulfur, and iron and aluminum sulfates. Because of their relatively low cost and proven effectiveness, gypsum, sulfur, and sulfuric acid are the most common amendments for reclamation.

1. Gypsum

Gypsum is the most commonly used amendment for sodic soil reclamation and for reducing the harmful effects of high-sodium irrigation waters because of its solubility, low cost, and availability. Gypsum added to a sodic soil can cause permeability changes by increasing electrolyte concentration and by cation-exchange effects (Loveday, 1976; Keren and Shainberg, 1981; Keren et al., 1983). In order to obtain permanent improvement, however, the cation-exchange effect is the important one. Potentially achievable electrolyte concentration, when the soil solution saturated in respect to gypsum, ranges from 15 to 133 mol m^{-3} as the ESP of the soil varies from 0 to 40 (Oster, 1982).

Water penetration and storage benefits due to gypsum have been described for a variety of soils throughout the world. In a ponding experiment on a brown sodic clay soil, McIntyre et al. (1982) found that without gypsum, 292 mm of water infiltrated in 379 days of ponding, wetting the profile to 2.1 m, while with 10 t ha^{-1} of gypsum, 605 mm of water infiltrated in 145 days, enough of which passed beyond 2 m to raise the groundwater level. The increased leaching of soluble salts, resulting from the extra profile drainage after gypsum application, has been documented on many occasions. McIntyre et al. (1982) also found that the depth of leaching of Cl^-, beyond which accumulation occurred, was 1.0 m when gypsum was absent and 2.8 m in the presence of it. Data from this experiment clearly demonstrated the presence of a region of low HC in the upper part of the profile which was ameliorated by the addition of gypsum.

Gypsum added to a sodic soil can initiate permeability increases due to both electrolyte concentration and cation-exchange effects (Loveday, 1976). The relative significance effect of the electrolyte and the cation exchange on HC was estimated by comparing the effects of gypsum and $CaCl_2$ in equivalent amounts on the HC of some soils from Israel (Fig. 5) (Shainberg et al., 1982). The chemically stable noncalcareous (Golan) soil was very sensitive to the type of Ca amendment added. Complete sealing of the soil eventually took place in the $CaCl_2$ treatments, but high hydraulic conductivity was maintained in the gypsum treatment. Since Na replacement was similar for both amendments (when applied in equivalent amounts), the differences were attributed to an electrolyte effect. The soil, which does not have the potential to release appreciable amounts of additional salt, is sensitive to even low concentrations of Na on the exchange complex. Thus, sustained release of electrolytes by gypsum particles is essential if one is to maintain high hydraulic conductivity. Conversely, for the calcareous (Nahal-Oz) soil, no difference between the two amendments was observed. The $CaCO_3$ in the Nahal-Oz soil dissolves readily during leaching with distilled water, so the release of electrolytes from gypsum particles is not beneficial (Sec. III.A.3).

The relative significance of the two effects is of interest for several reasons. If the electrolyte effect is sufficiently great to prevent dispersion and swelling of soil clays, the surface application of gypsum may be worthwhile. In this case the amount of gypsum required depends on the amount of high-quality water applied and the rate of gypsum dissolution. It is somewhat independent of the amount of exchangeable Na in the soil profile. Conversely, in soils where the electrolyte concentration effect is insignificant and the main effect results from cation exchange (high ESP level), then the amount of gypsum required depends on the amount of exchangeable Na in a selected depth of soil.

The amount of exchangeable sodium to be replaced during reclamation depends on the initial exchangeable sodium fraction ($E_{\mathrm{Na\,i}}$), the cation-exchange capacity (CEC, $\mathrm{mol_c\ Mg^{-1}}$), soil bulk density (ρ_b, $\mathrm{Mg\ m^{-3}}$), the desired final exchangeable sodium fraction ($E_{\mathrm{Na\,f}}$), and the depth of soil to be reclaimed (L, m). Once the above parameters have been determined, the amount of exchangeable Na to be replaced per unit of land area (Q_{Na}, $\mathrm{mol_c\ ha^{-1}}$) can be calculated from

$$Q_{\mathrm{Na}} = 10^4 L \rho_b (\mathrm{CEC})(E_{\mathrm{Na\,i}} - E_{\mathrm{Na\,f}}) \qquad (1)$$

The $E_{\mathrm{Na\,f}}$ value depends on the response of the soil in terms of its physical conditions.

The amount of gypsum needed to reclaim a sodic soil—the ''gypsum requirement'' (Richards, 1954)—can be calculated from

$$\mathrm{GR} = (8.61 \times 10^{-5}) Q_{\mathrm{Na}} \qquad (2)$$

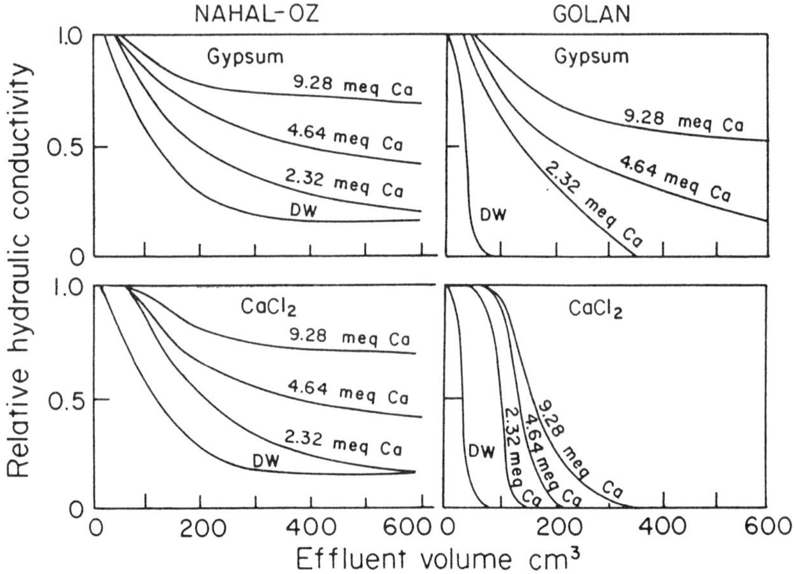

Fig. 5 The effect of gypsum and calcium chloride on the hydraulic conductivity of Nahal-Oz and Golan soils. (From Shainberg et al., 1982.)

The efficiency and rate of exchange, namely, the percentage of applied Ca which exchanges for adsorbed Na, varies with ESP, being much greater at high ESP values (Chaudhry and Warkentin, 1968). Removal of Na at ESP levels below 10 is slow, and part of the applied Ca displaces exchangeable Mg so that the efficiency declines to about 30% (Loveday, 1976; Greene and Ford, 1983). Efficiency may also be low (20–40%) in fine-textured soils, because of the slowness of exchange of Na inside the structural elements (Marin et al., 1982).

Generally, soil-water penetration is too low to allow reclamation of sodic soils to be attained in a single leaching. For example, a 50-cm depth of water applied for leaching can only dissolve about 12 Mg/ha of gypsum (reaching a saturation solution in respect to gypsum). Larger gypsum applications, therefore, will not be effective unless soil-water penetration is sufficient for larger water applications. Thus, sodic soil normally can only be reclaimed to a shallow depth the first year (assuming 500 mm of rainfall a year), but this will often permit a shallow-rooted crop to be grown following leaching. Subsequently, annual amendment applications and leaching volumes can be applied to reclaim the entire profile over a period of years.

In evaluating sodic soil reclamation, equilibrium chemistry and piston movement of soil solution are usually assumed (Oster and Frenkel, 1980). The soil solution in the gypsum layer, however, reaches equilibrium (in the presence of excess gypsum) only when the contact time between an elemental volume of solution and gypsum fragment surface is sufficiently long or when the surface area of the gypsum fragments per volume of water is sufficiently large. During irrigation or rainfall, the contact time can be short or the surface area of the gypsum fragments per unit volume of water can be small (low gypsum content in soil or large gypsum fragment size); thus the percolated solution can be undersaturated with respect to gypsum. The benefit of gypsum for reclaiming sodic soils depends not only on the infiltration characteristics of the soil but also on the gypsum dissolution properties. Some of the factors that influence the dissolution rate are the surface area of gypsum fragments (Fig. 4), soil-water velocity during leaching, and the electrolyte composition of the soil solution (Kemper et al., 1975; Keren and O'Connor, 1982).

The time-dependent dissolution of gypsum is

$$\frac{dC}{dt} = K(C_s - C) \tag{3}$$

where

dC/dt = net rate of dissolution
K = dissolution coefficient
C_s, C = solution concentration at saturation and at a particular time, respectively

Equation (3) is a first-order kinetic model with the term $C_s - C$ being the chemical concentration gradient. In accordance with Kemper et al. (1975), Keren and O'Connor (1982) concluded that the rate of dissolution of gypsum is strongly influenced by both gypsum content and soil-water velocity. Increasing solution velocity increased the dissolution rate coefficient but decreased the contact time between gypsum and the flowing solution; the net effect was a decreasing dissolution rate with increasing soil-water velocity.

Integrating Eq. (3) yields

$$-\ln\left(1 - \frac{C}{C_s}\right) = Kt \tag{4}$$

and a plot of $\ln(1 - C/C_s)$ versus t should yield a straight line.

The time t was calculated from Eq. (6). Since the thickness of the film around the gypsum fragments is changing with the soil-water velocity, the dissolution rate coefficient for a given surface of gypsum fragment is also changing. Thus, the lines obtained from Eq. (4) are not linear under such conditions. Keren and

O'Connor (1982) reported that the left side of Eq. (4) is empirically related linearly to the square root of time (t_c),

$$-\ln\left(1 - \frac{C}{C_s}\right) = \alpha t_c^{1/2} + \beta \tag{5}$$

where α and β are the slope and intercept, respectively, and t_c is the time for an increment of solution to leave a given soil depth:

$$t_c = \frac{L}{V} \tag{6}$$

where

L = length of layer of soil-gypsum mixture
V = soil-water velocity ($V = v/\theta$, where v = soil-water flux and θ = soil porosity)

Combining Eqs. (5) and (4) gives

$$K = \frac{1}{t_c}(\alpha t_c^{1/2} + \beta) \tag{7}$$

Introducing Eq. (6) in (7) yields

$$K = \left(\frac{V}{L}\right)^{1/2}\left[\alpha + \left(\frac{V}{L}\right)^{1/2}\beta\right] \tag{8}$$

The dissolution rate coefficient as a function of water flow velocity can be evaluated for a given surface area of gypsum. Since the permeability of a sodic soil is low, the dissolution rate is relatively high during the initial stage of reclamation. During the reclamation process, however, the permeability increases and the dissolution rate decreases. Thus, the soil regulates the gypsum dissolution rate during reclamation. Where it is possible to control the water flow velocity in soils (e.g., sprinkler irrigation), reduced velocities which give increased contact time between water and gypsum fragments are to be preferred for increased rate of gypsum dissolution and greater efficiency of exchange (Keren and O'Connor, 1982).

Surface area of gypsum fragments has also been invoked to account for differences in infiltration rates of soils resulting from mined and PG (Keren and Shainberg, 1980). The latter, with a bulk density half of the former, dissolved 10 times as fast. Another factor affecting dissolution rate is the coating of $CaCO_3$ on gypsum fragments in calcareous soils (Keren and Kauschansky, 1981).

In estimating the amount of gypsum to be applied, little attention has been directed to the possibility that incorporating a large amount of gypsum powder into the soil may cause temporary reduction in HC. The influence of gypsum fragments and content on the HC of soils before the gypsum has dissolved was studied for loamy soil and sandy loam soil (Keren et al., 1980). Because of the limited solubility of gypsum, excess small particles of gypsum may block the conducting pores of the soil and reducing its HC in the early stage of the dissolution process. It was concluded that the effect of small particles of gypsum in diminishing the HC of soils should be considered when recommendations are made based on the sodicity of the soil, using large amounts of gypsum as an amendment (higher than 2% of the soil by weight).

Tanji and Deveral (1984) described representative modeling approaches for the reclamation of sodic soils. The models include many of the known chemical reactions occurring in natural soils as a result of $CaCO_3$ and $CaSO_4$ dissolution, salt leaching, and exchange reactions. These models were subdivided into chromatographic and miscible displacement models. The models of Dutt and Terkeltoub (1972) and Tanji et al. (1972) are examples of chromatographic models. Both models assume that local equilibrium occurs between the plates, for chemical processes within the solution phase, between the solution and exchanger phases, and between the solution and soil mineral phases. The models compute monovalent-divalent exchange using the Gapon and/or Davis equations. Solubilities of gypsum and $CaCO_3$ are considered in the models. The models of Dutt and Terkeltoub (1972) and Tanji et al. (1972) have been field-tested with data from sodic soil reclamation leaching plots. The differences between measured and computed results were found to be no more than the horizontal variations typically found in salt-affected lands. Both models provide powerful tools for quantitative predictions of water and gypsum required to reclaim soil profiles to prescribed levels of salinity and ESP. However, assumptions must be made about the desired level of reclamation, such as the level to which ESP must be reduced, and the depth of soil required by the ESP reduction. The soil surface, being very susceptible to sodicity, may tolerate only very low ESP (<3), whereas at depths below the cultivated layer, ESP >10 may be tolerable.

2. *Acids and Sulfur*

These amendments are of value for ameliorating soils containing $CaCO_3$ with which the acid interacts to form gypsum (using H_2SO_4) or calcium chloride (using HCl). Sulfur requires an initial phase of microbiological oxidation to produce H_2SO_4.

In some areas, the availability of acid waste products arising from mining and industrial activities is increasing markedly. Use of these waste products as soil amendments may provide a safe means of disposal (Miyamoto et al., 1975b), and could greatly influence the economics of any amendment proposal.

Being highly corrosive, neither H_2SO_4 nor SO_2 should be added to water which flows through metal or concrete irrigation systems. Miyamoto et al.

(1975a) claim that the concentrated acid, sprinkled directly onto the soil surface, has advantages in better distribution, less destruction of soil aggregates, and more efficient leaching of salts. Another technique which has been used commercially (Tisdale, 1970) is direct application by chiseling into the soil in bands about 45 cm apart.

For elemental S the dust problem has been a serious disadvantage, but this can be overcome by using conventional fluid fertilizer equipment to apply water suspensions containing 55% to 60% S (Thorup, 1972).

Regarding the effectiveness of acidulents, some inconsistent results have been reported from time to time, but generally H_2SO_4 has been the most effective of them. In studies involving packed soil columns Yahia et al. (1975) and Prather et al. (1978) found H_2SO_4 to be more effective than gypsum in increasing water penetration into calcareous sodic soils. The former found that the rate of water penetration increased with increasing rates of acid up to about 15 t ha^{-1}, but thereafter decreased. The latter showed that for equivalent leachate volumes, H_2SO_4 treatment removed much greater quantities of exchangeable Na than did gypsum.

With regard to the amounts of H_2SO_4 and leaching water required, the experimental results of Miyamoto et al. (1975a) show that the same depth of water (as of soil to be treated), equivalent to at least two soil pore volumes, leaches 70% to 80% of the soluble salts originally in the soil together with those formed by the acid treatment. The amount of acid required to obtain a desired ESP may be calculated from equations using CEC and concentrations of Na, Ca, and Mg in the saturation extract made with the water to be used for leaching (Miyamoto et al., 1975a).

Other amendments such as sulfur, pyrites, and polysulfides, which must first be oxidized by soil microorganisms, will be slower acting than H_2SO_4. Their effectiveness in field experiments has been variable, perhaps being related to the presence or absence of appropriate microbial populations.

3. *Calcium Chloride and Calcium Carbonate*

Although $CaCl_2$ is generally too expensive to compete with other amendments, there are situations where it is available as an industrial waste product and could therefore be considered for reclamation. Because of its solubility it gives, initially, high electrolyte levels and high water intake rates, which make it a more efficient amendment than gypsum for high ESP soils (Alperovitch and Shainberg, 1973; Prather et al., 1978). With time, however, the slow dissolution of gypsum can be more significant than $CaCl_2$ in maintaining the electrolyte effect, especially for soils which do not contain minerals able to release electrolytes into the soil solution (Shainberg et al., 1982). Prather et al. (1978) have suggested that for high-ESP soils with low water intake rates, a combination of

$CaCl_2$ and gypsum might give quicker and more effective reclamation than gypsum alone.

Successful leaching reclamation without amendments has frequently been demonstrated (Jury et al., 1979). Reclamation under such conditions takes place when the drainage through the soil profile is good, an adequate leaching water is present, and a soil source of Ca exists (e.g., calcium carbonate). The rate at which $CaCO_3$ dissolution in water approaches equilibrium is dependent upon a number of factors, which include the surface area–solution volume ratio, the ionic composition of the solution, the ion composition of the adsorbed phase, affinity of the clay minerals to cations, the temperature, and the local partial pressure of CO_2. Amrhein et al. (1985) concluded that in soils the kinetics of $CaCO_3$ dissolution is not a simple diffusion-controlled or first-order reaction. In these calcareous systems, the transfer of atmospheric CO_2 to solution was an important rate-limiting step in the kinetics of dissolution.

$CaCO_3$ particles (isolated from the soils) in water give ion activity product (IAP) values which were expected for calcite (Suarez and Rhoades, 1982). On the contrary, a supersaturation of calcite was observed in solution extracts from calcareous soils (Levy, 1981). The supersaturation of calcite in soil is related to the presence of silicates in the soil, which is more soluble than calcite, and it is not related to the unstable $CaCO_3$ phases (Suarez and Rhoades, 1982). Plummer et al. (1979) concluded that there are several uncertainties in modeling the kinetics of carbonate chemistry: at low pH the rate depends significantly on the thermodynamic transport constant for H^+ which is not well defined, and reaction site density and controls on pH at the surface interface between the crystal and the bulk solution are also not well understood. Moreover, kinetics of precipitation and dissolution of $CaCO_3$ have not yet been modeled for soil-water systems.

Soil $CaCO_3$ may be dissolved slowly to contribute Ca, especially in the reclamation of saline sodic soils in which its solubility is enhanced (Oster, 1982). However, it has generally been considered of doubtful value to add $CaCO_3$ to nonsaline sodic soils because its dissolution rate is too slow to provide much Ca for exchange, unless an acid or acid former is applied concurrently. Sodic soils containing $CaCO_3$ are common in the arid and semiarid regions of the world. The beneficial effect of $CaCO_3$ in preventing HC reduction of sodic soil (low ESP levels) exposed to rainwater is its potential for dissolving and maintaining the soil-solution concentration levels above the flocculation value of the soil clays (Alperovitch et al., 1981) and reducing ESP. However, the beneficial effect of $CaCO_3$ in preventing clay dispersion may particularly manifest itself at the soil surface. Calcareous soils with moderate ESP levels maintain reasonable physical properties through most of the profile but remain susceptible to dispersion near the surface. This is because the soil solution electrolyte concentration near the soil surface may be insufficient to maintain physical structure during raindrop impact. Under such conditions an application of soil amendment on the soil surface is necessary to keep the infiltration rate sufficiently high.

B. Successive Dilutions of High-Salt Water

The benefit of an amendment for reclaiming saline-sodic soils depends primarily on the infiltration characteristics of the soil. Successive dilutions of saline water may be applied when (1) the soil physical conditions have deteriorated and HC of the soil is so low that the time required for reclamation or the amount of amendment required is excessive, or (2) if a sodic soil is to be leached with a water so low in salinity that water infiltration decreases adversely. The method involves application of successive dilutions of a high-salt water containing divalent cations (Reeve and Bower, 1960). This principle was later successfully tested in experiments, and a monogram was furnished for predicting the reclamation at various stages of leaching (Reeve and Doering 1966b). In the early phase, the high salinity of the water applied prevents clay dispersion and induces flocculation of the soil colloids. Simultaneously, the Ca ions in the perculating water decrease the sodicity level by exchange reaction. On dilution with high-quality water, the SAR of the irrigation water is reduced by the square root of the dilution factor. To ensure exchange equilibrium to the desired depth in each step of dilution, the total equivalent surface depth of applied water for reclamation should be about nine times the depth of soil to be reclaimed (Reeve and Doering, 1966a,b). The method is particularly effective for soils with the expanding clay type that have extremely low HC. A theoretical analysis of the use of high-salt water in reclaiming sodic soils by the addition of a constant quantity of Ca^{2+} in every step in successive dilution was reported by Misopolinos (1985).

Intermittent ponding required less water than continuous ponding to achieve the same degree of leaching (Miller et al., 1965), and the sprinkling irrigation system is more efficient than other methods at removing salt from small pores in the soil profile (Nielsen et al., 1966b).

In a field study with highly sodic clay loam soil, Reeve and Doering (1966b) found that the IR for saturation gypsum solution was only 7.2×10^{-2} mm h^{-1}. With such a low IR, it was estimated that seven years of continuous leaching would be required to reclaim the profile to a depth of 0.90 m. However, the profile was reclaimed in three days with a series of $CaCl_2$ solutions, starting with a concentration of 300 mol m^{-3}, and following with solutions of 150, 75, 38, 19, and 6 mol m^{-3}. The average IR for the $CaCl_2$ solutions was about 150 times the IR for the gypsum solution. This method was used successfully to reclaim a sodic soil under humid environment, where HC and IR were increased from 30% to over 100% (Rahman et al., 1974).

The successful application of this method requires a balance between keeping the electrolyte concentration high to reduce reclamation time and keeping the electrolyte concentration low to reduce the amount of amendment required. A practical technique is to apply only two-thirds of the solution depth required for exchange equilibria with three dilutions of $CaCl_2$, for example, starting with a solution of 250 mol m^{-3} $CaCl_2$ and following with 100 and 50 mol m^{-3}. This is followed by a gypsum application to satisfy the remaining exchange requirement, and is completed by leaching with about a 0.3-m depth of water per meter depth of soil to be reclaimed. This final step is essential to leach the saline solutions below the root zone.

REFERENCES

Alperovitch, N., and Shainberg, I. (1973). Reclamation of alkali soils with $CaCl_2$ solutions. In *Physical Aspects of Soil, Water and Salts in Ecosystems* (A. Hadas et al., eds.), Springer, Berlin, pp. 431–440.

Alperovitch, N., Shainberg, I., and Keren, R. (1981). Specific effect of magnesium and hydraulic conductivity of sodic soil. *Clays Clay Miner. 33*: 443–450.

Amrhein, C., Jurinak, J. J., and Moore, W. M. (1985). Kinetics of calcite dissolution as affected by carbon dioxide partial pressure. *Soil Sci. Soc. Am. J. 49*: 1393–1398.

Arora, H. W., and Coleman, N. T. (1979). The influence of electrolytes concentration on flocculation of clay suspensions. *Soil Sci. 127*: 134–138.

Ben-Hur, M., and Letey, J. (1989). Effect of polysaccharides, clay dispersion and impact energy on water infiltration. *Soil Sci. Soc. Am. J. 53*: 233–238.

Chaudhry, G. H., and Warkentin, B. P. (1968). Studies on exchange of sodium from soils by leaching with calcium sulfate. *Soil Sci. 105*: 190–197.

Chiang, S. C., Radcliffe, D. E., Miller, W. P., and Newman, K. D. (1987). Hydraulic conductivity of three southeastern soils as affected by sodium, electrolyte concentration and pH. *Soil Sci. Soc. Am. J. 51*: 1293–1299.

Dutt, G. R., and Terkeltoub, R. W. (1972). Prediction of gypsum and leaching requirements for sodium-affected soil. *Soil Sci. 114*: 93–103.

Emerson, W. W., and Bakker, A. C. (1973). The comparative effects of exchangeable calcium, magnesium and sodium on some physical properties of red-brown earth subsoils. The spontaneous dispersion of aggregates in water. *Aust. J. Soil Res. 11*: 151–157.

Felhendler, R., Shainberg, I., and Frenkel, H. (1974). Dispersion and hydraulic conductivity of soils in mixed solution. *Trans. 10th Int. Cong. of Soil Sci.*, Moscow, I, pp. 103–112.

Frenkel, H., Goertzen, J. O., and Rhoades, J. D. (1978). Effect of clay type and content, exchangeable sodium percentage, and electrolyte concentration on clay dispersion and soil hydraulic conductivity. *Soil Sci. Soc. Am. J. 42*: 32–39.

Goldberg, S., and Glaubing, R. A. (1987). Effect of saturating cation, pH and aluminum and iron oxide on the flocculation of kaolinite and montmorillonite. *Clays Clay Miner. 35*: 220–227.

Greene, R. S. B., and Ford, G. W. (1985). The effect of gypsum on cation exchange in two red duplex soils. *Aust. J. Soil Res. 23*: 61–74.

Helalia, A. M., and Letey, J. (1988). Cationic polymer effects on infiltration rates with a rainfall simulator. *Soil Sci. Soc. Am. J. 52*: 247–250.

Hillel, D., and Gardner, W. R. (1970). Transient infiltration into crust topped profiles. *Soil Sci. 109*: 69–70.

Jury, W. A., Jarrell, W. M., and Devitt, D. (1979). Reclamation of saline-sodic soils by leaching. *Soil Sci. Soc. Am. J. 43*: 1100–1106.

Kemper, W. D., Olsen, J., and DeMooy, C. J. (1975). Dissolution rate of gypsum in flowing water. *Soil Sci. Soc. Am. J. 39*: 458–463.

Keren, R. (1989). Water-drop kinetic energy effect on water infiltration in calcium and magnesium soils. *Soil Sci. Soc. Am. J. 53*: 1624–1628.

Keren, R. (1990). Water-drop kinetic energy effect on infiltration in sodium-calcium-magnesium soils. *Soil Sci. Soc. Am. J. 54*: 983–987.

Keren, R. (1991). Specific effect of magnesium on soil erosion and water infiltration. *Soil Sci. Soc. Am. J. 55*: 783–787.

Keren, R., and Kauschansky, P. (1981). Coating of calcium carbonate on gypsum particle surfaces. *Soil Sci. Soc. Am. J. 45*: 1242–1244.

Keren, R., Kreit, J. F., and Shainberg, I. (1980). Influence of size of gypsum particles on the hydraulic conductivity of soils. *Soil Sci. 130*: 113–117.

Keren, R., and O'Connor, G. A. (1982). Gypsum dissolution and sodic soil reclamation as affected by water flow velocity. *Soil Sci. Soc. Am. J. 46*: 726–732.

Keren, R., and Shainberg, I. (1981). The efficiency of industrial and mined gypsum in reclamation of a sodic soil-rate of dissolution. *Soil Sci. Soc. Am. J. 45*: 103–107.

Keren, R., Shainberg, I., Frenkel, H., and Kalo, Y. (1983). The effect of exchangeable sodium and gypsum on surface runoff from loess soil. *Soil Sci. Soc. Am. J. 47*: 1001-1004.

Keren, R., and Singer, M. J. (1988). Effect of low electrolyte concentration on hydraulic conductivity of Na/Ca-montmorillonite-sand system. *Soil Sci. Soc. Am. J. 52*: 368–373.

Levy, R. (1981). Effect of dissolution of alumosilicates and carbonates on ionic activity products of calcium carbonate in soil extracts. *Soil Sci. Soc. Am. J. 45*: 250–255.

Loveday, J. (1976). Relative significance of electrolyte and cation exchange effects when gypsum is applied to a sodic clay soil. *Aust. J. Soil Res. 14*: 361–371.

Low, P. F. (1980). The swelling of clay II: Montmorillonites. *Soil Sci. Soc. Am. J. 44*: 667–676.

Manin, M., Pissarra, A., and Van Hoorn, J. W. (1982). Drainage and desalinization of heavy clay soil in Portugal. *Agric. Water Manag. 5*: 227–240.

McIntyre, D. S. (1979). Exchangeable sodium, subplasticity and hydraulic conductivity of some Australian soils. *Aust. J. Soil Res. 17*: 115–120.

McIntyre, D. S., Loveday, J., and Watson, C. L. (1982). Field studies of water and salt movement in an irrigated swelling clay soil. I: Infiltration during ponding. II: Profile hydrology during ponding. III: Salt movement during ponding. *Aust. J. Soil Res. 20*: 81–90, 91–99, 101–105.

McNeal, B. L., and Coleman, N. T. (1966). Effect of solution composition on soil hydraulic conductivity. *Soil Sci. Soc. Am. Proc. 30*: 308–312.

McNeal, B. L., Layfield, D. A., Norvell, W. A., and Rhoades, J. D. (1968). Factors influencing hydraulic conductivity of soils in the presence of mixed salt solutions. *Soil Sci. Soc. Am. J. 32*: 187–190.

Miller, R. J., Biggar, J. W., and Nielsen, D. R. (1965). Chloride displacement in Panoche clay loam in relation to water movement and distribution. *Water Resour. Res. 1*: 63–73.

Misopolinos, N. D. (1985). A new concept for reclaiming sodic soils with high-salt water. *Soil Sci. 140*: 69–74.

Miyamoto, S., Prather, R. J., and Stroelhein, J. L. (1975a). Sulfuric acid and leaching requirements for reclaiming sodium-affected calcareous soils. *Plant Soil 3*: 573–585.

Miyamoto, S., Ryan, J., and Stroehlein, J. L. (1975b). Potentially beneficial uses of sulfuric acid in south-western agriculture. *J. Environ. Qual. 4*: 431–437.

Morin, J., Keren, R., Benyamini, Y., Ben-Hur, M., and Shainberg, I. (1989). Water infiltration as affected by soil crust and moisture profile. *Soil Sci. 148*: 53–59.

Nielsen, D. R., Biggar, J. W., and Luthin, J. N. (1966). Desalinization of soils under controlled unsaturated flow conditions. *6th Int. Congr. Comm. on Irrig. and Drainage*, New Delhi. pp. 19.15–19.24.

Oster, J. D. (1982). Gypsum usage in irrigated agriculture: A review. *Fertil. Res. 3*: 73–89.

Oster, J. D., and Frenkel, H. (1980). The chemistry of the reclamation of sodic soils with gypsum and lime. *Soil Sci. Soc. Am. J. 44*: 41–45.

Oster, J. D., and Schroer, F. W. (1979). Infiltration as influenced by irrigation water quality. *Soil Sci. Soc. Am. J. 43*: 444–447.

Oster, J. D., Shainberg, I., and Wood, J. D. (1980). Flocculation value and gel structure of Na/Ca montmorillonite and illite suspension. *Soil Sci. Soc. Am. J. 44*: 955–959.

Plummer, L. N., Parkhurst, D. L., and Wigley, T. M. L. (1979). Critical review of the kinetics of calcite dissolution and precipitation. In *Chemical Modeling in Aqueous Systems* (E. A. Jenne, ed.), *Am. Chem. Soc. Symp. Ser. 93*: 537–573.

Prather, R. J., Goertzen, J. O., Rhoades, J. D., and Frenkel, H. (1978). Efficient amendment use in sodic soil reclamation. *Soil Sci. Soc. Am. J. 42*: 782–786.

Pupisky, H., and Shainberg, I. (1979). Salt effects on the hydraulic conductivity of sandy soil. *Soil Sci. Soc. Am. J. 43*: 429–433.

Quirk, J. P., and Schofield, R. K. (1955). The effect of electrolyte concentration on soil permeability. *J. Soil Sci. 6*: 163–178.

Rahman, M. A., Hiler, E. A., and Runkles, J. R. (1974). High electrolyte water for reclaiming slowly permeable soils. *Trans. ASAE 17*: 129–133.

Reeve, R. E., and Bower, C. A. (1960). Use of high-salt waters as a flocculant and source of divalent cations for reclaiming sodic soils. *Soil Sci. 90*: 139–144.

Reeve, R. C., and Doering, E. J. (1966a). The high salt-water dilution method for reclaiming sodic soils. *Soil Sci. Soc. Am. Proc. 39*: 498–504.

Reeve, R. C., and Doering, E. J. (1966b). Field comparison of the high salt-water dilution method and conventional methods for reclaiming sodic soils. *6th Int. Comm. on Irrig. and Drainage*, New Delhi, R. I. Questions 19, pp. 19.1–19.14.

Rengasamy, P., Greene, R. S. B., Ford, G. W., and Mehanni, A. H. (1984). Identification of dispersive behavior and the management of real brown earths. *Aust. J. Soil Res. 22*: 413–432.

Richards, L. A. (1954). Diagnosis and improvement of saline and alkali soils. *U.S. Department of Agriculture, Handbook 60.*

Russo, D., and Bresler, E. (1977a). Analysis of the saturated-unsaturated hydraulic conductivity in a mixed Na/Ca soil system. *Soil Sci. Soc. Am. J. 41*: 706–710.

Russo, D., and Bresler, E. (1977b). Effect of mixed Na/Ca solutions on the hydraulic properties of unsaturated soils. *Soil Sci. Soc. Am. J. 41*: 713–717.

Shainberg, I., Bresler, E., and Klausner, Y. (1971). Studies on Na/Ca montmorillonite systems. I: The swelling pressure. *Soil Sci. 111*: 214–219.

Shainberg, I., Keren, R., and Frenkel, H. (1982). Response of sodic soils to gypsum and calcium chloride application. *Soil Sci. Soc. Am. J. 46*: 113–117.

Shainberg, I., and Letey, J. (1984). Response of soils to sodic and saline conditions. *Hilgardia 52*: 1–57.

Shainberg, I., Warrington, D., and Regasamay. (1990). Effect of soil conditioners and gypsum application on rain infiltration and erosion. *J. Soil Sci. 149*: 301–307.

Shaviv, A., Ravina, I., and Zaslavsky, D. (1985). Surface application of anionic surface conditioners to reduce crust formation. Assessment of soil surface sealing and crusting. Proc. Symp. Gent, Belgium, pp. 286–293.

Singer, M. J., Janitzky, P., and Blackard, J. (1982). The influence of exchangeable sodium percentage on soil erodibility. *Soil Sci. Soc. Am. J. 46*: 117–121.

Suarez, D. L., and Rhoades, J. D. (1982). The apparent solubility of calcium carbonate in soils. *Soil Sci. Soc. Am. J. 46*: 716–722.

Tanji, K. K., and Deveral, S. J. (1984). Simulation modeling for reclamation of sodic soils. In *Soil Salinity under Irrigation* (Shainberg and Shalhevet, eds.), Springer-Verlag, New York, pp. 238–251.

Tanji, K. K., Doneen, L. D., Ferry, G. V., and Ayers, R. S. (1972). Computer simulation analysis on reclamation of salt-affected soil in San Joaquin Valley, California. *Soil Sci. Soc. Am. J. 36*: 127–133.

Thorup, J. T. (1972). Soil sulphur application. *Sulphur Inst. J. 8*: 16–17.

Tisdale, S. L. (1970). The use of sulphur compounds in irrigated arid-land agriculture. *Sulphur Inst. J. 6*: 2–7.

Van der Merwe, A. J., and Burger, R. du Toit. (1969). The influence of exchangeable cations on certain physical properties of a saline-alkali soil. *Agrochemophysica 1*: 63–66.

Yahia, T. A., Miyamoto, S., and Stroelhein, J. L. (1975). Effect of surface applied sulfuric acid on water penetration into dry calcareous and sodic soils. *Soil Sci. Soc. Am. J. 39*: 1201–1204.

Yousaf, M., Ali, O. M., and Rhoades, J. P. (1987). Clay dispersion and hydraulic conductivity of some salt-affected arid land soils. *Soil Sci. Soc. Am. J. 51*: 905–907.

16

Impact of Crop Rotation and Land Management on Soil Erosion and Rehabilitation

Jacob Amir
Gilat Experiment Station, Agricultural Research Organization, Negev, Israel

I. INTRODUCTION

It is generally accepted that soil erosion and water runoff affect the water and nutrient availability for crops; thus management that alleviates these constraints will have a profound impact on crop production (Pierce, 1991). Crop rotation (C.R.), universally used as a major tool in soil conservation, is defined as a management system in which crops are grown in sequence on the same field and repeated cyclically. It may include a short cycle of two phases (e.g., C-S (corn-soybean)) or more (e.g., C-S-W (corn-soybean-wheat)). A three-phase C.R. can occur in three different sequences: C-S-W, W-C,S, or S-C-W or more.

In modern agriculture, short-term economic considerations generally dictate the management practices to be implemented, and these are sometimes contrary to soil conservation requirements. However, a sustainable crop production system must not only be suited to the prevailing socioeconomic conditions but should also ensure the preservation of the natural resources.

The practices of C.R., ley farming, and no-tillage management are examples of such an approach in modern agriculture. Besides the need for preventive conservation measures to control erosion and runoff, technologies are needed to improve productivity after the soil has been damaged. Two of the key crop management practices used today to prevent or reduce erosion were evolved in ancient times. Roman farmers discovered the value of legumes and an alternate fallow-crop rotation to increase or stabilize yields (Cato, 234 B.C.; Varo, 116 B.C.). The beneficial effect of legumes in crop production was discovered ca. 2838 B.C. in China, and they are still used today (Hymowitz, 1970). Where these ancient innovations have been ignored, the results have been a loss in productivity and soil degradation. This was especially so in the early cultivation of new lands in America, Australia, Asia, and Africa (Bennett, 1955; Lal and Stewart, 1990; McWilliam, 1990).

Soil rehabilitation following severe soil erosion in humid and semiarid regions is less complicated than in arid marginal regions since it is possible to incorporate beneficial plants into the C.R. system. The problem is more complex in arid marginal regions because the option of using C.R. for rehabilitation and prevention of soil erosion is virtually nonexistent. Under conditions of severe water limitation, runoff and wind erosion can be serious due to the scarcity of natural vegetation cover and low-organic-matter content in arid soils (Stewart et al., 1991).

Government legislation is crucial in preserving soil and water resources. The United States agricultural policy, formulated in 1985, is an excellent example of soil conservation legislation introduced to withdraw marginal lands from cultivation and to promote permanent grass coverage of the soil (Scaling, 1990). Such a policy is appropriate for developed countries, but not necessarily so for developing countries, where the only alternative is to develop inexpensive technologies to cope with their soil and water conservation problems.

The purpose of this chapter is to review the state of knowledge concerning the impact of the C.R. system on prevention and alleviation of soil erosion in agriculture.

II. GEOGRAPHIC AND SOCIOECONOMIC CONSTRAINTS ON CROP ROTATION

Cropping sequences and rotations intended to prevent soil deterioration and promote soil rehabilitation are dependent on geographical environment. The major environmental factor involved is the region's water availability, which almost completely dictates the crop and its management selection. However, other climatic, edapic, and socioeconomic factors are also involved and have to be taken into account.

One conventional method of classifying the cropping system is dividing it into extensive (low-input) and intensive (high-input) agriculture. Intensive agriculture practiced in high-rainfall or irrigated regions allows full use of the range of C.R. options. Extensive agriculture is practiced in lower rainfall and nonirrigated regions where C.R. options are minimal. In the marginal semiarid and arid areas the options are fallow-crop rotations, continuous cropping, and cropping with grazing of aftermaths and grazing of natural pasture (McWilliam, 1990; Amir et al., 1991a).

In modern agriculture, price fluctuation in the world market seriously affects the choice of cropping system. For example, relative prices of wool and grain have a strong effect on the C.R. system in Australia (McWilliam, 1990). Highly subsidized cereal production in Europe has replaced conventional C.R. with cereal monoculture (Hanley and Ridgman, 1978; Power and Follett, 1987). Where there is high demand for straw for forage, fuel, and construction, as in Asia and Africa, soil conservation methods based on plant residues and mulching are not practical (Unger, 1990). In many countries, millet (*Penisetum* sp.) and sorghum (*Sorghum* sp.) are the staple diets; therefore a conservation strategy must cope with this basic regional requirement (Little et al., 1987). Although forages, oil crops, and pulses are considered excellent break crops in cereal monoculture, they can be cultivated only in areas with over 500 mm of rainfull (Al-Fakhry, 1990). In arid regions with less than 250 mm of rainfall, a fallow-crop rotation with small-livestock grazing, aftermaths, and natural pasture may be the only possible economic options.

No single agronomic remedy exists for alleviating and preventing soil erosion and water loss. It is imperative to find the cropping system and agroengineering solution best suited to each region in order to achieve conservation and to maximize economic returns.

III. WATER LIMITATIONS IN CROPPING SYSTEM

Crop yields are in the first instance dependent on water availability. Where water availability is via either precipitation or irrigation, and other growth conditions are optimal, C.R. is the best method for improving and/or sustaining yields. Under these conditions, the standard conservational measures such as permanent plant coverage, water runoff, and drainage control are sufficient (Follett and Stewart, 1985).

In regions where the water availability falls below the crop's minimal requirement as dictated by evapotranspiration, it is possible to compensate by implementing special cultivation strategies which will improve the water availability in the dry environment. Runoff and wind erosion are severe in problems in semiarid and arid regions, so both water and soil conservation measures must be employed. Since the beginning of the century, intensive research has been conducted in the Great Plains of the United States, in Australia, and other countries to resolve problems of erosion and water loss (Greb, 1979; MacLennan, 1990; Unger, 1990). The specific effects of different cropping systems are discussed in a later section; therefore, only several general principles pertaining to soil erosion and rehabilitation will be considered here.

There are two basic cropping system approaches for reclaiming areas affected by runoff and wind erosion. The first is based on preserving vegetation or residue cover wherever possible. This can be achieved by introducing a pasture into the cropping sequences (one to four years) or introducing cover crops (legumes, oil crops, etc.), rather than allowing the soil to remain bare. This method is more suitable for humid areas. Soil erosion caused by runoff and its prevention is presented in Table 1. The data presented was measured over a period of 14 years in Missouri on a loam soil with a gentle slope of 3.6% (Miller, 1936). A second approach modifies the soil-water availability for plants by reducing evaporation, increasing infiltration, and minimizing water loss. This method is used in dry environments where legumes and sown pastures cannot be grown due to the water and climatic limitations (Greb, 1979; Unger, 1990).

IV. CROP ROTATION AND EROSION IN HUMID AREAS

Constraints on crop production in humid regions are primarily radiation, temperature, nutrition, sanitation, erosion, and waterlogging. In the humid tropics, where slash-and-burn agriculture is practiced, the cleared land produces economic yields for approximately three years but declines thereafter because of soil erosion and leaching of nutrients. Applying fertilizer may maintain yield levels for five consecutive years. For longer periods of productivity, e.g., 15 consecutive years, a relatively new management system called "alley cropping" has been introduced (Lal, 1991). This system includes rows of leguminous shrubs between rows of the crop. The shrubs are pruned at knee level and the branches used for mulching between crop rows. This system controls erosion, runoff, and nutrient leaching but is recommended only in fertile areas where C.R. is practiced, Table 2 (Lal, 1991; Szott et al., 1991).

Table 1 Average of 14 Years Measurements of Runoff and Erosion, Missouri Experiment Station, Columbia[a]

Cropping system or cultural treatment	Average annual erosion per acre in tons	Percentage of total rainfall running off the land
Bare, cultivated, no crop	41.0	30
Continuous corn	19.7	29
Continuous wheat	10.1	23
Rotation–corn, wheat, clover	2.7	14
Continuous bluegrass	0.3	12

[a]Soil type was Shelby loam; length of slope was $90^3/_4$ ft; degree of slope was 3.68%.
Source: Miller (1936).

Table 2 Alley Cropping Effects on Runoff and Soil Erosion under Maize-Cowpea Rotation Measured in 1984

	Runoff		
Treatment	Millimeters	Percentage of rainfall	Soil erosion (t ha^{-1} per year)
Plow-till	232	17.1	14.9
No-till	6	0.4	0.03
Leucaena, 4 m	10	0.7	0.2
Leucaena, 2 m	13	1.0	0.1
Gliricidia, 4 m	20	1.5	1.7
Gliricidia, 2 m	38	2.8	3.3

Source: Lal (1991).

In the humid temperate zone, in contrast to the unstable ecosystem of the tropics, cereal yield can be maintained through monoculture in heavy and fertile soil for more than 150 years by an annual supply of fertilizers or manure, Fig. 1 (Jenkinson, 1991). The relative high stability of soil organic matter in the topsoil on unfertilized and fertilized plots in Rothamsted's classic experiment provides convincing evidence regarding the vital role of organic matter in sustainable crop production, Fig. 2.

The poor soil sanitation, typical to the monoculture cropping system in semi-arid and arid zones, did not affect the Rothamsted experiment, suggesting an equilibrium of some kind between soil pathogens and their antagonists, or a possible control of soil pathogens by frost. Soil erosion appears not to be a problem in the Rothamsted published reports. The long-term experiments in North America provided another excellent source of valuable information regarding fertility, erosion, and C.R. practices. The Sanborn field at Columbia in Missouri was established in 1888 on a fine silt loam. Erosion of the top soil was analyzed after 75 years of cropping. The surface soil of most plots was modified by erosion. Table 3 shows topsoil thickness in plots under different crop managements. The greatest erosion was found in continuous corn and the lowest in continuous timothy (*Phleum pratense*) plots. This finding indicates that, on a slope of 0.5 to 3.0%, the continuous corn plots lost all topsoil within 100 years. The six-year rotation (C-O (oat)–W-C-T (timothy))-T lost only about half of the topsoil during the same period (Gantzer et al., 1991). C.R. alone, without fertilizer supplements, did not prevent yield decline. Yields of continuous wheat (C.W.), untreated with manure or fertilizer, declined due to disease and weed infestation. When topsoil has been removed by erosion and the remaining soil has lower water-retention capacity, adding nutrients is of limited value in restoring soil productivity (Mitchell et al., 1991)

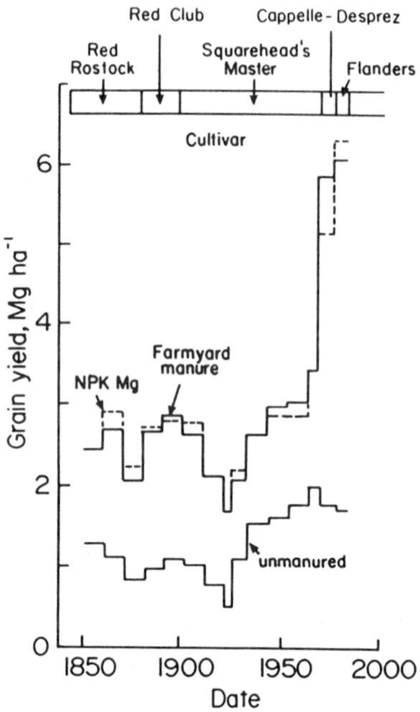

Fig. 1 Yields on three plots of the Broadbalk Winter Wheat Experiment, 1852 to 1986: unmanured, inorganic fertilizers and farmyard manure. (From Jenkinson, 1991.)

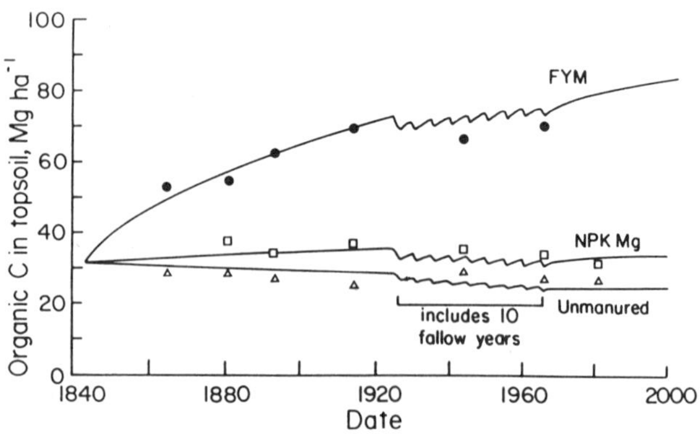

Fig. 2 Organic C in the top 23 cm of soil from three plots of the Broadbalk Winter Wheat Experiment, 1843 to 1980: unmanured, inorganic fertilizers and farmyard manure. The symbols show the observations; lines show the fit to the data given by the Rothamsted organic matter turnover model. (From Jenkinson, 1991.)

Table 3 Topsoil Thickness (depth to Bt horizon) and Percent Clay in the Ap Horizon Measured in Selected Plots from Sanborn Field, 1988

		Topsoil		Clay	
Management	Plot no.	Mean (cm)	SD (cm)	Mean (%)	SD (%)
Corn, unfertilized	17	17.8	2.3	26.8	2.5
Corn, manured	18	21.3	1.0	30.0	2.4
Mean		19.6	2.2	28.4	2.8
Six-yr rotation, unfertilized	13	31.0	2.0	14.6	2.0
Six-yr rotation, manured	19	28.5	3.9	18.4	2.0
Six-yr rotation, fertilized	20	33.5	4.7	17.6	2.0
Mean		31.0	4.0	16.8	2.0
Timothy, unfertilized	23	45.0	4.9	16.9	0.6
Timothy, manured	22	43.5	4.6	16.7	0.9
Mean		44.3	4.4	16.7	0.7

Source: Gantzer et al. (1991).

In intensive modern agriculture, including high-input crops like potato, tomato, and other horticultural crops, prolonged C.R. cycles of four to eight years are superimposed by the severe sanitation problems (Cook, 1984). In this type of enterprise, the problem of excess water causing erosion, runoff, and leaching is common. This is the main source of contamination of rivers, lakes, and wells with nitrates, pesticides, and herbicides (Schepers, 1988). Standard soil conservation methods such as contour planting, cover crop/residues, and conservative use of fertilizer and pesticide will suffice to maintain high yields and fulfill conservation requirements under this condition.

V. ALTERNATE CROP-FALLOW PRACTICE

Soil and water conservation managements are especially important to low-input dryland agriculture with 200- to 500-mm precipitation. Annual cropping and intensive tillage were the methods used in dryland farming until the turn of the century. This highly exploitative management led to rapid degradation of soil structure, decline of soil organic matter, acceleration of soil erosion, and, subsequently, a significant decline in yields (Bennett, 1955). To overcome the yield decline, two new management systems were developed in the United States and Australia—use of fertilizer or manure to restore depleted nutrients, and adaptation of the fallow-crop cropping system to alleviate the water and nutrient shortages (Greb, 1979; McWilliam, 1990). However, these solutions proved to be inadequate in maintaining long-term productivity, and cereal yields could not be sustained. The fallow system was continually improved by intensive research and accumulated farming experience. Results published by Greb (1979) reveal a dramatic rise in water conservation, from 19 to 27% (in the 1916–1960 period) to 33 to 40% in recent years in the central Great Plains.

Maximal water-use efficiency (WUE) published for wheat grain production after fallow was 3.5 kg ha^{-1} mm^{-1} in the U.S. Great Plains, 11.4 kg ha^{-1} mm^{-1} in South Australia, 8.4 kg ha^{-1} mm^{-1} in Syria, and 15.0 kg ha^{-1} mm^{-1} in Israel (French, 1978; Greb, 1979; Cooper et al., 1987; Amir et al., 1991a). The extremely high WUE in the northern Negev of Israel was measured after fallow with no stored soil water from previous years (dry fallow). In the same experiment similar WUE was achieved with a 10-year C.W. cropping system by using a soil-applied biocide metham-sodium, suggesting that the dry-fallow effect is caused by improved soil sanitation (Amir et al., 1991a,b).

Reports from Australia suggest that the beneficial effect from fallow rotations can be ascribed to soil-stored water at sowing in 59% of the cases, while 41% of yield increases could be attributed to control of soilborne pathogens or additional soil N release (French, 1978). The higher wheat yields after dry fallow were obtained in a long-term experiment in Israel when the N was not a limiting factor in the experiment, indicating that sanitation was the main beneficial factor (Amir et al., 1991b).

The root system in dry fallow was compared with that in C.W., and the results showed in root density in dry fallow, which was almost double that of the C.W. treatment, Fig. 3. The difference in root density explains the more efficient WUE of the dry fallow wheat system (F.W.), by a higher transpiration: evapotranspiration ratio relative to C.W. (Amir et al., 1990).

When rainfall was above 300 mm, differences between the above systems gradually disappeared. This surprising result is explained by a measured linear increase in transpiration : evapotranspiration ratio in C.W. management up to 300 mm of rainfall. Above 300 mm the transpiration: evapotranspiration ratio remained constant at 0.55 and similar to the dry F.W. system (Amir et al., 1990). These results led to a hypothesis that the use of surface residues evaporation could be reduced and subsequently soil available water increased (Amir and Sinclair, 1991).

VI. LEY FARMING

In Australia after the cultivation of C.W. on soils that had previously been covered with native vegetation yields declined by 43% in 30 cropping years. From the beginning of the 20th century until 1945, yields recovered due to the use of phosphorus fertilizer and the introduction of bare fallow management for improving the water, N supply, and weed control. In regards to soil conservation, this system was not stable in the long-term due to soil degradation and loss of

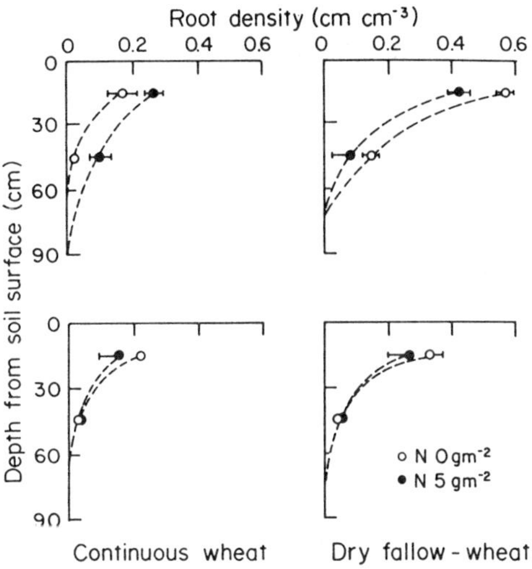

Fig. 3 Root density at anthesis (upper) and at maturity (lower), of continuous wheat and dry-fallow/wheat managements; at two levels of applied N. Bars represent the SE of means. (From Amir et al., 1991.)

organic matter. The remedy to this problem was ley farming utilization (Sims, 1977; McWilliam, 1990).

The self-regenerating legumes (*Trifolium* sp. and *Medicago* sp.) in rotation with wheat resulted in significant improvement of wheat yields and animal production (Davidson, 1981). The positive effect of legume ley farming has been attributed to its ability to control soil erosion, fixate atmospheric N, break the pathogen cycle, and improve soil structure (McWilliam, 1990). International price fluctuation had a marked impact on ley farming in Australia. The increased demand for wheat and declining prices of N fertilizer have shifted production toward C.W. cropping (McWilliam, 1990). The introduction of ley farming in other Mediterranean dryland areas has not been very successful. Improved *Trifolium* and *Medicago,* although successful in Australia, have been less valuable when reintroduced to their native regions, mainly because of disease, ruderal annual grass competition, pest problems, and possibly lower rainfall (Eyal, Benjamin, and Tadmore, 1975).

VII. THE MONOCULTURE CROPPING SYSTEM

Raising a crop on the same land continuously is referred to as monoculture. This practice is widespread in the United States, Europe, Australia, and Asia. About half of the wheat and sorghum in the United States is produced by monoculture (Power and Follett, 1987). The outcome of monoculture was the development of specialized machinery, professional skill, and, often, maximal farm income. An influential factor in the trend toward continuous cropping is, perhaps, government policy, which affects farmers' decision making. Monoculture has led to the exclusion of the grazing animal and the development of feeding animals in feedlots.

The major drawback of monoculture from the conservation perspective is the danger of topsoil degradation by erosion. Erosion changes the physical and hydrological properties of the soil, and as a result crop yields are usually lower regardless of the amount of fertilizer applied (Mitchell et al., 1991). Another significant limitation of monoculture stems from pest, weed, and disease infes-

Table 4 Influence of Crop Rotation on the Yield Response of Wheat to Soil Fumigation in Cultivated (tilled) Fields in Eastern Washington[a]

Previous crops[b]	Exp.	Check (t ha^{-1})	Fumigated (t ha^{-1})	Increase (%)
Continuous wheat				
WW/WW/WW	1	3.2	6.5	105
WW/WW/WW	2	3.6	6.2	72
L/WW/SW	3	4.1	4.8	14
WW/WW/WW	4	1.7	3.5	106
WW/WW/WW	5	3.6	5.5	53
Average				70
Two-year rotations				
P/WW/P	1	4.9	6.8	36
L/WW/L	2	6.9	8.3	20
P/WW/P	3	6.2	7.2	16
P/WW/P	4	5.5	6.2	13
L/WW/L	5	4.8	6.0	26
Average				22
Three-year rotations				
WW/SB/P	1	7.5	8.4	12
WW/SB/L	2	9.6	10.0	4
WW/SB/P	3	7.5	8.6	15
WW/SB/P	4	6.1	6.1	2
WW/SB/P	5	7.0	7.2	3
Average				7

[a]Representative trials from about 50 experiments conducted over a 15-yr period.
[b]WW = winter wheat; SW = spring wheat; L = lentil; P = pea; SB = spring barley.
Source: Gantzer et al. (1991).

tation, Table 4 (Cook, 1990). When the water supply is adequate, the alternative to monoculture is introduction of a break crop. Productivity can be restored to eroded soil, especially if a forage crop, such as legume or grass, is included in the sequential order. In extremely dry areas, where the bare fallow crop is the sole alternative to continuous cropping, the only conservation measures against erosion are stubble mulch or no-tillage management (Smika and Unger, 1986).

Fallow is an inefficient system because only 30–50% of the land is cropped annually and the soil is exposed to erosion during the 6- to 18-month fallow period. In Mediterranean areas receiving about 250-mm annual rainfall, monoculture is the most feasible soil cover to prevent or reduce erosion. However, the principle limitation to monoculture is the sanitation problem, such as nematodes and root-rot diseases (Amir et al., 1991b). However, these problems can be overcome by the new conservation tillage technologies (Amir and Sinclair, 1991).

VIII. CONSERVATIVE TILLAGE, NO-TILLAGE MANAGEMENT

Conservative tillage is defined as tillage retaining at least 30% crop cover residue on the soil surface after planting. The development and practice of conservation tillage has been reviewed intensively (McCalla and Army, 1961; Phillips and Phillips, 1984; Allmaras et al., 1991). Crop residues placed on the surface or incorporated into the soil help prevent erosion and crusting while improving soil aggregation, infiltration, and water retention in the soil. Stubble mulch tillage is an important method of controlling wind erosion in the United States, and the data presented in Table 5 clearly show the superiority of this system in controlling wind erosion when compared to clean fallow. The results also illustrate the effect of wind erosion on different soil types; light soils erode more easily than heavy soils (Chephill and Woodruff, 1957). Duley and Russel (1939) studied the effect of tillage treatment and the amount of straw mulch on water conservation, Table 6. In their study the best treatment was 450 g m^{-2} of straw on the surface, which resulted in the soil storage of 54.3% of the 455-mm precipitation which reached a depth of 180 cm (Smika and Unger, 1986).

The data in Table 7 illustrate the progress achieved in water conservation in Colorado (U.S.A.). The results show a significant rise in fallow efficiency, from 19% in the period 1916–1930 to 40% in the 1980s, mainly due to stubble mulch, herbicides, and no-tillage managements (Greb, 1979).

The latest development in conservation tillage is no-tillage (N.T.) management, which allows direct seeding into stubble by use of special machinery and herbicides. The main advantages of this new technology are better erosion control by standing and flattened residue. Soil surface is left intact, lowering significantly organic matter oxidation and improving crop yield from a reduction

Table 5 Effectiveness of Stubble Mulch and Clean Fallow in Controlling Wind Erosion in Western Nebraska, 1957

Soil type	Kind of fallow[a]	Wind erodibility ($t\ ac^{-1}$)[b]	Straw mulch ($lb\ ac^{-1}$)
Keith loam	Stubble mulch	0.4	1,370
Do.	Clean fallow	9.8	425
Very fine sandy loam	Stubble mulch	.3	3,645
Do.	Clean fallow	16.5	800
Rosebud, very fine sandy loam	Stubble mulch	2.3	1,680
Do.	Clean fallow	22.7	480
Keith loam	Stubble mulch	0.1	3,780
Do.	Clean fallow	1.8	735

[a]Stubble mulch tillage accomplished with 30-in sweeps as required to control weeds. Clean fallow accomplished by chopping stubble, plowing 6 in deep, chiseling, spring-toothing twice, and rod-weeding twice.
[b]Portable tunnel data adjusted to annual soil loss for field 160 rods long.
Source: Chepill and Woodruff (1957).

Table 6 Effect of Straw and Tillage Treatments on Soil Water Storage, 23 April to 9 September 1938, at Lincoln, Nebraska

Treatment	Precipitation stored[a]		Depth of water penetration (m)
	mm	%	
Straw, 4.5 $t\ ha^{-1}$, on surface	247	54.3	1.8
Straw, 4.5 $t\ ha^{-1}$, disked in	176	38.7	1.5
Straw, 4.5 $t\ ha^{-1}$, plowed in	155	34.1	1.5
No straw, disked	89	19.6	1.2
No straw, plowed	94	20.7	1.2
Decayed straw, 2 $t\ ha^{-1}$, plowed in	79	17.4	1.2
Basin listed	126	27.7	1.5

[a]Precipitation totaled 455 mm.
Source: Smika and Unger (1986), adapted from Duley and Russel, 1939.

Table 7 Progress in Fallow Systems and Wheat Yields, U.S. Central Plains Research Station, Akron, Colo, Mado

Years	Changes in fallow systems	Average annual precipitation (in)	Drought years (No)	Fallow water storage[a] (in)	Fallow efficiency (%)	Wheat yield (Bu ac^{-1})	Water-use efficiency[b] (Bu ac^{-1})
1916–30	Maximum tillage; plow harrow (dust mulch)	17.3	1	4.0	19	15.9	0.46
1931–45	Conventional tillage; shallow disk, rodweeder	15.8	5	4.4	24	17.3	0.54
1946–60	Improved conventional tillage, begin stubble mulch 1957	16.4	3	5.4	27	25.7	0.78
1961–75	Stubble mulch; begin minimum tillage with herbicides (1969)	15.3	4	6.2	33	32.2	1.05
1976–90	Projected estimate. Minimum tillage; begin no-till 1983	16.2[c]	3	7.2	40	40.0	1.23

[a]Based on 14 months fallow, mid-July to second mid-September.
[b]Assuming two years precipitation per crop in a wheat-fallow system.
[c]Assuming average precipitation from 1976 to 1990.
Source: Greb (1979).

in soil evaporation and improved soil structure (Phillips and Phillips, 1984). Potential benefits of N.T. on crop yields depend on soil type, precipitation amount and distribution, soil drainage, and alleviation of the negative aspects of the N.T. system, namely poor sanitation and weed infestation (Dick et al., 1991). Evaporation is an important parameter in the water budget of field crops, especially in dry areas as it has a negative effect on crop production. For many crops, evaporation accounts for 30% to 50% of the total evapotranspiration, depending on crop cover (Unger and Parker, 1976; Amir et al., 1990). N.T. farming has a significant effect on the first stage of evaporation (energy limited evaporation), as illustrated in Fig. 4 (Unger and Parker, 1976). N.T. reduces evaporation over a certain length of time, depending on the amount of residue, evaporative demand, depth of water penetration, and the extent of competition between the evaporation and transpiration processes (Phillips et al., 1980; Allmaras et al., 1991). In dry environments below 250-mm annual rainfall, 400 g m^{-2} of straw coverage substantially reduced the evaporation and, consequently, increased the transpiration by almost 100%. This shift in the evaporation : transpiration ratio and its impact on grain yield was computer simulated (Amir and Sinclair, 1991) and validated in field experiments in the northern Negev in Israel, Table 8. The significant finding, following long-term experiments in the Negev desert, was that the 17-year C.W. with the straw coverage treatment produced

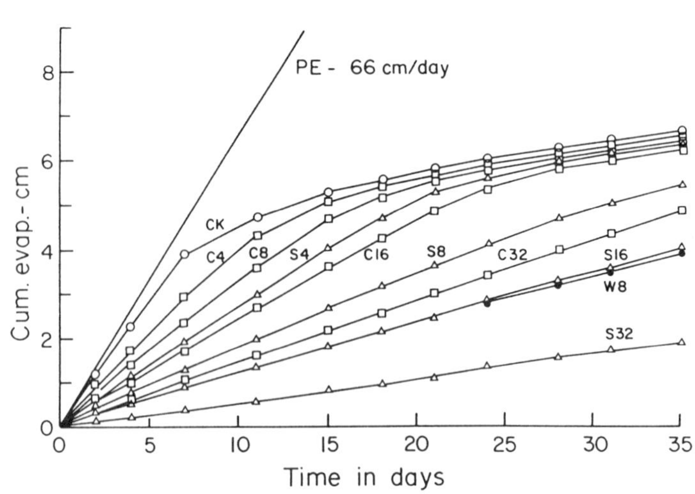

Fig. 4 Effects of residue treatments on cumulative evaporation at 0.66 cm day^{-1} potential evaporation. CK = check, C = cotton, S = grain sorghum, W = wheat. Numbers after letters designate metric tons of applied residues per hectare. (From Unger and Parker, 1976.)

Table 8 Straw Mulch Effect in Continuous Wheat and Wheat-Fallow at the Gilat Experiment Station, Negev Desert, Israel (1989–90)[a]

Managements	Grain yield ($g\ m^{-2}$)[b]		
		Straw on surface	
	No straw	(400 $g\ m^{-2}$)	(800 $g\ m^{-2}$)
Continuous wheat (17 years)	139a	229b	222b
Wheat-fallow[c]	288a	360b	338b

[a]Based on total precipitation of 243 mm.
[b]Row values followed by the same letter are not significantly different at the 5% level (Duncan's multiple test).
[c]Dry fallow without soil stored water at sowing.
Source: Amir (unpublished data 1990).

a yield equal to the "dry" fallow wheat C.R. The "dry" fallow lacked stored soil water at sowing (Amir et al., 1991a), and the 17-year wheat monoculture was heavily infested with nematodes and root-rot disease (Amir et al., 1991b). From the data published from the northern Negev, it is clear that fallowing's beneficial effect on yield through upgrading soil sanitation gradually diminished when rainfall exceeded 300 mm (Amir et al., 1990). Similarly, in a below-300-mm rainfall year, straw-covered C.W. treatment lowered evaporation and significantly raised transpiration, and yields were similar to those in above-300-mm rainfall threshold, successfully bypassing the soil sanitation effect, Table 8.

The beneficial effect which residue coverage has on water budget for crops opens up the possibility for crop diversification, e.g., introducing a nongraminae crop into a rotation sequence. If other restricting factors such as sanitation, weeds, allelopathy, and drainage problems do not interfere, this residue management can compete economically with the F.W. system in even drier areas. In arid environments, where erosion potential can be high due to topography, low organic residue, and strong winds, crop residue management can be employed as a promising soil reclamation method.

In a humid environment in northern Illinois, Hayes and Kimberlin (1978) investigated the effects of crop residues and six-year crop rotation: C-C-W-H-H-H. By using the universal soil loss equation (USLE) formulated by Wishmeier and Smith (1978), they calculated the crop management factor (*C*-factor). For three years of hay, the *C*-factor per day, calculated from residue management, crop cover, C.R., and erosion effects, was 0.00001, and the soil erosion was 0.004 soil loss, compared to 1.0 for bare fallow. For wheat, the *C*-factor per day was 0.0018–0.0160, depending on the crop stage, and the erosion hazard was 0.03–0.8. For the second corn in sequence the *C*-factor ranged from 0.0005 to 0.0022, and the erosion hazard from 0.2 to 0.48, depending on the crop, Table 9. Crop production due to the exclusion of pasture and forage crops, is a major cause of increased erosion and pollution in the last decade. Studies indicate that the risk of erosion and runoff exists even on flat land and gentle slopes where sheet erosion exceeds 1.1 kg m^{-2} yr^{-2} soil loss tolerance. Hence, the value of conservation methods including C.R. should be viewed in a much wider context.

IX. MODELING CROP ROTATION EFFECT

Research involving C.R. has been considered in the past as a fertility factor, especially N. Legume in the rotation was the main N source for nonsymbiotic crops. It is generally accepted that the N availability is a major growth factor affecting crop production. However, today this element can be easily applied with inexpensive fertilizers. Even with a high rate of N availability, crops in monoculture yield less than in C.R., Fig. 5. Thus, there is benefit of C.R. in addition to the N availability (Baldock et al., 1981; Hesterman et al., 1987).

Table 9 *C*-Factor Computations for a Crop Sequence of Corn, Two Years; Wheat, One Year; and Hay, Three Years[a]

1 Crop	2 Dates	3 Crop stage[b]	4 R-E index[c]	5 Soil loss compared to fallow[d]	6 (Column 4 × column 5) crop management factor crop stage period	7 Crop management factor: rate per day
Corn	4/10–5/1	F	0.04	0.08	0.0032	0.0016
	5/1–6/1	1	0.10	0.25	0.0250	0.00083
	6/1–7/1	2	0.18	0.17	0.0306	0.00102
	7/1–10/10	3	0.55	0.10	0.0550	0.00055
	10/10–4/10	4	0.13	0.11	0.0143	0.00008
Corn	4/10–5/1	F	0.04	0.25	0.0100	0.00050
	5/1–6/1	1	0.10	0.48	0.0480	0.00160
	6/1–7/1	2	0.18	0.37	0.0666	0.0022
	7/1	3	0.52	0.30	0.1040	0.00116
Wheat	10/1–11/1	1	0.06	0.80	0.0480	0.00160
	11/1–4/10	2	0.10	0.50	0.0500	0.00036
	4/10–7/10	3	0.41	0.15	0.0615	0.00068
	7/10–9/10	4	0.37	0.03	0.0111	0.00018
Hay	9/10–4/10	—	0.22	0.004	0.009	0.00000
Hay (1st)	4/10–4/10	—	1.00	0.004	0.0040	0.00001
Hay (2nd)	4/10–4/10	—	1.00	0.004	0.0040	0.00001
Hay (3rd)	4/10–4/10	—	1.00	0.004	0.0040	0.00001

Sum of column 6 = 0.5402 + 6 yr = 0.0900/yr *C*-factor for CCWHHH = 0.090.

[a]Corn is conventionally tilled, corn residue is left following first year corn. Second year corn is harvested as silage. Wheat straw is harvested. Crop yields are high. Northern Illinois.

[b]F = between plowing and planting, 1 = from planting to 30 days after, 2 = 30 to 60 days after, 3 = 60 days after to harvest, 4 = from crop harvest to plowing for next crop (after wheat that is seeded it ends 2 months after harvest). No period 4 after 2nd year corn when followed by winter wheat.

[c]Percent of the erosion producing rainfall that occurs during a crop stage period + 100.

[d]Percent of soil loss that will occur during this crop stage period when growing this crop with this management compared to continuous fallow (continuous fallow is void of any vegetation and is considered 100% soil loss) + 100.

Source: Wischmeier and Smith (1978).

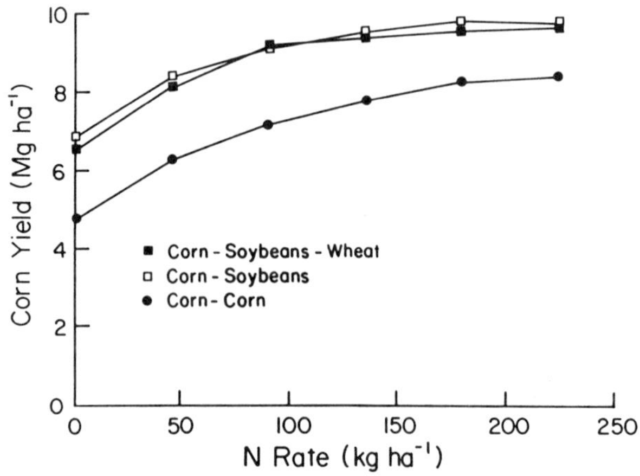

Fig. 5 Corn yield in three crop rotations as influenced by N application rate on a Webster clay loam (fine loamy, mixed, mesic Typic Haplaquoll) at Waseca, Minnesota, averaged for 1975 to 1986. (From Pierce and Rice, 1988.)

The analysis of C.R. effect, other than N, is a difficult research investigation because too many possible factors are involved, e.g., (1) reduction in pests and diseases, (2) better control of weeds and reduction of allelopathy, (3) improvement of soil properties, (4) better utilization of residual soil-stored water, and (5) alleviated erosion effects. The major growth factors in crop production have been quantified and used in computer-simulated crop growth models (Hank and Ritchie, 1991). The direct effects of C.R. on plant growth, yields, soil erosion, and rehabilitation, however, have not yet been simulated. The impact of C.R. on soil erosion is expressed by the *C*-factor in the USLE. The *C*-factor takes into account crop residue cover, roughness of surface, and crop residues buried in the topsoil layer. The erosion productivity import calculator (EPIC), which has been used to simulate infiltration, runoff, soil water, and plant growth, performed well in humid and semiarid conditions (Williams and Renard, 1985; Steiner et al., 1987). EPIC requires calibration of coefficients for each location when addressing the C.R. effect on yields.

A prerequisite for simulation of C.R. impact on soil and water conservation is the quantification of the major C.R. effects, namely, sanitation effect, nutritional benefits, improvement in soil properties, and water regime. It is possible after studying the effects of each factor to incorporate them as subcomponents in the existing crop models. In a recently published wheat model by Amir and Sinclair (1991a) for unlimited and limited growth conditions, an attempt was made to simulate some of the above components (Amir and Sinclair, 1991a,b).

The database for the development and validation of the model came from an arid environment where the fallows were ''dry'' without soil-stored water at sowing, and the fallow N benefit was neutralized by four N treatments in C.W. system. By eliminating the water and N fallow benefits, it is possible to quantify the fallow sanitation effects. The direct sanitation effects of the rotation were on the transpiration : evapotranspiration ratio, due to normal and healthy root systems in the F.W. rotation (Amir et al., 1990). In the model, the growth process was inhibited as a logistic function of fraction of transpirable soil water (Amir and Sinclair, 1991b). However, it is still necessary to modify this function to account for the effect of pathogens that damage root systems in C.W. management. A new series of experiments are now being implemented to evaluate new relationships for C.W. management. The new function will simulate the C.W. effect by reducing the transpiration amount available to plants, and thus the leaf-area index, dry-matter production and finally the grain field. The simulation of the damaged root system on plant water regimes does not at this stage take into account the specific damage caused by nematodes, diseases, and pests.

Recently, some model versions were published which evaluated specific biotic stresses on crop growth, such as weeds, insects, nematodes, and diseases (Duncan and Ferris, 1983; Szmedra et al., 1988; Retta et al., 1991). Some of the models showed reasonable prediction of biotic stresses on some aspects of crop performance and can be used in the C.R. effect model.

In conclusion, we are in an early stage of C.R. effect modeling. Further predicative capabilities should be considered as an important research task.

X. CONCLUSION AND FUTURE NEEDS

Crop rotation benefits that maintain high productivity and conserve soil and water resources have long been recognized. Soil erosion and water runoff are major causes where there has been rapid soil deterioration and yield decline of cultivated soils. Soil degradation has intensified in the last decades by the trend toward monoculture, less C.R., conventional tillage and bare fallow. Since the turn of the century strategies to cope with erosion and runoff damage have been developed by conservationists. However, C.R. has proved to be the most effective of all the soil conservation strategies examined. Plant ground cover during the erosion period reduces the kinetic energy of the raindrops, increases the infiltration rate, and by the plants' continual transpiration, new soil volume is available; thus more rainfall can be absorbed. The introduction of a long-term permanent grass-legume pasture into severely eroded soils is the most effective reclamation method known. Ley farming, using self-regenerating legume species in the C.R., is successful in Australia, being not only a conservation tool but also a profitable agricultural system. Ley farming is now being investigated in ecosystems similar to those in Australia.

Alternatives to the ley farming conservation system are the crop residue options, minimum tillage, and N.T. These strategies were adopted successfully in a humid and semiarid conditions and are now being investigated in more arid areas. Residue cover has reduced erosion, runoff, and water shortage effects, but weed infestation and sanitation problems remain.

A complete solution appears to be stubble mulch technology combined with C.R. The success of stubble mulch management without C.R. will depend on the development of effective new herbicides and disease tolerance/resistance of plant varieties. Each method used to control erosion and runoff fits a specific situation; however, there are situations where combination of the methods is necessary. After the topsoil has been either severely damaged or lost completely by erosion, rehabilitation can be a long process that will involve a succession of methods that include promotion of pastures that eventually, with soil improvement, can be replaced by C.R. and conservation tillage.

Soil and water conservation practices are crucial for long-term sustainable agriculture. To ensure that these practices are implemented, there must be positive cooperation and interaction between research, regional planning, extension service, and government legislation.

Future research on C.R. effect on soil erosion and water conservation must include the following considerations:

1. Due to the complexities of the interaction between the factors involved in the C.R. effect, additional research is needed through long-term multifactorial experiments at different sites. The database from these experiments will provide the information needed for understanding and quantifying the processes involved that can then be used for the development of models to predict yields by computer simulation.
2. Considerable effort must be directed toward quantification of the sanitation and allelopathy effect involved in the C.R. effect.
3. Improved understanding of the evaporation : transpiration ratio under field conditions in different C.R. systems.
4. Research on control of weeds under stubble mulch management conditions.
5. Breeding of improved crop varieties for stubble mulch systems tolerant/resistant to pests and diseases.

REFERENCES

Al-Fakhry, A. K. (1990). Studies on fallow and the possible alternatives in rainfed region of North Iraq. In *Challenges in Dryland Agriculture* (P. W. Unger, T. V. Sneed, W. R. Jordan, and R. Jensen, eds.), Texas Agricultural Experiment Station, College Station, Texas, pp. 147–149.

Allmaras, R. R., Langdale, G. W., Unger, P. W., Dowdy, R. H., and Van Doren, D. M. (1991). Adopting conservation tillage and associated planting systems. In *Soil Management for Sustainability* (R. Lal and F. J. Pierce, eds.), Soil and Water Conservation Society, Ankeny, Iowa, pp. 53–83.

Amir, J., Krikun, J., Orion, D., Putter, J., and Klitman, S. (1991a). Wheat production in an arid environment. 1: Water-use efficiency, as affected by management practices. *Field Crops Res. 27*: 351–364.

Amir, J., Krikun, J., Orion, D., Putter, J., and Klitman, S. (1991b). Wheat production in an arid environment. 2: Role of soil pathogens. *Field Crops Res. 27*: 365–376.

Amir, J., Krikun J., and Putter J. (1990). Wheat production in an arid environment. In *Challenges in Dryland Agriculture* (P. W. Unger, T. V. Sneed, W. R. Jordan, and R. Jensen, eds.), Texas Agricultural Experiment Station, College Station, Texas, pp. 183–186.

Amir, J., and Sinclair, T. R. (1991a). A model of the temperature and solar-radiation effects on spring wheat growth and yield. *Field Crops Res. 28*: 47–58.

Amir, J., and Sinclair, T. R. (1991b). A model of water limitation on spring wheat growth and yield. *Field Crops Res. 28*: 59–69.

Baldock, J. O., Higgs, R. H., Paulson, W. H., Jackobs, J. A., and Shrader, W. D. (1981). Legume and mineral nitrogen effects on crop yields in several crop sequences in the upper Mississippi valley. *Agron. J. 73*: 885–890.

Bennett, H. H. (1955). *Elements of Soil Conservation*. McGraw-Hill, New York.

Cato, P. (149–234 B.C.). *De Agricultura*.

Chephill, W. S., and Woodruff, N. P. (1957). *U.S. Department of Agriculture Res. Serv.*, Manhattan, Kansas.

Cook, R. J. (1984). Root health: Importance and relationship to farming practices. In *Organic Farming: Current Technology and Its Role in a Sustainable Agriculture*, Am. Soc. Agron., Madison, Wisconsin, pp. 111–127.

Cook, R. J. (1990). Diseases caused by root-infecting pathogens in dryland agriculture. In *Dryland Agriculture Strategies for Sustainability*, Vol. 13, Advances in Soil Science (R. P. Singh, J. F. Parr, and B. A. Stewart, eds.), Springer-Verlag, New York, pp. 215–239.

Cooper, P. J. M., Gregory, P. J., Keatinge, J. D. H., and Brown, S. C. (1987). Effect of fertilizer, variety and location on barley production under rainfed conditions in northern Syria. *Field Crops Res 16*: 67–84.

Davidson, B. R. (1981). *European Farming in Australia: An Economic History of Australian Farming*. Elsevier, Amsterdam.

Dick, W. A., McCoy, E. L., Edwards, W. M., and Lal, R. (1991). Continuous application of no-tillage to Ohio soils. *Agron. J. 83*: 65–75.

Duncan, L. W., and Ferris. (1983). Validation of a model for prediction of host damage by two nematode species. *J. Nematol. 15*: 227–234.

Eyal, E., Banjamin, R. W., and Tadmore, N. H. (1975). Sheep production on seeded legumes, planted shrubs and dryland grain in a semi-arid region of Israel *J. Range Management 28*: 101–107.

Follett, R. F., and Stewart, B. A. (1985). *Soil Erosion and Crop Productivity*. Soil Science Society of America, Madison, Wisconsin.

French, R. J. (1978). The effect of fallowing on the field of wheat. II: The effect on grain yield. *Aust. Aust. J. Agric. Res. 29*: 669–684.

Gantzer, G. J., Anderson, S. H., Thompson, A. L., and Brown, J. R. (1991). Evaluation of soil loss after 100 years of soil and crop management. *Agron. J. 83*: 74–77.

Greb, B. (1979). U.S.D.A. Reducing drought effects on croplands in the West-Central Great Plains. *Agriculture Information Bulletin*, No. 420, U.S. Government Printing Office, Washington, DC.

Hank, J., and Ritchie, J. T. (Eds.). (1991). Modeling plant and soil systems. *Agron. Monograph*, No. 31, Am. Soc. Agron., Madison, Wisconsin.

Hanley, F., and Ridgman, W. J. (1978). Some effects of growing winter wheat continuously. *J. Agric. Sci. Camb. 90*: 517–521.

Hayes, W. A., and Kimberlin, L. W. (1978). A guide for determining crop residue for water erosion control. In *Crop Residue Management Systems.* (W. R. Oschwald, ed.), *Am. Soc. of Agron.*, Special Pub. No. 31, Madison, Wisconsin, pp. 35–48.

Hesterman, O. B., Russelle, M. P., Sheaffer, C. C., and Heichel, G. H. (1987). Nitrogen utilization from fertilizer and legume residues in legume-corn rotation. *Agron. J. 79*: 726–731.

Hymowitz, T. (1970). *Elements of Soil Conservation.* McGraw-Hill, New York.

Jenkinson, D. S. (1991). The Rothamsted long-term experiments: Are they still of use. *Agron. J. 83*: 2–10.

Lal, R. (1991). Soil surface management in the tropics. In *Advances in Soil Science*, Vol. 15 (B. A. Stewart, ed.), Springer-Verlag, New York, pp. 91–137.

Lal, R., and Stewart, B. A. (1990). Agroforestry for soil management in tropics. In *Soil Degradation—A Global Threat* (R. Lal and B. A. Stewart, eds.), Springer-Verlag, New York, pp. XIII-XVII.

Little, P. D., Horowitz, M. M., and Nyerges, A. E. (1987). Land at risk in the third world. In *Local Level Perspectives.* Westview Press, Boulder, Colorado.

MacLennan, H. S. (1990). Adoption of trash retention farming techniques in Victoria. In *Challenges in Dryland Agriculture* (P. W. Unger, T. V. Sneed, W. R. Jordan, and R. Jensen, eds.), Texas Agricultural Experiment Station, College Station, Texas, p. 83.

McCalla, T. M., and Army, T. J. (1961). Stable mulch farming. *Adv. Agron. 13*: 125–196.

McWilliam, J. R. (1990). Striving for sustainability in dryland farming: The Australian experience. In *Challenges in Dryland Agriculture* (P. W. Unger, T. V. Sneed, W. R. Jordan, and R. Jensen, eds.), Texas Agricultural Experiment Station, College Station, Texas, pp. 45–49.

Miller, M. F. (1936). *Missouri Agricultural Station Bulletin*, No. 366.

Mitchell, C. C., Westerman, R. L., Brown, J. R., and Peck, T. R. (1991). Overview of long-term agronomic research. *Agron. J. 83*: 24–29.

Phillips, R. E., Blevis, R. L., Thomas, G. W., Froje, W. W., and Phillips, S. H. (1980). No-tillage agriculture. *Science 208*: 1108–1113.

Phillips, S. H., and Phillips, R. E. (1984). *No-Tillage Agriculture Principle and Practices.* Van Nostrand Reinhold, New York.

Pierce, F. J. (1991). Erosion productivity impact prediction. In *Soil Management for Sustainability* (R. Lal and F. J. Pierce, eds.), Soil and Water Conservation Society, Iowa, pp. 35–52.

Pierce, F. J., and Rice, C. W. (1988). Crop rotation and its impact on efficiency of water and nitrogen use. In *Cropping Strategies for Efficient Use of Water and Nitrogen* (W. L. Hargrove, ed.), *Am. Soc. Agron.*, Special Pub. No. 51, Madison, Wisconsin, pp. 21–40.

Power, J. F., and Follett, R. F. (1987). Monoculture. *Sci. Am. 256*: 57–64.

Retta, A., Vanderlip, R. L., Higgins, R. A., Moshier, L. J., and Ferjerherm, A. M. (1991). Sustainability of corn growth models for incorporation of weed and insect stresses. *Agron. J. 83*: 757–765.

Scaling, W. (1990). Getting dryland resource management systems on the ground. In *Challenges in Dryland Agriculture* (P. W. Unger, T. V. Sneed, W. R. Jordan, and R. Jenson, eds.), Texas Agricultural Experiment Station, College Station, Texas, pp. 36–37.

Schepers, S. (1988). Role of cropping system in environmental quality: groundwater nitrogen. In *Cropping Strategies for Efficient Use of Water and Nitrogen* (W. L. Hargrove, ed.), *Am. Soc. Agron.* Special Pub. No. 51, Madison, Wisconsin, pp. 167–178.

Sims, H. J. (1977). Cultivation and fallowing practices. In *Soil Factors in Crop Production in a Semi-arid Environment* (J. S. Russell and E. C. Greacen, eds.), University of Queensland Press, St. Lucia, pp. 243–261.

Smika, D. E., and Unger, P. W. (1986). Effect of surface residue on soil water storage. In *Advances in Soil Science*, Vol. 5 (B. A. Stewart, ed.), Springer-Verlag, New York, pp. 111–138.

Steiner, J. L., Williams, J. R., and Jones, O. R. (1987). Evaluation of the EPIC simulation model using a dryland wheat-sorghum-fallow crop rotation. *Agron. J. 79*: 732–738.

Stewart, B. A., Lal, R., and El-Swaify, A. (1991). Sustaining the resource base of an expanding world agriculture. In *Soil Management for Sustainability* (R. Lal and F. J. Pierce, eds.), Soil and Water Conservation Society, Ankeny, Iowa, pp. 125–144.

Szmedra, P. I., Wetztein, M. E., and McClendon, R. W. (1988). *Research Bulletin 353*. University of Georgia, College of Agriculture, Experiment Station, Athens, Georgia.

Szott, L. T., Palm, C. A., and Sanchez, P. A. (1991). Agroforestry in arid soils of the humid tropics. *Adv. Agron. 45*: 275–300.

Unger, P. W., and Parker, J. J. (1976). Evaporation reduction from soil with wheat, sorghum, and cotton residues. *Soil Sci. Soc. Am. J. 40*: 938–942.

Unger, P. W. (1990). Residue management for dryland farming. In *Challenges in Dryland Agriculture* (P. W. Unger, T. V. Sneed, W. R. Jordan, and R. Jensen, eds.), Texas Agricultural Experiment Station, Texas, pp. 483–489.

Varo, T. (27–116 B.C.). *De re Rustica*.

Williams, J. R., and Renard, K. G. (1985). *Soil Erosion and Crop Productivity* (R. F. Follett and B. A. Stewart, eds.), *Am. Soc. Agron.*, Madison, Wisconsin, pp. 67–103.

Wischmeier, W. H., and Smith, D. D. (1978). *Agric. Handbook*, No. 537, U.S. Department of Agriculture, Washington, DC.

Index

Aggregates
breakdown (disintegration), 67
stability, 54
Alkali, 317, 339
Antecedent moisture, 66–67, 80, 86, 95, 133, 182
Aspect, 95

Boron, 330

Calcium carbonate (Ca), 52–54, 354, 368–370
Canopy (cover crops), 107, 109, 250
Chemical dispersion, 3
Clay minerals, 3, 44
illite, 319
kaolinite, 318–319
montmorilonite, 318–319
Concentrated flow, 72
Consecutive rainstorms, 14–16
Conservation methods, 239–263
Cover management factor (C), 176–182
Crop rotation, 375–385
Crusting (*see* Surface sealing)

Deflocculation, 80, 85
Demixing, 319
Deposition (*see* Particles deposition)
Depression storage (*see* Surface storage)
Detachment of soil
by raindrops, 81, 127
by water flow, 128
Diffuse double layer, 318
Dispersion, 134, 318, 358

Electrolyte concentration (EC), 12–14, 16–17, 80, 316, 354
Erodibility, K factor, 64, 67, 85, 108, 133, 143, 175, 239
Erosion forms
  channels, 305–310
  gullies (*see* Gullying)
  interrill, 71–72, 107, 126, 134, 136
  rill, 72, 107, 126, 153
Erosivity (R factor), 26, 33–35, 63, 173–175
Exchangeable cations, 4
  magnesium percentage, 8, 356
  potassium percentage, 9
  sodium percentage (ESP), 6–8, 16–17, 133, 316, 354

Flocculation value, 322
Furrows
  diking, 253–256
  graded, 259

Geotextyles, 303
GIS, 159, 303
Gullying
  ephemeral, 155, 302
  upland, 156
  valley floor, 157–159
Gypsum, 14, 251, 268–282, 324, 335, 338, 358, 362–367

Infiltration, 1–2, 18, 215, 227–229, 356

Land
  imprinting, 261
[Land]
  leveling, 257
  smoothing, 258
Lee farming, 382–380

Models
  of soil detachment, 142
  of soil erosion and runoff, 138
    AGNPS, 159, 198
    ANSWERS, 159, 198
    CLIGEN, 235
    CREAMS, 170, 198, 234
    EPIC, 234
    KINEROS, 198, 233
    MUSLE, 198–199, 205
    RUSLE, 170, 173
    SPUR, 198, 234
    SWRRB, 198, 234
    USLE, 77–79, 170–186, 205
    WEPP, 137, 140–142, 159, 186–199, 205, 233–244
Mulch, 107, 109–117, 132, 249

Organic matter, 45–49
Overland flow velocity, 84, 131
Oxides (Al, Fe), 51

Particle deposition, 130
Particle transportation, 239
Peak discharge, 208, 229–231
Ph, 324
Phosphogypsum (*see* Gypsum)
Polymers, 282–295
  polyacrylamide (PAM), 251, 283, 362
  polysaccharide, 283, 362

Raindrops
erosivity (*see* Erosivity)
impact, 63, 81, 127, 129–130
kinetic energy, 10–12, 16, 26, 63, 66
shape, 63
size, 62, 63, 68, 130
splash, 62, 65, 67, 71–72
Rainfall
analysis, 26–33
intensity, 26, 66, 130
properties, 9
volume, 35–39
Rainstorm probability analysis, 24–26
Runoff, 203, 206
Runoff estimation methods
curve number (CN), 207
kinematic wave mode, 154, 221–227

Sealing (*see* Surface sealing)
Sediment (*see also* Particles)
load (yield), 72, 84
transportability, 134
transport by flow (wash, sheet), 128–129
transport by splash (air splash), 128
Shear strength (*see* Soil shear strength)
Sodium adsorption ratio (SAR), 320, 354
Slope (slope factor, S, L, topographic factor), 131, 277–281
leeward, 96–101
length, 61, 79, 89, 159, 175
steepness (gradient), 61, 69, 79, 93, 175
windward, 97–101

Soil
amendments (*see* Soil stabilizers)
characteristics (properties), 41–44, 67
conditioners (*see* Soil stabilizers)
conservation methods (*see* Conservation methods)
crusting (*see* Surface sealing)
dispersion (*see* Dispersion)
salinity, 316, 326–330
sealing (*see* Surface sealing)
shear strength, 63–64, 68, 79, 127, 133, 154, 159
slacking, 71, 80, 85
sodicity (*see also* ESP), 49–51, 80, 316, 332
splash (*see* Detachment, Raindrop splash)
stabilizers (chemical additives, 250, 267–296
swelling, 71, 134, 318, 358
texture, 42–44
Stripcropping, 258
Surface
cover (*see* Mulch)
drying, 70
roughness (microrelief), 94, 132, 160
sealing, crusting, 3, 17, 69–70, 134, 269–277
storage (water ponding depth, depression storage, water flow depth), 68, 91–94, 220

Terraces
bench, 260
discontinuous parallel, 261
graded, 259
intermittent, 260
Tillages
contour, 251–253

[Tillages]
mulch, 241–243
no, 244, 249, 385
reduced, 243–244
Time to concentration, 211
Topographic factor (*see* Slope factor)
Water
erosion, 125–126
ponding depth (*see* Surface storage)
Water erosion prediction project (WEPP), 137, 140–142, 159, 186–199, 205, 233–244